Molecular Interventions in Lifestyle-Related Diseases

OXIDATIVE STRESS AND DISEASE

Series Editors

LESTER PACKER, PH.D.
ENRIQUE CADENAS, M.D., PH.D.

University of Southern California School of Pharmacy
Los Angeles, California

1. Oxidative Stress in Cancer, AIDS, and Neurodegenerative Diseases, *edited by Luc Montagnier, René Olivier, and Catherine Pasquier*
2. Understanding the Process of Aging: The Roles of Mitochondria, Free Radicals, and Antioxidants, *edited by Enrique Cadenas and Lester Packer*
3. Redox Regulation of Cell Signaling and Its Clinical Application, *edited by Lester Packer and Junji Yodoi*
4. Antioxidants in Diabetes Management, *edited by Lester Packer, Peter Rösen, Hans J. Tritschler, George L. King, and Angelo Azzi*
5. Free Radicals in Brain Pathophysiology, *edited by Giuseppe Poli, Enrique Cadenas, and Lester Packer*
6. Nutraceuticals in Health and Disease Prevention, *edited by Klaus Krämer, Peter-Paul Hoppe, and Lester Packer*
7. Environmental Stressors in Health and Disease, *edited by Jürgen Fuchs and Lester Packer*
8. Handbook of Antioxidants: Second Edition, Revised and Expanded, *edited by Enrique Cadenas and Lester Packer*
9. Flavonoids in Health and Disease: Second Edition, Revised and Expanded, *edited by Catherine A. Rice-Evans and Lester Packer*
10. Redox–Genome Interactions in Health and Disease, *edited by Jürgen Fuchs, Maurizio Podda, and Lester Packer*
11. Thiamine: Catalytic Mechanisms in Normal and Disease States, *edited by Frank Jordan and Mulchand S. Patel*
12. Phytochemicals in Health and Disease, *edited by Yongping Bao and Roger Fenwick*
13. Carotenoids in Health and Disease, *edited by Norman I. Krinsky, Susan T. Mayne, and Helmut Sies*

Molecular Interventions in Lifestyle-Related Diseases

edited by

Midori Hiramatsu
Toshikazu Yoshikawa
Lester Packer

CRC Press
Taylor & Francis Group
Boca Raton London New York

CRC Press is an imprint of the
Taylor & Francis Group, an **informa** business
A TAYLOR & FRANCIS BOOK

CRC Press
Taylor & Francis Group
6000 Broken Sound Parkway NW, Suite 300
Boca Raton, FL 33487-2742

First issued in paperback 2019

© 2006 by Taylor & Francis Group, LLC
CRC Press is an imprint of Taylor & Francis Group, an Informa business

No claim to original U.S. Government works

ISBN-13: 978-0-8247-2958-5 (hbk)
ISBN-13: 978-0-367-39168-3 (pbk)
Library of Congress Card Number 2005051435

Library of Congress Cataloging-in-Publication Data

Molecular interventions in lifestyle-related diseases / edited by Midori Hiramatsu, Toshikazu
    Yoshikawa, Lester Packer.
            p. cm. -- (Oxidative stress and disease)
    Includes bibliographical references and index.
    ISBN 0-8247-2958-7 (alk. paper)
    1. Oxidative stress--Molecular aspects. 2. Oxidative stress--Prevention. 3. Antioxidants--
Therapeutic use. 4. Lifestyles--Health aspects. I. Hiramatsu, Midori. II. Yoshikawa, Toshikazu. III.
Packer, Lester. IV. Series.

RB170.M658 2005
616.3'9--dc22   2005051435

Visit the Taylor & Francis Web site at
http://www.taylorandfrancis.com

and the CRC Press Web site at
http://www.crcpress.com

Series Introduction

Oxygen is a dangerous friend. Through evolution, oxygen — itself a free radical — was chosen as the terminal electron acceptor for respiration. The two unpaired electrons of oxygen spin in the same direction; thus, oxygen is a biradical. Other oxygen-derived free radicals, such as superoxide anion or hydroxyl radicals, formed during metabolism or by ionizing radiation, are stronger oxidants, i.e., endowed with a higher chemical reactivity. Oxygen-derived free radicals are generated during oxidative metabolism and energy production in the body and are involved in regulation of signal transduction and gene expression, activation of receptors and nuclear transcription factors, oxidative damage to cell components, the antimicrobial and cytotoxic action of immune system cells, neutrophils and macrophages, as well as in aging and age-related degenerative diseases. Overwhelming evidence indicates that oxidative stress can lead to cell and tissue injury. However, the same free radicals that are generated during oxidative stress are produced during normal metabolism and, as a corollary, are involved in both human health and disease.

In addition to reactive oxygen species, research on reactive nitrogen species has been gathering momentum to develop an area of enormous importance in biology and medicine. Nitric oxide or nitrogen monoxide (NO) is a free radical generated by nitric oxide synthase (NOS). This enzyme modulates physiological responses in the circulation such as vasodilation (eNOS) or signaling in the brain (nNOS). However, during inflammation, a third isoenzyme is induced, iNOS, resulting in the overproduction of NO and causing damage to targeted infectious organisms and to healthy tissues in the vicinity. More worrisome, however, is the fact that NO can react with superoxide anion to yield a strong oxidant, peroxynitrite. Oxidation of lipids, proteins, and DNA by peroxynitrite increases the likelihood of tissue injury.

Both reactive oxygen and nitrogen species are involved in the redox regulation of cell functions. Oxidative stress is increasingly viewed as a major upstream component in the signaling cascade involved in inflammatory responses and stimulation of adhesion molecule and chemoattractant production. Hydrogen peroxide decomposes in the presence of transition metals to the highly reactive hydroxyl radical, which by two major reactions — hydrogen abstraction and addition — accounts for most of the oxidative damage to proteins, lipids, sugars, and nucleic acids. Hydrogen peroxide is also an important signaling molecule that, among others, can activate NF-κB, an important transcription factor involved in inflammatory responses. At low concentrations, hydrogen peroxide regulates cell signaling and stimulates cell proliferation; at higher concentrations, it triggers apoptosis and, at even higher levels, necrosis.

Virtually all diseases thus far examined involve free radicals. In most cases, free radicals are secondary to the disease process, but in some instances free radicals are causal. Thus, there is a delicate balance between oxidants and antioxidants in health and disease. Their proper balance is essential for ensuring healthy aging.

The term *oxidative stress* indicates that the antioxidant status of cells and tissues is altered by exposure to oxidants. The redox status is thus dependent on the degree to which the cell's components are in the oxidized state. In general, the reducing environment inside cells helps to prevent oxidative damage. In this reducing environment, disulfide bonds (S—S) do not spontaneously form because sulfhydryl groups are maintained in the reduced state (SH), thus preventing protein misfolding or aggregation. This reducing environment is maintained by oxidative metabolism and by the action of antioxidant enzymes and substances, such as glutathione, thioredoxin, vitamins E and C, and enzymes such as superoxide dismutases, catalase, and the selenium-dependent glutathione reductase and glutathione and thioredoxin hydroperoxidases, which serve to remove reactive oxygen species (hydroperoxides).

Changes in the redox status and depletion of antioxidants occur during oxidative stress. The thiol redox status is a useful index of oxidative stress mainly because metabolism and NADPH-dependent enzymes maintain cell glutathione (GSH) almost completely in its reduced state. Oxidized glutathione (glutathione disulfide, GSSG) accumulates under conditions of oxidant exposure, and this changes the ratio GSSG/GSH; an increased ratio is usually taken as indicating oxidative stress. Other oxidative stress indicators are ratios of redox couples such as NADPH/NADP, NADH/NAD, thioredoxin$_{reduced}$/thioredoxin$_{oxidized}$, dihydrolipoic acid/α-lipoic acid, and lactate/pyruvate. Changes in these ratios affect the energy status of the cell, largely determined by the ratio ATP/ADP + AMP. Many tissues contain large amounts of glutathione, 2 to 4 mM in erythrocytes or neural tissues and up to 8 mM in hepatic tissues. Reactive oxygen and nitrogen species can oxidize glutathione, thus lowering the levels of the most abundant nonprotein thiol, sometimes designated as the cell's primary preventive antioxidant.

Current hypotheses favor the idea that lowering oxidative stress can have a health benefit. Free radicals can be overproduced or the natural antioxidant system defenses weakened, first resulting in oxidative stress, and then leading to oxidative injury and disease. Examples of this process include heart disease, cancer, and neurodegenerative disorders. Oxidation of human low-density lipoproteins is considered an early step in the progression and eventual development of atherosclerosis, thus leading to cardiovascular disease. Oxidative DNA damage may initiate carcinogenesis. Environmental sources of reactive oxygen species are also important in relation to oxidative stress and disease. A few examples: UV radiation, ozone, cigarette smoke, and others are significant sources of oxidative stress.

Compelling support for the involvement of free radicals in disease development originates from epidemiological studies showing that an enhanced antioxidant status is associated with reduced risk of several diseases. Vitamins C and E and prevention of cardiovascular disease are notable examples. Elevated antioxidant status is also associated with decreased incidence of cataracts, cancer, and neurodegenerataive disorders. Some recent reports have suggested an inverse correlation between antioxidant status and the occurrence of rheumatoid arthritis and diabetes mellitus. Indeed, the number of indications in which antioxidants may be useful in the prevention or treatment of disease is increasing.

Oxidative stress, rather than being the primary cause of disease, is more often a secondary complication in many disorders. Oxidative stress diseases include inflammatory bowel diseases, retinal ischemia, cardiovascular disease and restenosis, AIDS, adult respiratory distress syndrome, and neurodegenerative diseases such as stroke, Parkinson's disease, and Alzheimer's disease. Such indications may prove amenable to antioxidant treatment (in combination with conventional therapies) because there is a clear involvement of oxidative injury in these disorders.

In this series of books, the importance of oxidative stress and disease associated with organ systems of the body is highlighted by exploring the scientific evidence and the medical applications of this knowledge. The series also highlights the major natural antioxidant enzymes and antioxidant substances such as vitamins E, A, and C, flavonoids, polyphenols, carotenoids, lipoic acid, coenzyme Q_{10}, carnitine, and other micronutrients present in food and beverages. Oxidative stress is an underlying factor in health and disease. More and more evidence indicates that a proper balance between oxidants and antioxidants is involved in maintaining health and longevity and that altering this balance in favor of oxidants may result in pathophysiological responses causing functional disorders and disease. This series is intended for researchers in the basic biomedical sciences and clinicians. The potential of such knowledge for healthy aging and disease prevention warrants further knowledge about how oxidants and antioxidants modulate cell and tissue function.

Lester Packer
Enrique Cadenas
Series Editors

Preface

This book is a treatise aimed at providing insights on lifestyle factors relevant for health-related issues and is a response to the concept of lifestyle-related diseases advanced by the Japanese Ministry of Health, Labor, and Welfare in 1996. Factors that affect lifestyle-related diseases include — among others — diet, alcohol consumption, cigarette smoking, and stress. The recognition of the occurrence of risk factors and their identification is an important step in overcoming lifestyle-related diseases and building a healthy lifestyle.

At a molecular level, lifestyle factors are encompassed by the concepts of oxidative stress, those underlying optimum diet, and the cellular effects elicited by micronutrients and antioxidants. This book is divided into three main sections that address the molecular basis of free radicals and lifestyle-related diseases and preventive/therapeutic approaches including nutraceuticals, functional foods, and pharmacological interventions. Each contains several chapters addressing critical molecular mechanisms, therapeutic interventions, and other issues of relevance to human health.

We thank the authors for their outstanding contributions to assemble this treatise.

Midori Hiramatsu
Toshikazu Yoshikawa
Lester Packer

About the Editors

Midori Hiramatsu is a professor at Tohoku University of Community Service and Science, and its graduate school. She has studied pharmacology at Kobe College of Pharmacy (Bachelor of Pharmacy in 1972). She has served as a research fellow and an instructor in the Department of Neurochemistry, Okayama University Medical School, where she obtained her Ph.D. in 1983, as a research general at the Laboratory of Yamagata Technopolis Foundation, as Research Director at the Division of Medical Science, Institute for Life Support Technology, Yamagata Technopolis Foundation, and as the head of the Media Center at Tohoku University of Community Service and Science.

Toshikazu Yoshikawa is Professor and Chairman, Inflammation and Immunology, Graduate School of Medical Science, Kyoto Prefectural University of Medicine, and in addition holds the posts of Professor in the Department of Internal Medicine, the Department of Kampo Medicine, the Department of Biomedical Safety Science, and the Department of Medical Proteomics.

He obtained his training at Kyoto Prefectural University of Medicine, graduating with a Ph.D. degree in 1983, and has held posts as visiting professor at Lousiana State University (1984–1985) and at the University of Tokyo.

He is currently serving as editor of the journal *BioFactors* and as president of the Japanese Society of Lipid Peroxide and Free Radical Research, and the Society for Free Radical Research International.

Lester Packer received a Ph.D. in microbiology and biochemistry from Yale University and was Professor and Senior Researcher at the University of California at Berkeley for 40 years. Currently he is Adjunct Professor, Molecular Pharmacology and Toxicology, School of Pharmacy, University of Southern California Health Science Center, Los Angeles. His research interests include the molecular, cellular, and physiological aspects of oxidants and antioxidants in biological systems.

In addition to his membership in numerous professional societies, Dr. Packer serves on editorial advisory boards for scientific journals related to biochemistry, antioxidant metabolism and nutrition. He has an unmatched scientific record on antioxidants and micronutrients, considered key components in achieving healthy aging.

Dr. Packer has been president of the Society of Free Radical Research International, founder and president of the Oxygen Club of California, and vice president of UNESCO — the United Nations Global Network on Molecular and Cell Biology. Dr. Packer is the recipient of numerous scientific awards including three honorary doctoral degrees. He has published more than 800 scientific papers and more than 100 books on many aspects of antioxidants and health, including standard references such as *Vitamin E in Health and Disease*, *Vitamin C in Health and Disease*, *The Handbook of Natural Antioxidants*, *Understanding the Process of Aging: The Roles of Mitochondria, Free Radicals, and Antioxidants*, and *Carotenoids and Retinoids: Molecular Aspects and Health Issues*. *The Antioxidant Miracle*, published in 1999 is a book for non-scientists.

Contributors

S. Adhikari
Bhabha Atomic Research Centre
Mumbai, India

Bharat B. Aggarwal
M.D. Anderson Cancer Center
University of Texas
Houston, Texas

Tamar Amit
Technion–Israel Institute of
  Technology
Haifa, Israel

I.F.F. Benzie
The Hong Kong Polytechnic
  University
Hong Kong, China

C.K. Chan
The Hong Kong Polytechnic
  University
Hong Kong, China

Kyung-Joo Cho
Korea Advanced Institute of Science
  and Technology
Daejeon, South Korea

S.W. Choi
The Hong Kong Polytechnic
  University
Hong Kong, China

An-Sik Chung
Korea Advanced Institute of Science
  and Technology
Daejeon, South Korea

Dipak K. Das
University of Connecticut School of
  Medicine
Farmington, Connecticut

T.P.A. Devasagayam
Bhabha Atomic Research Centre
Mumbai, India

Kelly L. Drew
University of Alaska
Fairbanks, Alaska

Manuchair Ebadi
University of North Dakota
Grand Forks, North Dakota

Hesham El ReFaey
University of North Dakota
Grand Forks, North Dakota

Yoshihiko Hatano
NUPALS
Niigata, Japan

Shingo Hayakawa
University of Yamanashi
Kofu, Japan

David Heber
UCLA Center for Human Nutrition
Los Angeles, California

Yoshihiro Higuchi
Kanazawa University
Kanazawa, Japan

Keisuke Hirai
Takeda Chemical Industries
Osaka, Japan

Midori Hiramatsu
Tohoku University of Community
  Service and Science
Yamagata, Japan

Yuko Hirose
University of Yamanashi
Kofu, Japan

Haruyo Ichikawa
M.D. Anderson Cancer Center
University of Texas
Houston, Texas

Takashi Ichiyanagi
NUPALS
Niigata, Japan

Yasuyuki Ishii
AIST
Osaka, Japan

K.K. Janardhanan
Amala Cancer Research Centre
Kerala, India

James A. Joseph
Tufts University
Boston, Massachusetts

Kan Kanamori
Toyama University
Toyama, Japan

Takao Kaneyuki
Kurasiki Sakuyo University
Kurasiki, Japan

Yong-Chul Kim
Kyoto University
Kyoto, Japan

Norihiko Kondo
Kyoto University
Kyoto, Japan

Tetsuya Konishi
NUPALS
Niigata, Japan

B. Lakshmi
Amala Cancer Research Centre
Kerala, India

Hyoung-gon Lee
Case Western Reserve University
Cleveland, Ohio

Silvia Mandel
Technion–Israel Institute of
  Technology
Haifa, Israel

Toshiki Masumizu
JEOL Ltd.
Tokyo, Japan

Hiroshi Masutani
Kyoto University
Kyoto, Japan

Seiichi Matsugo
University of Yamanashi
Kofu, Japan

Yoshiyuki Matsuo
Kyoto University
Kyoto, Japan

Nilanjana Maulik
University of Connecticut School of
  Medicine
Farmington, Connecticut

Chie Mihara
Toyama University
Toyama, Japan

K.P. Mishra
Bhabha Atomic Research Centre
Mumbai, India

Yuji Mizukami
Kanazawa University
Kanazawa, Japan

Hadi Moini
University of Southern California
Los Angeles, California

Akitane Mori
Okayama University
Okayama, Japan

Yuji Naito
Kyoto Prefectural University of
  Medicine
Kyoto, Japan

Hajime Nakamura
Kyoto University Hospital
Kyoto, Japan

Etsuo Niki
National Institute of Advanced
  Industrial Science and Technology
Ikeda, Japan

Yumiko Nishinaka
Kyoto University
Kyoto, Japan

Akihiko Nunomura
Asahikawa Medical College
Asahikawa, Japan

Lester Packer
University of Southern California
Los Angeles, California

George Perry
Case Western Reserve University
Cleveland, Ohio

Junko Sakakura
AIST
Osaka, Japan

Kazuo Sasaki
Toyama University
Toyama, Japan

Navindra P. Seeram
UCLA
Los Angeles, California

Sushil Sharma
University of North Dakota
Grand Forks, North Dakota

Shaik Shavali
University of North Dakota
Grand Forks, North Dakota

Shishir Shishodia
M.D. Anderson Cancer Center
University of Texas
Houston, Texas

Mark A. Smith
Case Western Reserve University
Cleveland, Ohio

Aoi Son
Kyoto University
Kyoto, Japan

Y.T. Szeto
The Hong Kong Polytechnic
  University
Hong Kong, China

Atsushi Takeda
Tohoku University
Sendai, Japan

Hideji Tanii
Kanazawa University
Kanazawa, Japan

J.C. Tilak
Bhabha Atomic Research Centre
Mumbai, India

Yuzo Uchida
University of Yamanashi
Kofu, Japan

Kazuhiko Uchiyama
Kyoto Prefectural University of
  Medicine
Kyoto, Japan

Sawitri Wanpen
University of North Dakota
Grand Forks, North Dakota

Toshiyuki Washizu
University of Yamanashi
Kofu, Japan

Orly Weinreb
Technion–Israel Institute of
  Technology
Haifa, Israel

Yorihiro Yamamoto
Tokyo University of Technology
Tokyo, Japan

Fumihiko Yasui
University of Yamanashi
Kofu, Japan

Junji Yodoi
Kyoto University
Kyoto, Japan

Isao Yokoi
Oita University
Oita, Japan

Tanihiro Yoshimoto
Kanazawa University
Kanazawa, Japan

Toshikazu Yoshikawa
Kyoto Prefectural University of
  Medicine
Kyoto, Japan

Moussa B.H. Youdim
Technion–Israel Institute of
  Technology
Haifa, Israel

Baolu Zhao
Institute of Biophysics
Beijing, China

Xiongwei Zhu
Case Western Reserve University
Cleveland, Ohio

Table of Contents

SECTION II Brain Diseases, and Free Radicals and Their Protection

SECTION III Nutraceuticals, Functional Foods, Micronutrients, and Pharmacological Interventions

Section I

Lifestyle-Related Diseases, and Free Radicals and Their Protection

1 Redox Regulation as an Underlying Factor in Health and Disease

Junji Yodoi
Kyoto University, Kyoto, Japan

Norihiko Kondo
Kyoto University, Kyoto, Japan

Yasuyuki Ishii
AIST, Osaka, Japan

Junko Sakakura
AIST, Osaka, Japan

Hajime Nakamura
Kyoto University Hospital, Kyoto, Japan

CONTENTS

1.1 INTRODUCTION

Thioredoxin (TRX) is a small ubiquitous protein containing a conserved active site, -Cys-Gly-Pro-Cys, and plays a variety of redox-related roles in organisms ranging from *Escherichia coli* to humans.[1] Human TRX was originally cloned as a soluble factor produced by human T-cell leukemia virus type I (HTLV-I)-transformed ATL2 cells.[2,3] Human TRX has critical roles in many cellular functions, including cell activation, differentiation, proliferation, and apoptosis, via

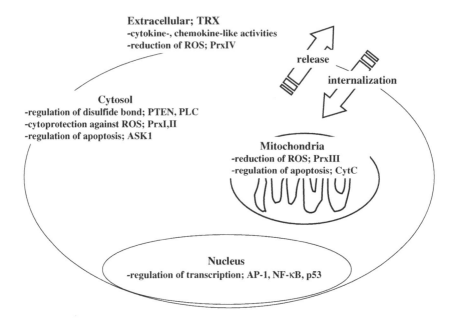

FIGURE 1.1 Redox regulation of cell signaling by TRX.

regulation of redox status[4] (Figure 1.1). In our studies, we produced TRX-overexpressing transgenic (TG) mice and investigated the role of TRX in stress response. TRX-TG mice are more resistant to a variety of oxidative stresses compared with wild-type C57BL/6 mice (Figure 1.2).

1.2   DISCUSSION

TRX is also released from various types of mammalian cells despite the absence of a typical secretory signal sequence[5,6] and has been shown to have several cytokine- and chemokine-like activities.[4,7] However, the mechanism of TRX release is still unknown. To better understand the mechanism, we generated the Jurkat stable transfectant cells expressing wild-type TRX and a mutant TRX (C32S/C35S), which has two cysteines displaced by serine in the active site of TRX.[8] We tested whether the redox-active site of TRX is involved in the release of TRX in response to oxidative stress, e.g., by exposure to H_2O_2 (Figure 1.3A). The wild-type TRX was rapidly released from the Jurkat transfectant expressing wild-type TRX but not from the mutant TRX C32S/C35S. The result suggested that the intact sequence of a redox-active site (-CGPC-) in the TRX protein is indispensable for its active release from T lymphocytes by oxidative stress. We also found that this TRX release is regulated by extracellular TRX (Figure 1.3B). TRX release by exposure to H_2O_2 was inhibited in a dose-dependent manner by the presence of rTRX-WT. The result suggested the presence of target molecules

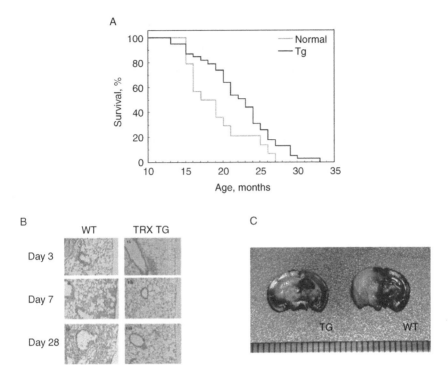

FIGURE 1.2 Characterization of TRX-TG mice against oxidative stress. A: TRX TG mice exhibit extended life span.

on the plasma membrane, which sense the concentration of TRX in extracellular fraction. Next, we studied the physiological functions of extracellular TRX. The pretreatment with rTRX suppressed H_2O_2-induced apoptosis. As shown in Figure 1.3C, extracellular rTRX attenuated ROS production and ROS-mediated apoptosis via caspase-3 activation. Based on these results, we speculated that the release of TRX from T lymphocytes is regulated by a negative feedback loop that senses the concentration of TRX in and/or outside the cells for the purpose of maintaining the physiological condition of the cells.

We attempted to identify the target molecules on the plasma membrane for extracellular TRX, which might be involved in the signal transduction and its entry into the cells (in submission). We proposed that the reduced form of TRX binds to its target molecules through its hydrophobic residues, that nucleophilic attack by the thiolate of Cys32 in the TRX molecule results in the formation of a transient mixed disulfide bond with the target molecule,[9] and that this reaction will be followed by a nucleophilic attack of the deprotonated Cys35, resulting in the generation of TRX-S2 and reduction of disulfide in the target molecule. Based on this hypothesis, we anticipated that the transient formation of the disulfide bond between Cys32 in TRX and a thiol group in target molecules may be

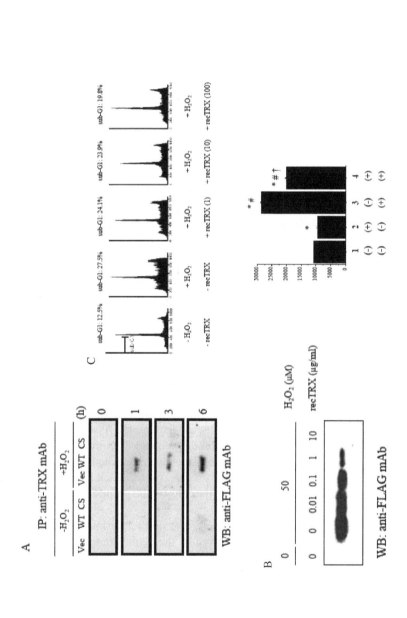

FIGURE 1.3 Redox-sensing release and regulation of apoptosis signaling. A: Redox-active site in TRX is indispensable for its release by H_2O_2. B: Inhibition of TRX release by the addition of extracellular TRX in a dose-dependent manner. C: Suppression of H_2O_2-induced apoptosis by the addition of recombinant TRX.

Cell   TRX-WT   TRX-C32S/C35S   TRX-C35S

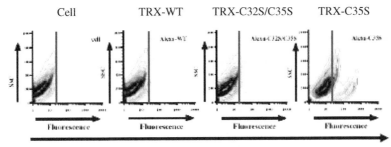

Fluorescence Intensity

FIGURE 1.4 Binding of recombinant TRXs to ATL2 cells.

involved in the binding of TRX to the cell surface. To test this possibility, we prepared a modified recombinant TRX in which Cys35, or both Cys32 and Cys35, were replaced with serine (i.e., TRX-C35S and TRX-C32S/C35S in the recent experiments) and investigated the effect of the replacement of Cys35 with serine, which would make the disulfide bond between TRX and target molecules persistent, on the binding of rTRX molecules to a T-cell line. As shown in Figure 1.4, Alexa-rTRX-C35S bound to the cells but not Alexa-rTRX-WT or Alexa-C32S/C35S. This binding of TRX-C35S to the cells was inhibited by the presence of dithiothreitol (DTT) and also by the presence of 10 to 5 50 folds of wild-type TRX. The results suggested that the binding of TRX-C35S to the cells was achieved with disulfide bonds, and TRX-WT and TRX-C35S may share a common target molecule on the cell surface. In the most labeled cells, Alexa-TRX-C35S was associated only with the plasma membrane by confocal analysis. In the remaining cells, the Alexa fluorescence label was homogeneously distributed in the cytosol. This observation — the incorporation of TRX-C35S into the cells — was confirmed by Western blotting analysis.

We assessed the binding/incorporation of TRX-C35S to HTLV-I-positive (ATL2 cells) and HTLV-I-negative (Jurkat T cells) T-cell lines. The result showed that Alexa-labeled TRX-C35S bound to ATL2 cells but not to HTLV-I-negative cell line cells (Jurkat T cells). The incorporation of TRX-C35S in ATL2 cells is also more enhanced than in Jurkat T cells. As expected, all HTLV-I-positive cell line cells employed in the experiment expressed CD25 (IL-2R alpha) chain on their cell surface but not the HTLV-I-negative cell line cells. CD25 (IL-2R alpha) chain is a cell activation marker and is known to be associated with activated lipid rafts fraction. These results suggested the possibility that the rapid internalization of TRX-C35S might be related to the lipid rafts microdomains in the plasma membranes of HTLV-I-positive T cells. To confirm this possibility, we analyzed the fractionation of Triton X-100 lysate of ATL2 cells and Jurkat T cells by floatation through OptiPrep gradients. The results suggested that a portion of TRX-C35S in ATL2 cells is associated with the lipid rafts microdomains, but not in Jurkat T cells.

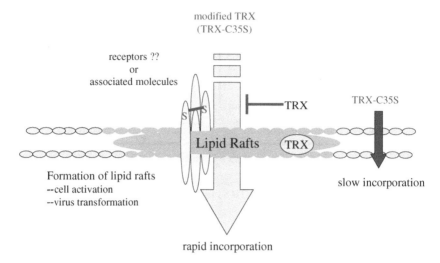

FIGURE 1.5 The scheme for the internalization mechanism of TRX-C35S via lipid rafts microdomain.

1.3 SUMMARY

Much evidence regarding the cytokine- and chemokine-like activities of TRX has accumulated in recent years. Although reports suggest the existence of target molecules or "receptor molecules" of TRX on cell surfaces, the molecules have not been identified. Our experiments showed marked binding/incorporation of TRX-C35S to HTLV-I-transformed T cells but not to HTLV-I-negative T cells, including the Jurkat T cells. We suggest that the receptors for TRX-C35S are associated with lipid rafts, which are conserved structures that play a role in many signaling processes involving receptors. As lipid rafts are induced by various stimuli, including virus transformation, it is hypothesized that the enhanced binding/incorporation of TRX-C35S to ATL2 cells, as compared to Jurkat T cells, is related to the lipid rafts formation. Because the binding/incorporation of TRX is dependent on the cysteine–cysteine interaction, expression of the target molecules or receptor molecules of TRX in lipid rafts fraction may be regulated by the redox status. As receptor signaling is under redox regulation, the enhanced binding and rapid internalization of a modified TRX may play a critical role in T-cell activation.

REFERENCES

1. Holmgren, A. Thioredoxin. *Annu. Rev. Biochem.* 54, 237–271 (1985).
2. Tagaya, Y. et al. ATL-derived factor (ADF), an IL-2 receptor/Tac inducer homologous to thioredoxin; possible involvement of dithiol-reduction in the IL-2 receptor induction. *EMBO J.* 8, 757–764 (1989).

3. Yodoi, J. and Uchiyama, T. Diseases associated with HTLV-I: virus, IL-2 receptor dysregulation, and redox regulation. *Immunol. Today.* 13, 405–11 (1992).
4. Nakamura, H., Nakamura, K., and Yodoi, J. Redox regulation of cellular activation. *Annu. Rev. Immunol.* 15, 351–369 (1997).
5. Rubartelli, A., Bajetto, A., Allavena, G., Wollman, E., and Sitia, R. Secretion of thioredoxin by normal and neoplastic cells through a leaderless secretory pathway. *J. Biol. Chem.* 267, 24161–24164 (1992).
6. Ericson, M. L., Horling, J., Wendel-Hansen, V., Holmgren, A., and Rosen, A. Secretion of thioredoxin after *in vitro* activation of human B cells. *Lymphokine Cytokine Res.* 11, 201–207 (1992).
7. Nakamura, H. et al. Circulating thioredoxin suppresses lipopolysaccharide-induced neutrophil chemotaxis. *Proc. Natl. Acad. Sci. USA.* 98, 15143–15148 (2001).
8. Kondo, N. et al. Redox-sensing release of human thioredoxin from T lymphocytes with negative feedback loops. *J. Immunol.* 172, 442–448 (2004).
9. Holmgren, A. Thioredoxin structure and mechanism: conformational changes on oxidation of the active-site sulfhydryls to a disulfide. *Structure.* 3, 239–243 (1995).

2 Diet and Lifestyle in Cancer Prevention

David Heber
UCLA Center for Human Nutrition, Los Angeles, California

CONTENTS

2.1 INTRODUCTION

There may have been many different environments in which ancient man found it possible to survive, but it was the rich diversity of the plant world that provided the background for the evolution of man. *Homo sapiens* (the species of all humans today) evolved over 50,000 years ago in response to a nutritional environment that was largely plant based[1] and which had been in existence for over 2 billion years, evolving, in turn, in response to predators and climate. The plant foods available were typically low in fat, high in fiber, and rich in the special preventive substances found only in plants (phytonutrients). This diverse diet provided all the elements needed for a long life, and we have evidence that ancient man knew of the medicinal properties of the plants as well. The discovery of agriculture 10,000 years ago meant that the biodiversity of the diet became narrowed, with an emphasis on those plants that could be easily cultivated. With the Industrial Revolution 200 years ago, many foods began to be processed for purposes such as economic gain, resulting in a further separation of man from his plant-based environment and further imbalances due to the refining and removal of the husks of grains such as rice or the refining of oils, resulting in the loss of vitamins.

Today we find ourselves faced with an epidemic of overnutrition and obesity in the industrialized world and a diet that comprises some foods that may promote heart disease and cancer. Although we derive more than enough calories from our foods, we have lost many of the natural preventive substances that provided the checks and balances in our bodies. Our modern diet is largely derived from animal sources and is higher in fat, lower in fiber, and poorer in micronutrients, especially antioxidants. Obesity, although often considered synonymous with overnutrition, is more accurately depicted as overnutrition of calories but undernutrition of many essential vitamins, minerals, and phytonutrients.

In most countries, obesity is associated with affluence and higher socioeconomic status, and it is thought to be aggravated by Western influences. For example, China, at one time, had one of the world's lowest rates of obesity, with only 2% of men and 6% of women being affected. However, the number of overweight individuals has sharply increased in the last decade, with 14% of men and 17% of women now classified as being overweight. The Chinese appear to be extremely sensitive to the consequences of carrying excess weight: even a modest increase is more likely to induce the development of type 2 diabetes, hypertension, and stroke than in either Americans or Western Europeans. More extreme examples of the increasing prevalence of obesity can be found in the Pacific Islands. In urban Samoa, an astounding 58% of men and 77% of women are obese. Even ethnic groups previously thought to be immune or less predisposed to obesity are now being affected. Japanese schoolchildren continue to have a very low rate of obesity, but the rate has doubled in recent years. Although our modern diet has virtually eliminated malnutrition in the majority of the population, there is an epidemic of obesity both in the United States and internationally. Countries such as Japan, with a traditional healthy plant-based diet, have noted an increase in obesity from 5% to 20% of the population in the last 30 years.

This increased incidence of obesity has been associated with an increased incidence of heart disease, breast cancer, prostate cancer, and colon cancer in comparison with populations eating a diet consisting of less meat and more fruits, vegetables, cereals, and whole grains. At the same time, successful primary treatment of many common forms of cancer has led to an increased population of cancer survivors. As a result, there is an opportunity to prevent the recurrence of cancer after primary treatment with dietary interventions in individual cases, even as public health measures that change diet and lifestyle in populations remain the mainstay of the primary prevention of cancer (see Figure 2.1).

2.2 OXIDANTS, ANTIOXIDANTS, AND PHYTOCHEMICAL EFFECTS BEYOND ANTIOXIDATION

The common forms of cancer, including breast, colon, and prostate cancer, are the result of genetic–environmental interactions. Although a small minority (5 to 10%) of patients have inherited forms of cancer due to alterations in the genes of the germ cell line, most cancers are an outcome of genetic changes at the somatic cell level. These genetic changes lead to unregulated growth through activation of growth-promoting genes (oncogenes) or inactivation of tumor suppressor

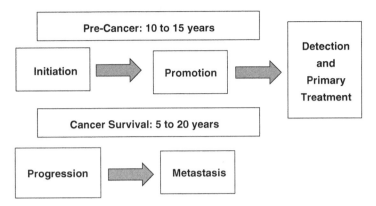

FIGURE 2.1 The two faces of cancer represent (1) primary prevention through early detection and public health measures affecting diet and lifestyle such as smoking cessation and (2) prevention of progression and metastasis in cancer survivors who have successfully completed primary treatment of their cancer.

genes. In some cases, oncogenes code for growth factor receptor proteins and growth factors. Reactive oxygen radicals are thought to damage biological structures and molecules, including lipids, protein, and DNA, and there is evidence that antioxidants can prevent this damage and that some phytochemicals can promote DNA repair.

The process of oxidation occurs in any oxygen-rich environment when substrates are exposed to ultraviolet light or heat. In order to understand the mechanisms by which antioxidants can prevent chronic diseases, it is necessary to understand the actions of free radicals and how antioxidants limit free radical reactions.[2] When free radicals attack polyunsaturated fats in the presence of oxygen, lipid peroxides are formed in a chain reaction that results in amplification of the original oxidative damage. For example, low density lipoprotein that is oxidized is more easily taken up by macrophages in the endothelial wall of blood vessels, promoting atherogenesis.[3] A DNA-based oxidation product (8-hydroxy-deoxyguanosine) can be detected in the urine of humans exposed to oxidant stresses, indicating the possibility that oxidation can alter genetic information. These oxidative processes have been associated with degenerative changes occurring with aging and the development of cardiovascular disease and cancer. On balance, a number of studies have shown that the various antioxidants act in concert in a cooperative system of antioxidant defense. Vitamin C quenches free radicals in aqueous systems and also regenerates cellular vitamin E, which helps to control lipid peroxidation.[4] Beta-carotene also traps free radicals in concert with vitamin E. The selenium-containing enzyme glutathione peroxidase destroys peroxides before they can damage cell membranes and interacts synergistically with vitamin E.[5] In a number of animal studies, the administration of antioxidants ameliorated damage from experimental oxidant stress. Furthermore, the antioxidant requirement in these studies was directly proportional to the increased tissue concentrations of free radicals. Although not proven, promotion of antioxidation through consumption of antioxidant-rich fruits and vegetables as well as dietary supplements may contribute to enhanced antioxidant defenses.

On the basis of population studies and animal studies, a number of governmental agencies including the National Research Council, the National Cancer Institute, and the U.S. Department of Agriculture have recommended that Americans eat five to nine servings of fruits and vegetables a day, in part to increase the intake of beneficial antioxidants. Recent surveys indicate that only a small fraction of the general population follows this advice, leading some nutrition authorities to recommend dietary supplementation or food fortification with antioxidants to reduce chronic disease incidence. Although there remains some uncertainty about the long-term effects of antioxidant supplementation, there is accumulating evidence that the practice of dietary supplementation may be beneficial.

For example, vitamin C has three proposed mechanisms of action in chemoprevention: (1) incorporation of ascorbic acid into the hyaluronidase inhibitor system, (2) prevention of the formation of nitrosamines in the gastrointestinal tract, and (3) inhibition of the potency of chemical carcinogens by enhancing the activity of the detoxifying cytochrome P-450 system. Beta-carotene and retinol have been shown *in vitro* and in animal studies to suppress carcinogenesis. Vitamin E in cell culture has been shown to reduce expression of certain oncogenes (H-ras and c-myc) in tumor cells and to inhibit cell proliferation in other cell lines. Selenium has been shown to prevent the metabolism of chemical carcinogens, which is necessary for their activation. Rats fed a diet rich in beta-carotene, vitamin C, or selenium were demonstrated to have a reduced incidence of carcinogen-induced pancreatic cancer.

Recent studies performed at the Sloane-Kettering Cancer Center in New York have shown that tumors concentrate vitamin C. Dehydroascorbate, the oxidized form of vitamin C, is taken up by the tumor cell, utilizing a glucose transporter. Once inside the cell, vitamin C is formed by reduction of dehydroascorbate, and the vitamin is trapped in the cell. This same phenomenon has been demonstrated in normal brain cells, which have a similar transport mechanism. This observation was interpreted in a *New York Times* editorial as indicating that tumor cells need vitamin C for growth and that vitamin C should not be taken by cancer patients. At about the same time, a study was done on rats at the University of North Carolina. In this study, rats with a carcinogen-induced brain tumor were either made vitamin E deficient or were fed a normal diet containing the usual amounts of vitamin E. When the tumors were treated with chemotherapy, the vitamin E-deficient animals had a better response to chemotherapy. Again, a *New York Times* editorial interpreted these results as indicating that cancer patients should not take any vitamin E supplements.

In general, there is uncertainty about the effects of vitamins during cancer therapy, with some studies showing protection of the host against oxidant damage to the heart from adriamycin, and none showing absolute interference with chemotherapy. However, many oncologists today advise cancer patients not to take vitamin supplements. Certainly, cancer survivors following completion of their therapy may take the usual doses of vitamin supplements until more research is done.

Although the antioxidant properties of vitamin C, vitamin E, and beta-carotene have been considered to be of primary importance,[6,7] other mechanisms may also be involved in their chemopreventive actions. These include effects on the immune system, on modulation of growth factors and growth factor receptors, and on intercellular interactions.[8] Epidemiologic studies provide evidence on the relationship of the intake of antioxidants to the incidence of lung cancer. In Finnish studies carried out over periods of 5 to 20 yr, there was an inverse relationship between serum levels of vitamin E and other antioxidants and the risk of cancer at several anatomic sites. In a study of 15,093 women aged 15 to 99 years and initially free of cancer, the disease was diagnosed in 313 women over an 8-year follow-up. Low serum levels of vitamin E strongly predicted the risk of epithelial cancers but demonstrated only slightly elevated risk of cancers in reproductive organs exposed to estrogens.[9] Therefore, whereas the effects of antioxidants can be demonstrated in animal studies, human studies are more complex, e.g., due to the known effects of obesity increasing estrogen levels in women and also other factors such as reduced fiber intake in the Western diet.

Sometimes smoking also appears to override the protective effects of antioxidant vitamins in some studies. A randomized double-blind prospective controlled trial carried out for 5 to 8 yr in 29,133 male smokers studied the effects of 50 mg synthetic vitamin E alone or in combination with 20 mg synthetic beta-carotene.[10] Among the 876 cases of lung cancer that developed, there was no evidence of decreased incidence in the men supplemented with vitamin E. There were 52 fewer cases of prostate cancer but, unexpectedly, the incidence of lung cancer was 18% higher and the mortality due to lung cancer was higher in the beta-carotene-supplemented group. This result runs counter to the body of experimental evidence. In a recent supplementation trial in China in which 29,584 adults aged 40 to 69 years were provided a variety of vitamin supplements, including vitamin A, riboflavin, vitamin C, and a combination of beta-carotene (15 mg), selenium (50 mcg), and vitamin E (30 mg), the latter combination of supplements was associated with a 13% decrease in cancer mortality. The major effect was on gastric cancer (21% reduction), which is the most common cancer in this population.

Clearly, much further research needs to be done on the effects of antioxidants, but the above studies indicate a potential protective effect on cancers of the aerodigestive system. In the presence of other major etiologic factors (e.g., estrogen for breast cancer or smoking for lung cancer), the beneficial effects of antioxidants are not demonstrable. In patients at increased risk of cancer, including the population of those who have undergone primary treatment of cancer, antioxidant supplementation is being recommended by many physicians.

Fruits, vegetables, spices, and some grains contribute thousands of phytochemicals to the human diet and many of these are absorbed into the body.

Although these are commonly considered antioxidants based on their ability to trap singlet oxygen, they have been demonstrated scientifically to have many functions beyond antioxidation. These phytochemicals can interact with the host to confer a cancer-preventive benefit by regulating the enzymes important for metabolizing xenobiotics and carcinogens, by modulating nuclear receptors and cellular signaling of proliferation and apoptosis, and by acting indirectly through antioxidant actions that reduce proliferation and protect DNA from damage.[11]

2.3 SPECIFIC ANTICANCER EFFECTS OF COLORFUL FRUITS AND VEGETABLES

Phytochemicals found in fruits and vegetables demonstrate synergistic and additive interactions through their effects on gene expression, antioxidation, and cytokine action. Fruits and vegetables are 10- to 20-fold less calorie-dense than grains, and provide higher amounts of dietary fiber, compared to refined grains. They also provide a balance of omega-3 and omega-6 fatty acids and a rich supply of micronutrients as well as a diverse array of phytochemicals. Nutrition authorities agree that a wide range and diversity of fruit and vegetable intake are desirable, but much work remains to be done on how to implement these recommendations.

Current NCI dietary recommendations emphasize increasing the daily consumption of fruits and vegetables from diverse sources such as citrus fruits, cruciferous vegetables, and green and yellow vegetables.[12]

This color coding has been further extended in a lay book titled *What Color Is Your Diet?* (Harper-Collins, 2001), to seven different groups based on their content of a primary phytochemical family for which there is evidence of cancer prevention potential (see Table 2.1).

Commonly eaten fruits and vegetables can be classified by color — red, red/purple, orange, orange/yellow, green, yellow/green, and white/green — depending on the specific chemicals that absorb light in the visible spectrum, thus, creating the different colors. Plant-eating animals naturally use color as an identifying marker of edible plant species. The changing color of ripening fruits and vegetables signify that they are at the peak of their taste and nutritive value. Many of the phytonutrients are actually the pigment molecules that lend ripe fruits and vegetables their distinctive hues. Only humans and a few primates have red–green color discriminating vision (trichromatic vision). All other mammals have dichromatic vision and are red–green color blind. It is believed that mankind evolved this type of vision to help in food selection with a plant-based diet.

Carotenoids are chemical compounds that absorb visible light and so determine that carrots are orange in color, tomatoes are red, and marigolds are yellow. Approximately 700 different carotenoids have been isolated from plants and animals. About 50 to 60 of these are present in a typical diet. These carotenoids are specifically broken down by the body, often during the process of absorption into the bloodstream from the small intestine.

TABLE 2.1

Color-Coded Fruits and Vegetables Organized by Main Phytonutrient and Benefits

Color	Fruits and Vegetables	Main Phytonutrient and Benefits
Red	Tomatoes, tomato soup, juices or sauces, pink grapefruit, watermelon	Lycopene is one of the most potent free radical scavengers in nature. It may reduce the risk of heart and lung disease, as well as prostate cancer.
Red/purple	Red grapes, blueberries, blackberries, cherries, plums, prunes, raspberries, strawberries, red apples	Anthocyanins are powerful antioxidants that strengthen skin and other tissues, tendons, and ligaments. They may help with age-related declines in mental function according to results of some recent experiments.
Orange	Apricots, acorn and winter squash, butternut and yellow squash, carrots, mangoes, cantaloupes, pumpkin, sweet potatoes	Alpha- and beta-carotene are carotenoids and very effective antioxidants. They protect against cancer by preventing oxidative damage, and promote vision by conversion to vitamin A.
Orange/yellow	Clementines, mandarin oranges, oranges and orange juice, peaches, pineapple and pineapple juice, nectarines, papayas, tangerines, tangelos	The rinds of citrus fruit contain limonene and other chemicals that have some anticancer effects. The vitamin C and flavonoids are included in a rich matrix in whole fruits.
Green	Broccoli and broccoli sprouts, bok choy, Brussels sprouts, cabbage, Chinese cabbage, kale	Sulforaphane, isothiocyanate, and indoles fight numerous diseases by eliminating toxic waste (carcinogens) from the body.
Yellow/green	Collard greens, green and yellow peppers, green beans, kale, beets and mustard greens, green peas, avocado, honeydew melon, yellow corn	Lutein and zeaxanthin are pigments that become concentrated in the retina where they help reduce the risk of cataracts and age-related macular degeneration.
White/green	Asparagus, celery, chives, endive, garlic, leeks, mushrooms, pearl onions, pears, shallots	Allyl sulfides are compounds that give garlic and onions their odor but also can promote blood vessel health. Quercetin is a flavonoid with anticancer potential as well.

The red group includes tomatoes, pink grapefruit, and watermelon, which contain lycopene. Lycopene is more readily available from cooked tomato products and juices than from whole tomatoes, and these products are the primary sources of lycopene in our diet. In practice, over 80% of the lycopene in the American diet comes from pasta sauce, tomato soup, tomato juice, and ketchup.

The red/purple group includes grapes, red wine, grape juice, prunes, cranberries, blueberries, blackberries, strawberries, and red apples. They contain anthocyanins, which are powerful antioxidants that may have a beneficial effect on cancer progression by inhibiting blood clot formation, angiogenesis, and cancer cell proliferation.

The orange group includes carrots, mangoes, apricots, cantaloupes, pumpkins, acorn squash, winter squash, and sweet potatoes. These are rich in alpha- and beta-carotenes. In this group, carrots provide about half of these carotenes in the U.S. diet, with additional significant contributions from tomato products.

The orange/yellow group includes orange juice, oranges and tangerines, peaches, papaya, and nectarines. These provide β-cryptoxanthin, a minor carotenoid, which accounts for only 0.03 mg of the 6 mg/d intake of all carotenoids by the average American. As a practical matter, 87% of cryptoxanthin comes from orange juice, oranges, and tangerines. Other fruits, providing smaller amounts, include peaches, papayas, and nectarines. These fruits obviously have other benefits and are a separate group, primarily, providing more diversity in the diet.

The green group includes broccoli, Brussels sprouts, cabbage, Chinese cabbage or bok choy, and kale. These contain sulforaphane, isothiocyanate, and indoles that induce the phase II enzyme systems involved in detoxification of carcinogens such as glutathione-S-transferases.

The yellow/green group includes spinach, collard, mustard or turnip greens, yellow corn, green peas, avocado, and honeydew melon. These provide lutein and zeaxanthin as well. These carotenoids concentrate in the eye and contribute to eye health. Low intakes have been associated with cataracts and age-related macular degeneration, the primary preventable cause of blindness in America. Lutein is also a potent antioxidant, which is synergistic in combination with lycopene (see text that follows).

The white/green group includes garlic, onions, celery, pears, white wine, endive, and chives. Plants in the onion family contain allicin that has been shown to have antitumor effects. Foods in this group are also rich sources of flavonoids including quercetin and kaempferol. Of all the antioxidants in fruits and vegetables, it is the flavonoids that are eaten in the largest quantity — up to 1 g/d. There are many flavonoid structures, and researchers in my laboratory are developing methods for determining the amount of flavonoids that are consumed on average, based on their breakdown products in the urine.

Often in the scientific and lay press a single phytochemical, such as lycopene from tomatoes, is featured for its antioxidant activities, whereas the beneficial effects actually result from the ingestion of foods which contain families of compounds that, in the case of tomatoes, include lycopene, phytoene, phytofluene, vitamin E, and vitamin C.[13] Tomato products, including soups, juices, pasta, and catsup, have received increased attention since Giovanucci et al.[14] reported that an increased dietary intake of lycopene was associated with a reduced risk of prostate cancer. Nonetheless, studies of lycopene exemplify the phytochemical functions beyond antioxidation that are manifested by many phytochemicals.

Lycopene has the highest antioxidant activity among all dietary carotenoids and is very efficient at quenching singlet oxygen and scavenging free radicals.[14] Although the antioxidant activity of lycopene in laboratory studies of multilamellar liposomes,[15,16] assayed by inhibition of formation of thiobarbituric acid-reactive substances, was in the following ranking: lycopene > alpha-tocopherol (vitamin E) > alpha-carotene > beta-cryptoxanthin > zeaxanthin = beta-carotene > lutein, mixtures of carotenoids were more effective than the single compounds. This synergistic effect was most pronounced when lycopene or lutein were present. The superior protection afforded by mixtures may be related to the specific positioning of different carotenoids in membranes.

Although lycopene is a carotenoid without provitamin-A activity, it has potent effects on prostate cancer cell proliferation alone, and in combination with alpha-tocopherol. In studies of the effects of lycopene and alpha-tocopherol on the growth of two different human prostate carcinoma cell lines (the androgen insensitive DU-145 and PC-3), it was found that lycopene alone was not a potent inhibitor of prostate carcinoma cell proliferation. However, the simultaneous addition of lycopene together with alpha-tocopherol, at physiological concentrations (less than 1 μM and 50 μM, respectively), resulted in a strong inhibitory effect on prostate carcinoma cell proliferation, which reached values close to 90%.[17] The effect of lycopene with alpha-tocopherol was synergistic but this was not seen with beta-tocopherol, ascorbic acid, and probucol. Furthermore, the combination of low concentrations of lycopene with 1,25-dihydroxyvitamin D3 synergistically inhibited cell proliferation and stimulated differentiation in the HL-60 promyelocytic leukemia cell line.[18] Lycopene treatment resulted in a concentration-dependent reduction in HL-60 cell growth as measured by [3H]thymidine incorporation and cell counting. This effect was accompanied by inhibition of cell cycle progression in the G0/G1 phase as measured by flow cytometry. Lycopene alone induced cell differentiation as measured by phorbol ester-dependent reduction of nitroblue tetrazolium and expression of the cell surface antigen CD14. Additional gene–nutrient interactions may also be involved in the observed effects of carotenoids. Studies using human and animal cells have identified a gene, connexin 43, whose expression is upregulated by chemopreventive carotenoids and which allows direct intercellular gap junctional communication (GJC). GJC is deficient in many human tumors and its restoration or upregulation is associated with decreased proliferation.[19]

It is feasible to study the effects of mixtures of phytochemicals on serum biomarkers as demonstrated by a recent clinical trial[20] in which a mixed intake of fruits and vegetables (500 g/d, meeting "five-a-day" guidelines, vs. 100 g/d) increased folate by 15% and decreased plasma homocysteine by 11% in 47 normal volunteers. The fruit- and vegetable-rich diet provided 13.3 mg carotenoids, 173 mg vitamin C, and 228 μg folate, compared to 2.9 mg carotenoids, 65 mg vitamin C, and 131 μg folate. Remarkably, plasma nutritional biomarkers were increased by

46% for lutein, 45% for beta-carotene, 64% for vitamin C, and 121% for alpha-carotene, demonstrating the feasibility of tracking such an intervention for cancer prevention.

2.4 FUTURE RESEARCH QUESTIONS

The key research questions that need to be answered are: (1) Can biomarkers of these phytochemical effects be used to follow changes in cancer cell progression, morphology, or biochemistry and be useful in cancer prevention studies? (2) Can methods be developed to simultaneously track the disposition of phytochemicals from fruits and vegetables, given the differences in volumes of distribution, pharmacokinetics of absorption, and the various pathways of metabolism and clearance of these substances? and (3) Can synergistic interactions of phytochemicals from fruits and vegetables be demonstrated *in vitro* and *in vivo* to provide the scientific substantiation for the popular recommendation[21] to eat a variety of colorful fruits and vegetables for their cancer-preventive benefits?

2.5 SUMMARY

Achieving the necessary breakthroughs in nutrition, physiology, and genetics will require an interdisciplinary team approaching the solutions to heart disease, cancer, and aging from a new perspective — human nutrition. The UCLA Center for Human Nutrition is dedicated to achieving this vision. Our center is approaching this challenge with a team of over 100 scientists drawn from many different fields of medicine and science.

The killer diseases, heart disease and cancer, and aging all result from an interaction of genes and environment, leading to changes in the functioning of the cells. Even though we know the entire human genome, we are only at the beginning of our journey to put an end to these diseases and extend the human lifespan. In the UCLA Center for Human Nutrition, researchers recently found that a gene coding for an enzyme implicated in activating colon cancer-causing chemicals present in smoked meat could be found in about 20 different versions in different individuals. Although the gene coded for the same enzyme (*N*-acetyl transferase) in each individual, there were differences in the structure of the gene, which defined different individuals in the same way as a fingerprint would. It was found that individuals with a certain fingerprint (called a polymorphism) for this particular gene would be protected from developing colon cancer, but only if they ate broccoli and other cruciferous vegetables. In individuals who did not eat these vegetables, there was no protection, regardless of the genes that each individual had in place. This discovery showed that for this gene, and possibly others, the information on how to react to the nutritional environment is encoded in our genes.

When cancers and precancer cells are examined, they are full of damaged genes. The process of cancer itself progresses through a series of mutations in the DNA of the cancer cell over a 10 to 15-year period. In both Israel and Japan, heavy smokers have less than half the incidence of lung cancer seen in the United States. Those who eat more fruits and vegetables appear to be the ones who are protected. The American people do not eat enough fruits and vegetables, cereals and grains, dietary fiber or, in many cases, enough high-quality protein. One out of three women and one out of two men will be diagnosed with some form of cancer during their lifetime. It is entirely possible that the high incidence of cancer in this country is due to the diet we eat or, rather, the one we do not eat.

High calories, high fat, lots of polyunsaturated fats, an epidemic of obesity, and smoking all put a stress on our cellular repair systems. As we age, our ability to rid our bodies of the abnormal cells that result from this damage wanes, and cancer can get a foothold. Food security is no longer a technical problem. We can grow enough to feed the whole world. Food security is a political and social problem. The impact of growing populations on the environment and the distribution of food have much more to do with wiping out malnutrition than science does. As the world's population continues to grow, the demand for food is expected to double in the next 30 years. Through advances in agricultural biotechnology, there are already prospects for producing food for people on the existing farmlands well into the future. Preserving the earth's environment and our internal environments will be the scientific challenge. This challenge will require better knowledge of how our genes interact with our diet. We already know something about these issues, and there is much we could already be doing that will improve our quality of life and hopefully extend our lifespan.

Our diets must be changed. Recommendations made in the late 1980s by the Surgeon General, the American Cancer Society, the American Heart Association, and every government agency concerned with nutrition have still not been put effectively into action. The University of California at Los Angeles (UCLA) Center for Human Nutrition developed a graphic nutrition plan called the California Cuisine Pyramid. This pyramid was designed by me and my colleagues to make easily understood the goals for eating fruits and vegetables for their special preventive nutrients (phytonutrients), eating whole grain cereals and breads to meet the fiber goal, eating plant-derived protein and some animal protein to meet specific protein goals, and adding taste enhancers, including natural oils from avocado, nuts, seeds, spices, and sweeteners to make the foods we eat for health more enjoyable. This pyramid, and the diets specific to it, are designed for preventing cancer and heart disease and go well beyond the U.S. Department of Agriculture's Food Guide Pyramid.

The so-called basic four food groups and other medical and specific diets popularized after World War II have succeeded in wiping out famine and hunger in what are now wealthy societies. This has had more of an impact in reducing deaths from tuberculosis, pneumonia, and other infectious diseases than all the antibiotics and other medicines combined. Nutrition and sanitation are public health strategies that are far more powerful than drugs and surgery. We have overdone it by creating an epidemic of obesity in the United States and throughout the world. Obesity is the result of an interaction of genes and environment and is a risk factor for heart disease and cancer.

Malnutrition adversely affects the immune system; malnourished children are more prone to infections and succumb more easily. When they are fed a nutritious diet, their immune systems revive, and they respond better to penicillin and other antibiotics. However, in the long term, there is a price to be paid for modern, convenient, cost-effective overnutrition. Overnutrition sends growth signals to every cell in the body, and this overgrowth leads to mutations in genes. Some of these lead to heart disease, cancer, or premature aging.

REFERENCES

1. Eaton SB, Konner M. Paleolithic nutrition. *New Engl J Med* 1985; 312: 283–289.
2. Harman D. Free radical theory of aging: the free radical diseases. *Age* 1984; 7: 111–131.
3. Rifici VA, Khachadurian AK. Dietary supplementation with vitamins C and E inhibits in vitro oxidation of lipoproteins. *J Am Coll Nutr* 1993; 12: 631–637.
4. Watson RR, Leonard TK. Selenium and vitamins A, E, and C: nutrients with cancer prevention properties. *J Am Diet Assoc* 1986; 86: 505–510.
5. Wefers H, Sies H. The protection by ascorbate and glutathione against microsomal lipid peroxidation is dependent on vitamin E. *Eur J Biochem* 1988; 174: 353–357.
6. Rimm EB, Stampfer MJ, Ascherio A, Giovannucci E, Colditz GA, Willett WC. Vitamin E consumption and the risk of coronary heart disease in men. *New Engl J Med* 1993; 328: 1450–1456.
7. Stampfer MJ, Hennekens CH, Manson JE, Colditz GA, Rosner B, Willett WC. Vitamin E consumption and the risk of coronary artery disease in women. *New Engl J Med* 1993; 328: 1444–1449.
8. Prasad K, Edwards-Prasad J, Kumar S, Meyers A. Vitamins regulate gene expression and induce differentiation and growth inhibition in cancer cells. *Arch Otolaryngol Head Neck Surg* 1993; 119: 1133–1140.
9. Knekt P. Serum vitamin E level and risk of female cancers. *Int J Epidemiol* 1988; 17: 281–286.
10. Alpha Tocopherol Beta Carotene Cancer Prevention Study Group. The effect of vitamin E and beta carotene on the incidence of lung cancer in chronic smokers. *New Engl J Med* 1994; 330: 1029–1035.
11. Blot WJ, Li J-Y, Taylor PR, Guo W, Dawsey S, Wang G-Q, Yang CS, Zheng S-F, Gail M, Li G-Y, Yu Y, Liu B-Q, Tangrea J, Sun Y-H, Liu F, Fraumeni JF, Zhang Y-H, Li B. Nutrition intervention trials in Linxiang, China: supplementation with specific vitamin/mineral combinations, cancer incidence, and disease-specific mortality in the general population. *J Natl Cancer Inst* 1993; 85: 1483–1492.
12. Steinmetz KA, Potter JD. Vegetables, fruits, and cancer. I. *Epidemiol Cancer Causes and Control* 1991; 2: 325–337.
13. Block G, Patterson B, Subar A. Fruit, vegetables, and cancer prevention. A review of the epidemiological evidence. *Nutr Cancer* 1992; 18: 1–29.
14. Giovanucci E, Ascherio A, Rimm EB, Stampfer MJ, Colditz GA, Willett VC. Intake of carotenoids and retinol in relation to risk of prostate cancer. *J Natl Cancer Inst* 1995; 87: 1767–1776.
15. Di Mascio P, Kaiser S, Sies H. Lycopene as the most efficient biological carotenoid singlet oxygen quencher. *Arch Biochem Biophys* 1989; 274: 532–538.

16. Stahl W, Junghans A, de Boer B, Driomina ES, Briviba K, Sies H. Carotenoid mixtures protect multilamellar liposomes against oxidative damage: synergistic effects of lycopene and lutein. *FEBS Lett* 1998; 427: 305–308.

17. Amir H, Karas M, Giat J, Danilenko M, Levy R, Yermiahu T, Levy J, Sharoni Y. Lycopene and 1,25-dihydroxyvitamin D3 cooperate in the inhibition of cell cycle progression and induction of differentiation in HL-60 leukemic cells. *Nutr Cancer* 1999; 33: 105–112.

18. Karas M, Amir H, Fishman D, Danilenko M, Segal S, Nahum A, Koifmann A, Giat Y, Levy J, Sharoni Y. Lycopene interferes with cell cycle progression and insulin-like growth factor I signaling in mammary cancer cells. *Nutr Cancer* 2000; 36: 101–111.

19. Stahl W, von Laar J, Martin HD, Emmerich T, Sies H. Stimulation of gap junctional communication: comparison of acyclo-retinoic acid and lycopene. *Arch Biochem Biophys* 2000; 373: 271–274.

20. Broekmans WM, Klopping-Ketelaars IA, Schuurman CR et al. Fruits and vegetables increase plasma carotenoids and vitamins and decrease homocysteine in humans. *J Nutr* 2000; 130: 1578–1583.

21. Heber D, Bowerman S. *What Color is Your Diet?* Harper-Collins/Regan Books, New York, 2001.

3 The Gene Regulatory Mechanism of Thioredoxin and Thioredoxin Binding Protein-2/VDUP1 in Cancer Prevention

Hiroshi Masutani, Yumiko Nishinaka, Yong-Chul Kim, and Junji Yodoi
Kyoto University, Kyoto, Japan

CONTENTS

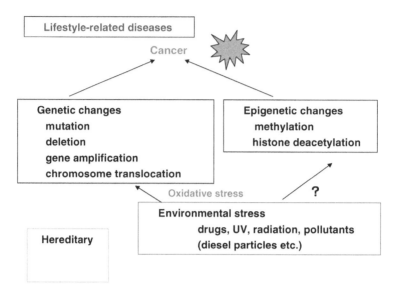

FIGURE 3.1 A simplified model of carcinogenesis caused by environmental stressors.

3.1 OVERVIEW

Cancer is an important lifestyle-related disease. An accumulation of changes in DNA is one cause of cancer. Environmental stressors such as infection, food additives, chemicals, and diesel exhaust particles cause various harmful changes, including DNA damage, leading to carcinogenesis (Figure 3.1). The host has multiple mechanisms to combat the effects of environmental stress, including immune responses, apoptosis, cell cycle control, and the induction of antioxidative enzymes. Thioredoxin is a vital component of the thiol-reducing system. The components of the thioredoxin system play important roles in protecting against oxidative stress. Expression of the thioredoxin gene is induced through the antioxidant responsive element (ARE) by Nrf2. Thioredoxin inducers may be useful for cancer prevention by reducing oxidative and DNA damage. We have identified thioredoxin-binding protein-2/vitamin-D3-upregulated protein 1 (TBP-2/VDUP1). We and others have shown that the expression of TBP-2 is lost in cancer cell lines and cancerous tissues, and ectopic overexpression of TBP-2 results in growth suppression. Thus, TBP-2 seems to play a role as an apparent suppressor oncogene. Use of the TBP-2 system may be another novel approach to cancer suppression.

3.2 THE THIOREDOXIN SYSTEM

The cellular reducing environment is provided by two mutually interconnected systems: the thioredoxin system and the glutathione system. Thioredoxin is a small protein with two redox-active cysteine residues in an active center (Cys-Gly-Pro-Cys-) and operates together with NADPH and thioredoxin reductase as an efficient reducing system for exposed protein disulfides.[1] Thioredoxin scavenges intracellular hydrogen peroxide in collaboration with a family of thioredoxin-dependent peroxidases, the peroxiredoxins.[2] Mammalian thioredoxin 2 is specifically expressed in mitochondria and is essential for cell survival.[3,4] Thus, the thioredoxin system is composed of several related molecules that form a network, maintain the cellular reducing environment, and protect cells from oxidative stress.

We identified and cloned a human thioredoxin[5] and have analyzed its role in the mammalian system. We have shown that the components of the thioredoxin system not only scavenge reactive oxygen species but also play a wide variety of cellular roles through protein–protein interactions.[6] Homozygous mutants carrying a targeted disruption of the thioredoxin gene died shortly after implantation, suggesting that thioredoxin expression is essential for cell survival and early development.[7] Thioredoxin transgenic mice display various phenotypes such as an elongated life span[8] and protection against ischemic injury,[9] acute lung failure,[10] diabetes mellitus,[11] and toxicity caused by environmental stressors.[12] Because oxidative stress has been implicated in these conditions, thioredoxin seems to play an important role in protection against oxidative-stress-associated diseases.

3.3 REGULATORY MECHANISM OF THIOREDOXIN GENE EXPRESSION

Thioredoxin is ubiquitously expressed in various tissues. The thioredoxin gene has several SP-1-binding sites in the gene regulatory region, characteristic of housekeeping genes.[13–15] However, we have often observed a cell-type-specific expression of thioredoxin, such as that in dendritic cells and activated macrophages.[16] In addition, thioredoxin expression is induced by a variety of stressors, including viral infection,[17,18] mitogens, phorbol myristate acetate (PMA), x-ray and ultraviolet irradiation,[19] hydrogen peroxide, and ischemic reperfusion.

Hemin induces thioredoxin expression at a transcriptional level.[20] Hemin is an oxidized form of heme, which is a prosthetic group of various important biomolecules such as hemoglobin. Hemin treatment is known to cause the differentiation of erythroid lineage cells and induce hemeoxygenase-1 gene expression. We demonstrated that hemin induces expression of the thioredoxin gene through the antioxidant responsive element (ARE).[21] Nrf2 was revealed to play a major regulatory role in the heme-mediated gene activation of thioredoxin.[21] We also showed that the expression of thioredoxin is induced by electrophile-targeting compounds such as tert-butylhydroquinone (tBHQ), which are known inducers of the ARE.[22]

Groups of redox-regulating enzymes such as γ-glutamylcysteine synthetase, NAD (P) H: quinone oxidoreductase, and glutathione S-transferase Ya contain the ARE/electrophile responsive element (EpRE). Cap 'n Collar transcription factors including NF-E2p45, Nrf1, and Nrf2 form heterodimers with small Maf proteins, binding to the ARE/EpRE/Maf recognition element (MARE). We propose that the redox regulation by the thioredoxin and glutathione systems, heme metabolism, and drug detoxification have interconnecting relationships.

Activation through the ARE of the thioredoxin gene seems to constitute a positive-feedback loop. Cap 'n Collar transcription factors such as Nrf2 and members of the Jun/Fos families of transcription factors have conserved cysteine residues, which are regulated by the redox status. Therefore, induction of thioredoxin may favor activation through the ARE.[22]

The idea of chemically preventing cancer has been widely accepted. Paul Thalaly showed that a substance from broccoli reduced cancer formation in a rat model.[23] Because oxidative DNA damage is one cause of cancer, the suppression of oxidative stress would seem beneficial for cancer prevention. In addition, the removal of harmful toxic substances through the facilitation of drug detoxification seems favorable for reducing oxidative stress and DNA damage. These results collectively indicate that thioredoxin inducers reduce oxidative and DNA damage, leading to cancer prevention.

3.4　THIOREDOXIN IN THE NERVOUS SYSTEM

Oxidative stress has also been implicated in the pathogenesis of a wide variety of neuronal diseases, including ischemic neuronal injury, Alzheimer's disease, and Parkinson's disease. Thioredoxin expression is upregulated by the nerve growth factor (NGF) through ERK and the CREB–cAMP responsive element (CRE) pathway in PC12 cells. Thioredoxin is necessary for NGF signaling through CRE, leading to c-*fos* expression, and also plays a critical role in NGF-mediated neurite outgrowth in PC12 cells.[24] Therefore, thioredoxin appears to be a neurotrophic cofactor, which augments the effect of NGF on neuronal differentiation and regeneration. Taking the cytoprotective effect of thioredoxin[25] into consideration, thioredoxin inducers seem to be beneficial for neuroregeneration and prevention of neurodegenerative disorders.

3.5　TBP-2

3.5.1　THIOREDOXIN-BINDING PROTEINS

One component of the thioredoxin system consists of thioredoxin-binding proteins. Ichijo et al. demonstrated that thioredoxin interacts with apoptosis signaling kinase 1 (ASK-1),[26,27] a mitogen-activated protein (MAP) kinase kinase kinase. We isolated TBP-2[28]/VDUP-1, which was originally reported as the product of a gene whose expression was upregulated in HL-60 cells stimulated with 1α, 25-dihydroxyvitamin D3.[33] The interaction of thioredoxin with TBP-2/VDUP-1 was observed *in vitro* and *in vivo*. Overexpression of TBP-2 suppressed both the reducing activity and expression of thioredoxin. Therefore, we reported TBP-2 as an apparent negative regulator of thioredoxin. Later, other groups reported the interaction between thioredoxin and this protein.[29,30] A reciprocal pattern of expression between thioredoxin and TBP-2 was also observed in T24 bladder carcinoma cells[31] and vascular smooth muscle cells.[32]

3.5.2 LOSS OF TBP-2 EXPRESSION IN HTLV-I-POSITIVE T-CELL LINES

Expression of TBP-2 mRNA was detected in HTLV-I-negative, but not in HTLV-I-positive T-cell lines. After infection with HTLV-I, T-cells are easily immortalized in the presence of IL-2 (IL-2-dependent cells), but some cells acquire the ability to proliferate in the absence of IL-2 (IL-2-independent cells). TBP-2 mRNA was not detected in IL-2-independent clones, but was detected in the IL-2-dependent cell lines. The loss of TBP-2 expression in the HTLV-I-positive T cells appears to occur during the transition from the IL-2-dependent growth phenotype to the IL-2-independent phenotype. Because a change in growth phenotype has been implicated in the progression of adult T-cell leukemia (ATL) development in an *in vitro* model,[33,34] the loss of TBP-2 expression may be a critical event associated with the leukemogenesis of ATL[35] (Figure 3.2). TBP-2 mRNA expression is downregulated in several tumors,[31,36] suggesting a close association between reduction and tumorigenesis. TBP-2 expression is also downregulated in melanoma metastasis.[37]

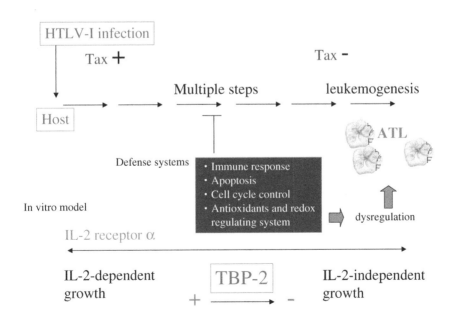

FIGURE 3.2 Model of the role of TBP-2 loss in ATL leukemogenesis.

3.5.3 GROWTH SUPPRESSIVE EFFECT OF **TBP-2**

TBP-2-transfected clones showed a marked reduction of cell proliferation compared to the control clones. The introduction of TBP-2 protein by TAT-mediated protein transfer resulted in retardation of cell growth.[35] Overexpression of TBP-2 was shown to lead to growth suppressive activity.[38,39] These results showed that TBP-2 is a growth suppressor. The mechanism by which TBP-2 suppresses growth is currently unclear, but several possibilities can be considered.

In the TBP-2-transfected clones, we observed a significant accumulation of cells in G1, accompanied by an increase in the levels of p16 and hypophosphorylated Rb.[35] The mechanism by which TBP-2 induces expression of p16 in these cells should be analyzed further. Thioredoxin has been demonstrated to have a variety of biological functions including cell-growth-promoting effects,[40–42] suggesting that inhibition of thioredoxin-reducing activity by TBP-2 contributes to the growth inhibitory effect of TBP-2. TBP-2 may interact with some other molecules related to the control of cell growth. We have found candidates for molecules interacting with TBP-2[39] (Masutani et al., unpublished data). Recently, it was reported that TBP-2 interacts with Fanconi anemia zinc finger and may be a component of a transcriptional repressor complex.[38]

3.5.4 ABNORMALITY OF **TBP-2** IN DISEASES

We have shown that TBP-2 has a suppressor-oncogene-like character. Our hypothetical model is outlined in Figure 3.3. TBP-2 expression is induced by various stimuli, including vitamin D3, high concentration of glucose, and treatment with HDAC inhibitors[31,43] (Masutani et al., in preparation). One strategy for cancer suppression may be the augmentation of TBP-2 expression, either through a reversal of the silencing of the TBP-2 gene or activation of the TBP-2 gene by inducers.

Interestingly, a report has suggested that loss of TBP-2/VDUP-1 contributes to hyperlipidemia in an animal model of the disease[44]: in mutant mice, production of CO_2 was decreased but the synthesis of ketone bodies was increased, suggesting that the altered redox status downregulates the citric acid cycle. To investigate the molecular basis underlining the loss of TBP-2 expression in cancer and hyperlipidemia, we are currently analyzing the molecular mechanism of the function of the TBP-2 molecule and in knockout mice.

3.6 SUMMARY AND FUTURE PERSPECTIVES

Oxidative stress is deeply involved in lifestyle-related diseases such as cancer. Enhancing the function of the thioredoxin system with thioredoxin inducers seems a suitable approach to cancer prevention. We are now testing micronutrients and extracts from various plants for thioredoxin-inducing activity, using a screening system based on our knowledge of the regulatory mechanism for the thioredoxin

Hypothetical model

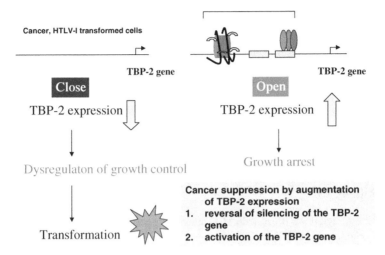

FIGURE 3.3 If the TBP-2 gene locus is open, the gene product is expressed, resulting in an induction of growth arrest. In HTLV-I-transformed cells or cancerous tissues, the TBP-2 gene is silenced, causing a loss of TBP-2 expression. This loss results in dysregulation of growth suppression and transformation.

gene. The study of TBP-2, however, has raised many questions. What is the mechanism for silencing the TBP-2 gene? How does the gene silencing occur in cancer tissues? What is the mechanism of growth suppression caused by TBP-2? How do thioredoxin and TBP-2 interact and regulate redox conditions in a physiological context? How is the response to vitamin D3 connected to the redox regulation? The answers to these questions will improve our understanding of redox regulation and stress signaling as well as the development of measures against cancer.

ACKNOWLEDGMENTS

We thank the members of our laboratory for helpful discussions, Y. Kanekiyo for secretarial help, and Y. Yamaguchi for technical assistance. This work was supported by a grant-in-aid for scientific research from the Ministry of Education, Science, and Culture of Japan and a grant-in-aid from the Research and Development Program for New Bio-industry Initiatives. Yumiko Nishinaka was supported by the New Energy and Industrial Technology Development Organization (NEDO) Fellowship Program. Yong-Chul Kim is a recipient of the Research Fellowship of the Japan Society for the Promotion of Science (JSPS).

ABBREVIATIONS

HTLV-I: Human T-cell leukemia virus type I
ATL: adult T-cell leukemia
TBP-2: thioredoxin binding protein-2
VDUP1: vitamin D3 up-regulated Protein-1
ARE: antioxidant responsive element
Nrf2: NF-E2 related factor 2
HDAC: histone deacetylase
NGF: nerve growth factor
CRE: cAMP responsive element
ASK-1: apoptosis signaling kinase 1

REFERENCES

1. Holmgren A, Björnstedt M. Thioredoxin and thioredoxin reductase. *Methods Enzymol* 1995; 252: 199–208.
2. Fujii J, Ikeda Y. Advances in our understanding of peroxiredoxin, a multifunctional, mammalian redox protein. *Redox Rep* 2002; 7: 123–130.
3. Spyrou G, Enmark E, Miranda VA, Gustafsson J. Cloning and expression of a novel mammalian thioredoxin. *J Biol Chem* 1997; 272: 2936–2941.
4. Tanaka T, Hosoi F, Yamaguchi-Iwai Y, Nakamura H, Masutani H, Ueda S, Nishiyama A, Takeda S, Wada H, Spyrou G, Yodoi J. Thioredoxin-2 (TRX-2) is an essential gene regulating mitochondria-dependent apoptosis. *EMBO J* 2002; 21: 1695–1703.
5. Tagaya Y, Mitsui A, Kondo N, Matsui H, Hamuro J, Brown N, Arai K, Yokota T, Wakasugi H, Yodoi J. ATL-derived factor (ADF), an IL-2 receptor/Tac inducer homologous to thioredoxin; possible involvement of dithiol-reduction in the IL-2 receptor induction. *EMBO J* 1989; 8: 757–764.
6. Masutani H, Yodoi J. Overview: thioredoxin. In Packer L, Ed. *Methods in Enzymology* Vol. 347. San Diego: Academic Press, 2002: 279–286.
7. Matsui M, Oshima M, Oshima H, Takaku K, Maruyama T, Yodoi J, Taketo MM. Early embryonic lethality caused by targeted disruption of the mouse thioredoxin gene. *Dev Biol* 1996; 178: 179–185.
8. Mitsui A, Hamuro J, Nakamura H, Kondo N, Hirabayashi Y, Ishizaki-Koizumi S, Hirakawa T, Inoue T, Yodoi J. Overexpression of human thioredoxin in transgenic mice controls oxidative stress and life span. *Antioxid Redox Signal* 2002; 4: 693–696.
9. Takagi Y, Mitsui A, Nishiyama A, Nozaki K, Sono H, Gon Y, Hashimoto N, Yodoi J. Overexpression of thioredoxin in transgenic mice attenuates focal ischemic brain damage. *Proc Natl Acad Sci USA* 1999; 96: 4131–4136.
10. Hoshino T, Nakamura H, Okamoto M, Kato S, Araya S, Nomiyama K, Oizumi K, Young HA, Aizawa H, Yodoi J. Redox-active protein thioredoxin prevents proinflammatory cytokine- or bleomycin-induced lung injury. *Am J Respir Crit Care Med* 2003; 168: 1075–1083.
11. Hotta M, Tashiro F, Ikegami H, Niwa H, Ogihara T, Yodoi J, Miyazaki J. Pancreatic beta cell-specific expression of thioredoxin, an antioxidative and antiapoptotic protein, prevents autoimmune and streptozotocin-induced diabetes. *J Exp Med* 1998; 188: 1445–1451.

12. Yoon BI, Hirabayashi Y, Kaneko T, Kodama Y, Kanno J, Yodoi J, Kim DY, Inoue T. Transgene expression of thioredoxin (TRX/ADF) protects against 2,3,7,8-tetrachlorodibenzo-p-dioxin (TCDD)-induced hematotoxicity. *Arch Environ Contam Toxicol* 2001; 41: 232–236.

13. Taniguchi Y, Taniguchi UY, Mori K, Yodoi J. A novel promoter sequence is involved in the oxidative stress-induced expression of the adult T-cell leukemia-derived factor (ADF)/human thioredoxin (Trx) gene. *Nucleic Acids Res* 1996; 24: 2746–2752.

14. Tonissen KF, Wells JR. Isolation and characterization of human thioredoxin-encoding genes. *Gene* 1991; 102: 221–228.

15. Wollman EE, d'Auriol L, Rimsky L, Shaw A, Jacquot JP, Wingfield P, Graber P, Dessarps F, Robin P, Galibert F, Bertoglio J, Fradelizi D. Cloning and expression of a cDNA for human thioredoxin. *J Biol Chem* 1988; 263: 15506-15512.

16. Masutani H, Naito M, Takahashi K, Hattori T, Koito A, Maeda Y, Takatsuki K, Go T, Nakamura H, Fujii S, Yoshida Y, Okuma M, Yodoi J. Dysregulation of adult T-cell leukemia-derived factor (ADF)/human thioredoxin in HIV infection: Loss of ADF high producer cells in lymphoid tissues of AIDS patients. *AIDS Res Hum Retroviruses* 1992; 8: 1707–1715.

17. Makino S, Masutani H, Maekawa N, Konishi I, Fujii S, Yamamoto R, Yodoi J. Adult T-cell leukaemia-derived factor/thioredoxin expression on the HTLV-I transformed T-cell lines: heterogeneous expression in ATL-2 cells. *Immunology* 1992; 76: 578–583.

18. Fujii S, Nanbu Y, Nonogaki H, Konishi I, Mori T, Masutani H, Yodoi J. Coexpression of adult T-cell leukemia-derived factor, a human thioredoxin homologue, and human papillomavirus DNA in neoplastic cervical squamous epithelium. *Cancer* 1991; 68: 1583–1591.

19. Wakita H, Yodoi J, Masutani H, Toda K, Takigawa M. Immunohistochemical distribution of adult T-cell leukemia-derived factor/thioredoxin in epithelial components of normal and pathologic human skin conditions. *J Invest Dermatol* 1992; 99: 101–107.

20. Leppa S, Pirkkala L, Chow SC, Eriksson JE, Sistonen L. Thioredoxin is transcriptionally induced upon activation of heat shock factor 2. *J Biol Chem* 1997; 272: 30400–30404.

21. Kim YC, Masutani H, Yamaguchi Y, Itoh K, Yamamoto M, Yodoi J. Hemin-induced activation of the thioredoxin gene by Nrf2: A differential regulation of the antioxidant responsive element (ARE) by switch of its binding factors. *J Biol Chem* 2001; 276: 18399–18406.

22. Kim Y-C, Yamaguchi Y, Kondo N, Masutani H, Yodoi J. Thioredoxin-dependent redox regulation of the antioxidant responsive element (ARE) in electrophile response. *Oncogene* 2003; 22: 1860–1865.

23. Fahey JW, Zhang Y, Talalay P. Broccoli sprouts: an exceptionally rich source of inducers of enzymes that protect against chemical carcinogens. *Proc Natl Acad Sci USA* 1997; 94: 10367–10372.

24. Bai J, Nakamura H, Kwon YW, Hattori I, Yamaguchi Y, Kim YC, Kondo N, Oka S, Ueda S, Masutani H, Yodoi J. Critical roles of thioredoxin in nerve growth factor-mediated signal transduction and neurite outgrowth in PC12 cells. *J Neurosci* 2003; 23: 503–509.

25. Masutani H, Bai J, Kim Y-C, Yodoi J. Thioredoxin as a neurotrophic cofactor and an important regulator of neuroprotection. *Mol Neurobiol* 2004; 29: 229–242.

26. Ichijo H, Nishida E, Irie K, ten Dijke P, Saitoh M, Moriguchi T, Takagi M, Matsumoto K, Miyazono K, Gotoh Y. Induction of apoptosis by ASK1, a mammalian MAPKKK that activates SAPK/JNK and p38 signaling pathways. *Science* 1997; 275: 90–94.

27. Saito M, Nishitoh H, Fujii M, Takeda K, Tobiume K, Sawada Y, Kawabata M, Miyazono K, Ichijyo H. Mammalian thioredoxin is a direct inhibitor of apoptosis signal-regulating kinase (ASK) 1. *EMBO J* 1998; 17: 2596–2606.

28. Nishiyama A, Matsui M, Iwata S, Hirota K, Masutani H, Nakamura H, Takagi Y, Sono H, Gon Y, Yodoi J. Identification of thioredoxin-binding protein-2/vitamin D(3) up-regulated protein 1 as a negative regulator of thioredoxin function and expression. *J Biol Chem* 1999; 274: 21645-21650.

29. Junn E, Han SH, Im JY, Yang Y, Cho EW, Um HD, Kim DK, Lee KW, Han PL, Rhee SG, Choi I. Vitamin D3 up-regulated protein 1 mediates oxidative stress via suppressing the thioredoxin function. *J Immunol* 2000; 164: 6287–6295.

30. Yamanaka H, Maehira F, Oshiro M, Asato T, Yanagawa Y, Takei H, Nakashima Y. A possible interaction of thioredoxin with VDUP1 in HeLa cells detected in a yeast two-hybrid system. *Biochem Biophys Res Commun* 2000; 271: 796–800.

31. Butler LM, Zhou X, Xu WS, Scher HI, Rifkind RA, Marks PA, Richon VM. The histone deacetylase inhibitor SAHA arrests cancer cell growth, up-regulates thioredoxin-binding protein-2, and down-regulates thioredoxin. *Proc Natl Acad Sci USA* 2002; 99: 11700–11705.

32. Schulze PC, De Keulenaer GW, Yoshioka J, Kassik KA, Lee RT. Vitamin D3-upregulated protein-1 (VDUP-1) regulates redox-dependent vascular smooth muscle cell proliferation through interaction with thioredoxin. *Circ Res* 2002; 91: 689–695.

33. Maeda M, Arima N, Daitoku Y, Kashihara M, Okamoto H, Uchiyama T, Shirono K, Matsuoka M, Hattori T, Takatsuki K, et al. Evidence for the interleukin-2 dependent expansion of leukemic cells in adult T cell leukemia. *Blood* 1987; 70: 1407–1411.

34. Uchiyama T, Yodoi J. *Adult T cell leukemia and related diseases:* Austin: R. G. Landes company, 1995: 1–139.

35. Nishinaka Y, Nishiyama A, Masutani H, Oka S, Ahsan MK, Nakayama Y, Ishii Y, Nakamura H, Maeda M, Yodoi J. Loss of thioredoxin binding protein-2/vitamin D3 up-regulated protein-1 (TBP-2/VDUP-1) in HTLV-I-dependent T-cell transformation: implications for ATL leukemogenesis. *Cancer Res* 2004; 64: 1287–1292.

36. Ikarashi M, Takahashi Y, Ishii Y, Nagata T, Asai S, Ishikawa K. Vitamin D3 up-regulated protein 1 (VDUP1) expression in gastrointestinal cancer and its relation to stage of disease. *Anticancer Res* 2002; 22: 4045–4048.

37. Goldberg SF, Miele ME, Hatta N, Takata M, Paquette-Straub C, Freedman LP, Welch DR. Melanoma metastasis suppression by chromosome 6: evidence for a pathway regulated by CRSP3 and TXNIP. *Cancer Res* 2003; 63: 432–440.

38. Han SH, Jeon JH, Ju HR, Jung U, Kim KY, Yoo HS, Lee YH, Song KS, Hwang HM, Na YS, Yang Y, Lee KN, Choi I. VDUP1 upregulated by TGF-beta1 and 1,25-dihydorxyvitamin D3 inhibits tumor cell growth by blocking cell-cycle progression. *Oncogene* 2003; 22: 4035–4046.

39. Nishinaka Y, Masutani H, Oka S, Matsuo Y, Yamaguchi Y, Nishio K, Ishii Y, Yodoi Y. Importin 1. (Rch 1) mediates nuclear translocation of thioredoxin-binding protein-2/vitamin D₃-up-regulated protein 1. *J Biol Chem* 2004; 279: 37559–37565.

40. Wakasugi N, Tagaya Y, Wakasugi H, Mitsui A, Maeda M, Yodoi J, Tursz T. Adult T-cell leukemia-derived factor/thioredoxin, produced by both human T-lympho-tropic virus type I- and Epstein-Barr virus-transformed lymphocytes, acts as an autocrine growth factor and synergizes with interleukin 1 and interleukin 2. *Proc Natl Acad Sci USA* 1990; 87: 8282–8286.

41. Tagaya Y, Wakasugi H, Masutani H, Nakamura H, Iwata S, Mitsui A, Fujii S, Wakasugi N, Tursz T, Yodoi J. Role of ATL-derived factor (ADF) in the normal and abnormal cellular activation: involvement of dithiol related reduction. *Mol Immunol* 1990; 27: 1279–1289.

42. Gallegos A, Gasdaska JR, Taylor CW, Paine-Murrieta GD, Goodman D, Gasdaska PY, Berggren M, Briehl MM, Powis G. Transfection with human thioredoxin increases cell proliferation and a dominant-negative mutant thioredoxin reverses the transformed phenotype of human breast cancer cells. *Cancer Res* 1996; 56: 5765–5770.

43. Takahashi Y, Nagata T, Ishii Y, Ikarashi M, Ishikawa K, Asai S. Up-regulation of vitamin D3 up-regulated protein 1 gene in response to 5-fluorouracil in colon carcinoma SW620. *Oncol Rep* 2002; 9: 75–79.

44. Bodnar JS, Chatterjee A, Castellani LW, Ross DA, Ohmen J, Cavalcoli J, Wu C, Dains KM, Catanese J, Chu M, Sheth SS, Charugundla K, Demant P, West DB, de Jong P, Lusis AJ. Positional cloning of the combined hyperlipidemia gene Hyplip1. *Nat Genet* 2002; 30: 110–116.

45. Oka S, Liu W, Masutani H, Hirata H, Shinkai Y, Yamada S, Yoshida T, Nakamura H, Yodoi J. Dysregulation of lipid catabolism in thioredoxin binding protein-2 (TBP-2)-deficient mice: a unique animal model of Reye Syndrome. *FASEB J* in press.

4 Coenzyme Q10, Free Radicals, and Heart Disease

Yorihiro Yamamoto
Tokyo University of Technology, Tokyo, Japan

CONTENTS

4.1 INTRODUCTION

Coenzyme Q10, an essential cofactor of mitochondrial ATP production, was discovered in 1957. Coenzyme Q10 is ubiquitously present in other biomembranes and even in lipoproteins. The reduced form of coenzyme Q10 (ubiquinol-10) is believed to work as an antioxidant in these locations. When human plasma was incubated under aerobic conditions, ubiquinol-10 depleted after the depletion of ascorbate, but α-tocopherol remained unchanged, indicating that ascorbate and ubiquinol-10 are frontline antioxidants against oxygen radicals. The redox status of plasma coenzyme Q10 is a good indicator of oxidative stress, oxidative stress being defined as a disturbance in the pro-oxidant–antioxidant balance in favor of the former. We, therefore, developed a simple and reliable method for the simultaneous detection of both oxidized and reduced forms of coenzyme Q10 and measured their plasma levels in nine patients who were treated with percutaneous transluminal coronary angioplasty (PTCA) after a heart attack. Plasma samples were collected on hospitalization and at various times (0, 4, 8, 12, 16, and 20 h; and 1, 2, 3, 4, and 7 d) after the PTCA. The percentage of the oxidized form of coenzyme Q10 in the total coenzyme Q10 (%CoQ10) before and immediately after PTCA were 9.9 ± 2.8 and 11.4 ± 2.0, respectively, reached a maximum (20 to 45) 1 or 2 d later, and decreased to 7.9 ± 2.7 at 7 d after PTCA, indicating an increase in oxidative stress in patients during coronary reperfusion.

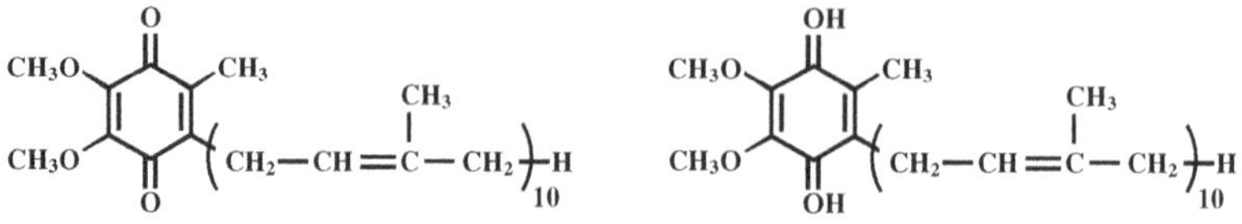

Ubiquinone-10 (CoQ-10)   Ubiquinol-10 (CoQH$_2$-10)

FIGURE 4.1 Chemical structure of coenzyme Q10.

4.2 COENZYME Q10

Fred Crane discovered a lipid-soluble, orange compound in bovine heart mito-chondria in 1957[1] and named the compound *coenzyme Q* because it is an essential cofactor of ATP production and possesses a quinone structure. At the same time, Morton also identified quinones, which are ubiquitously present in animals, plants, and even bacteria, and named them *ubiquinones*.[2] Soon, these two compounds were found to be identical, and Folkers et al. characterized the chemical structure.[3]

The length of isoprenoid side chains vary with the species. Humans have a coenzyme Q10 with 10 isoprenoid units. The oxidized and reduced forms of coenzyme Q10 are called *ubiquinone-10* and *ubiquinol-10*, respectively (Figure 4.1).

Humans biosynthesize coenzyme Q10, but its tissue concentration decreases with age.[4] In the human heart, for example, 30% and 50% reductions of coenzyme Q10 content were observed at ages of 40s and 80s, respectively, when compared with the value at 20s. Ubiquinol-10 is an important antioxidant for humans as described in the text that follows. For these reasons, taking supplements of coenzyme Q10 is popular in the U.S., European countries, and, lately, in Japan. About 100 t of coenzyme Q10 were produced exclusively by four Japanese companies in 2003.

4.3 ANTIOXIDANT ACTIVITY OF UBIQUINOL-10

In 1966, Mellors and Tappel showed that ubiquinol-6 was almost as effective as α-tocopherol in reducing 2,2-diphenyl-1-picrylhydrazyl, in inhibiting heme-catalyzed peroxidation of arachidonate emulsion, and in light-induced oxidation of mitochondria, indicating that ubiquinol is a good antioxidant.[5] In 1980, Takayanagi et al. showed that the inhibition of ADP-Fe(III)-induced oxidation of mitochondria was dependent on the concentration of ubiquinol controlled by exogenous reductants such as NAD(P)H and inhibitors of electron transport such as rotenone and antimycin A.[6] Furthermore, the administration of ubiquinone-10 protected rats from hepatic cellular damage caused by hepatic ischemia and reperfusion,[7] probably due to the inhibition of lipid peroxidation by ubiquinol-10 formed from exogenous ubiquinone-10 *in vivo*.[8] However, these pioneering works were generally overlooked until 1990, when the antioxidant activity of ubiquinol was reported independently by four groups.[9–12]

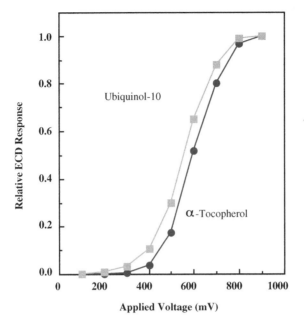

FIGURE 4.2 Hydrodynamic voltammograms of ubiquinol-10 and α-tocopherol measured by HPLC using 50 m*M* sodium perchlorate in methanol/tert-butyl alcohol (9/1, v/v) as the eluant.

Figure 4.2 shows that the oxidation potential of ubiquinol-10 is smaller than that of α-tocopherol, the best-known free radical scavenger, indicating that ubiquinol-10 should be a good antioxidant. We, therefore, compared the antioxidant activity of ubiquinol-10 with that of α-tocopherol under the condition of a constant rate of peroxyl radical formation using a free radical initiator. This is essential for a quantitative comparison. Ubiquinol-10 suppressed the oxidation of methyl linoleate in hexane, although its reactivity toward peroxyl radical was about 10 times less than that of α-tocopherol (Figure 4.3).[9] On the other hand, ubiquinol-10 inhibited the oxidation of phosphatidylcholine liposomal membranes as efficiently as α-tocopherol (Figure 4.4).[9] This may be ascribed to the fact that the ubisemiquinone radical, an electron oxidation product of ubiquinol, reacts with oxygen to give hydroperoxyl radicals, which continue the oxidation of methyl linoleate in hexane (Figure 4.3). In contrast, hydroperoxyl radicals migrate to the aqueous phase in liposomal oxidation and do not participate in lipid-phase oxidation (Figure 4.4). This is consistent with the stoichiometric number (about 1) of peroxyl radical trapped by ubiquinol-10.[9] When both ubiquinol-10 and α-tocopherol were present, ubiquinol-10 decreased first and α-tocopherol decreased only after all of ubiquinol-10 was consumed in both solution and liposomes (Figure 4.3 and Figure 4.4).[9] These results and ESR studies suggest that ubiquinol-10 regenerates α-tocopherol by reducing α-tocopheroxyl radical.[9,13] No significant antioxidant activity of ubiquinone-10 was observed (Figure 4.3, Figure 4.4, and Figure 4.5).[9]

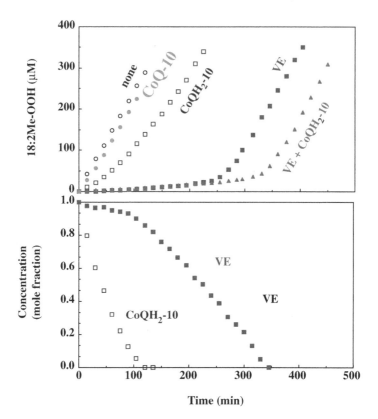

FIGURE 4.3 Formation of methyl linoleate hydroperoxide (18:2Me-OOH) during the aerobic oxidation of 453 mM methyl linoleate initiated with 0.2 mM 2,2′azobis(2,4-dimethylvaleronitrile) in the absence of antioxidant or the presence of 3 μM ubiquinol-10 (CoQH2-10) and/or 3 μM α-tocopherol (VE) in hexane at 37°C. The effect of addition of 3 μM ubiquinone-10 (CoQ-10) was also measured. The lower panel shows the decrease of CoQH2-10 and VE when both antioxidants were present in the system.

The aforementioned results suggest that ubiquinol-10 could be a frontline antioxidant against oxygen radicals *in vivo*. In fact, ubiquinol-10 decreased first when human plasma was oxidized with a lipid-soluble radical initiator, whereas ascorbate decreased first when oxidized with water-soluble radical initiator (Figure 4.6).[14] When human plasma was incubated at 37°C under aerobic conditions, ubiquinol-10 depleted after the depletion of ascorbate; however, no significant decay in α-tocopherol was seen (Figure 4.7).[15] A significant increase in cholesteryl ester hydroperoxides was observed after the depletion of ascorbate and ubiquinol-10 (Figure 4.7), indicating that α-tocopherol without ubiquinol-10 and ascorbate did not suppress the formation of lipid hydroperoxide. Bowry and Stocker showed that α-tocopheroxyl radicals act as a chain carrier in this type of lipid oxidation, and this oxidation process was called tocopherol-mediated

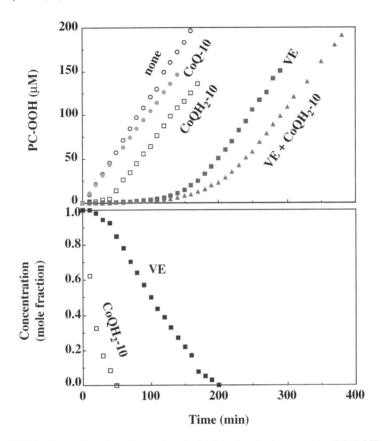

FIGURE 4.4 Formation of soybean phosphatidylcholine hydroperoxide (PC-OOH) during the aerobic oxidation of 5.2 m*M* soybean phosphatidylcholine multilamellar liposome dispersed in 0.1 *M* NaCl containing 100 μ*M* EDTA initiated with 1.0 m*M* 2,2'azobis(2,4-dimethylvaleronitrile) in the absence of antioxidant or the presence of 3 μ*M* ubiquinol-10 (CoQH2-10) and/or 3 μ*M* α-tocopherol (VE) in hexane at 37°C. The effect of addition of 3 μ*M* ubiquinone-10 (CoQ-10) was also measured. The lower panel shows the decrease of CoQH2-10 and VE when both antioxidants were present in the system.

peroxidation (TMP).[16] Ascorbate and ubiquinol-10 are the most important biological anti-TMP reagents.[15,17]

4.4 OXIDATIVE STRESS IN HEART DISEASE

The redox status of plasma coenzyme Q10 is a good indicator of oxidative stress because oxidative stress is a disturbance in the pro-oxidant–antioxidant balance in favor of the former, and ubiquinol-10 is a frontline antioxidant against oxidative stress, as described earlier. We, therefore, developed a simple and reliable method for the simultaneous detection of plasma ubiquinol-10 and ubiquinone-10.[18]

Roles of ubiquinol-10 in the oxidation of biological lipids (LH)

1. Scavenge peroxyl radicals (LOO•)
2. Reduce α-tocopheroxyl radical (VE•)
3. Transfer radicals in lipid phase into aqueous phase

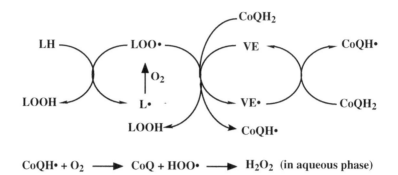

$$CoQH• + O_2 \longrightarrow CoQ + HOO• \longrightarrow H_2O_2 \text{ (in aqueous phase)}$$

FIGURE 4.5 Roles of ubiquinol-10 (CoQH2-10) in the oxidation of biological lipids (LH) in preventing the formation of lipid hydroperoxides (LOOH) and their reaction pathways.

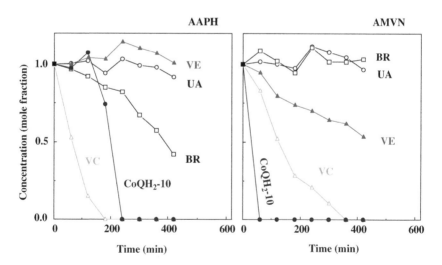

FIGURE 4.6 Changes in plasma antioxidant levels (mole fraction) during the aerobic oxidation of human plasma, initiated with a water-soluble initiator (10 mM 2,2′azobis(2-amidinopropane) dihydrochloride, AAPH) and a lipid-soluble initiator (10 mM 2,2′azobis(2,4-dimethylvaleronitrile), AMVN) at 37°C.

Figure 4.8 shows a HPLC system for the detection of plasma ubiquinol-10 and ubiquinone-10. By using this method we demonstrated an increase in oxidative stress in patients with hepatitis, cirrhosis, and hepatoma,[19] in LEC rats (animal

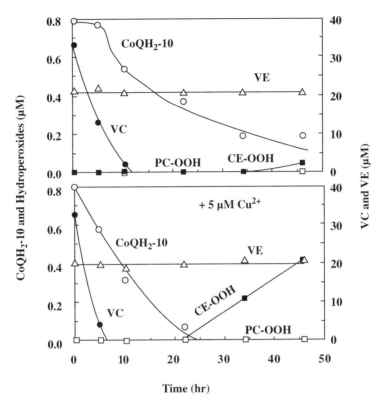

FIGURE 4.7 Changes in plasma antioxidant levels during the aerobic oxidation of human plasma in the absence (upper panel) or presence (lower panel) of 5 μM cupric chloride at 37°C.

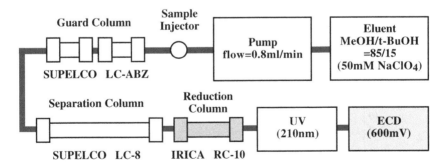

FIGURE 4.8 HPLC system for the detection of plasma ubiquinol-10 and ubiquinone-10.

model of liver cancer,[20] in newborn babies,[21] and in patients with Parkinson's disease.[22]

We also applied this method to patients with heart attacks. Nine patients (seven men and two women) with acute myocardial infarction were hospitalized at Health Insurance Naruto Hospital, Tokushima, and treated with percutaneous transluminal coronary angioplasty (PTCA). The average age (±S.D.) was 59.4 ± 15.1 yr. Blood was drawn from these patients and volunteers who acted as controls; sodium heparin was used as the anticoagulant. Plasma was separated by centrifugation at 1500 g for 10 min and was stored at 80°C until analysis.

Stored plasma (50 µl) was mixed vigorously with 250 µl of cold methanol and 500 µl of cold hexane in a 1.5-ml polypropylene tube. After centrifugation at 10,000 g for 3 min at 4°C, 5 µl of hexane layer (corresponding to 0.5 µl of plasma) was injected immediately and directly onto HPLC equipped with two guard columns (Type Supelguard LC-ABZ, 5 µm, 20 × 4.6 mm id, Supelco, Tokyo), an analytical column (Type Supelcosil LC-8, 5 µm, 250 × 4.6 mm id, Supelco), a reduction column (Type RC-10-1, Irica, Kyoto), and an amperometric electrochemical detector (ECD, Model Σ985, Irica). The oxidation potential for ECD was 600 mV on a glassy carbon electrode. The mobile phase consisted of 50 mM sodium perchlorate in methanol/tert-butyl alcohol (85/15, v/v) delivered at a flow rate of 0.8 ml/min.

Figure 4.9 shows the changes in percentage of the oxidized form of coenzyme Q10 in total coenzyme Q10 (%CoQ10) in patients treated with PTCA. Values of %CoQ10 before and immediately after PTCA were 9.9 ± 2.8 and 11.4 ± 2.0, respectively, reached a maximum (20 to 45%) 1 or 2 d later and decreased to 7.9 ± 2.7 at 7 d after PTCA. Elevation of urinary 8-epi-prostaglandin F_2, a marker

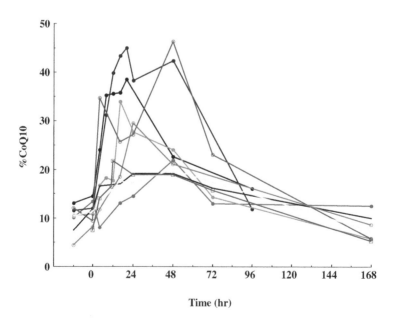

FIGURE 4.9 Changes in plasma ubiquinone-10 percentage (%CoQ10) in total amounts of ubiquinone-10 and ubiquinol-10 before and after the treatment of patients with percutaneous transluminal coronary angioplasty (PTCA) after a heart attack.

of nonenzymatic oxidation of lipids, in the first 6 h after PTCA for acute myocardial infarction was reported.[23] However, such dynamic changes in %CoQ10 for one week has not been reported by other researchers. These results strongly indicate an increase in oxidative stress in PTCA patients during coronary reperfusion. Antioxidant therapy should be beneficial to these patients.

ACKNOWLEDGMENTS

I thank the blood donors, Katusya Tamura (Health Insurance Naruto Hospital) for the help in collecting blood, and Satoshi Yamashita for measuring the %CoQ10.

REFERENCES

1. Crane FL, Hatefi Y, Lester RL, and Widmer C. Isolation of a quinone from beef heart mitochondria. *Biochim Biophys Acta* 1957; 25: 220–221.
2. Morton RA. Ubiquinone. *Nature* 1958; 182: 1764–1767.
3. Wolf DE, Hoffman CH, Trenner HR, Arison BH, Shunk CH, Linn BO, McPherson JF, and Folkers K. Coenzyme Q. I. Structure studies on the coenzyme Q group. *J Am Chem Soc* 1958; 80: 4753.
4. Kalen A, Appelkvist EL, and Dallner G. Age-related changes in the lipid compositions of rat and human tissues. *Lipids* 1989; 24: 579–584.
5. Mellors A and Tappel A. The inhibition of mitochondrial peroxidation by ubiquinone and ubiquinol. *J Biol Chem* 1966; 241: 4353–4356.
6. Takayanagi R, Takeshige K, and Minakami S. Lipid peroxidation and the reduction of ADP-Fe3+ chelate by NADH-ubiquinone reductase from bovine heart mitochondria. *Biochem J* 1980; 192: 853–860.
7. Marubayashi S, Dohi K, Ezaki H, Hayashi K, and Kawasaki T. Preservation of ischemic rat liver mitochondrial functions and liver viability with CoQ10. *Surgery* 1982; 91: 631–637.
8. Marubayashi S, Dohi KK, Yamada K, and Kawasaki T. Changes in the levels of endogenous coenzyme Q homologs, α-tocopherol, and glutathione in rat liver after hepatic ischemia and reperfusion, and the effect of pretreatment with coenzyme Q10. *Biochim Biophys Acta* 1984; 797: 1–9.
9. Yamamoto Y, Komuro E, and Niki E. Antioxidant activity of ubiquinol in solution and phosphatidylcholine liposome. *J Nutr Sci Vitaminol* 1990; 36: 505–511.
10. Kagan VE, Serbinova EA, Koynova GM, Kitanova SA, Tyurin VA, Stoytchev TS, Quinn PJ, and Packer L. Antioxidant action of ubiquinol homologues with different isoprenoid chain length in biomembranes. *Free Radic Biol Med* 1990; 8: 117–126.
11. Frei B, Kim MC, and Ames BN. Ubiquinol-10 is an effective lipid-soluble antioxidant at physiological concentrations. *Proc Natl Acad Sci USA* 1990; 87: 4879–4883.
12. Stocker R, Bowry VW, and Frei B. Ubiquinol-10 protects human low density lipoprotein more efficiently against lipid peroxidation than does α-tocopherol. *Proc Natl Acad Sci USA* 1991; 88: 1646–1650.
13. Mukai K, Kikuchi S, and Urano S. Stopped-flow kinetic study of the regeneration reaction of tocopheroxyl radical by reduced ubiquinone-10 in solution. *Biochim Biophys Acta* 1990; 1035: 77–82.

14. Yamamoto Y and Nagata Y. Oxidation of human plasma initiated with water- and lipid-soluble radical initiators. *J Oleo Sci* 1996; 45: 445–452.

15. Yamamoto Y, Kawamura M, Tatsuno K, Yamashita S, Niki E, and Naito C. Formation of lipid hydroperoxides in the cupric ion-induced oxidation of plasma and low density lipoprotein. In: *Oxidative Damage and Repair.* Davies KJA. Ed. Pergamon Press. New York. 1991: 287–291.

16. Bowry VW and Stocker R. Tocopherol-mediated peroxidation. The prooxidant effect of vitamin E on the radical-initiated oxidation of human low-density lipo-protein. *J Am Chem Soc* 1993; 115: 6029–6044.

17. Witting PK, Westerlund C, and Stocker R. A rapid and simple screening test for potential inhibitors of tocopherol-mediated peroxidation of LDL lipids. *J Lipid Res* 1996; 37: 853–867.

18. Yamashita S and Yamamoto Y. Simultaneous detection of ubiquinol and ubiquinone in human plasma as a marker of oxidative stress. *Anal Biochem* 1997; 250: 66–73.

19. Yamamoto Y, Yamashita S, Fujisawa A, Kokura S, and Yoshikawa T. Oxidative stress in patients with hepatitis, cirrhosis, and hepatoma evaluated by plasma antioxidants. *Biochem Biophys Res Commun* 1998; 247: 166–170.

20. Yamamoto Y, Sone H, Yamashita S, Nagata Y, Niikawa H, Hara K, and Nagao M. Oxidative stress in LEC rats evaluated by plasma antioxidants and free fatty acids. *J Trace Elem Exp Med* 1997; 10: 129–134.

21. Hara K, Yamashita S, Fujisawa A, Ogawa T, and Yamamoto Y. Oxidative stress in newborn infants with and without asphyxia as measured by plasma antioxidants and free fatty acids. *Biochem Biophys Res Commun* 1999; 257: 244–248.

22. Sohmiya M, Tanaka M, Tak NW, Yanagisawa M, Tanino Y, Suzuki Y, Okamoto K, and Yamamoto Y. Redox status of plasma coenzyme Q10 indicates elevated systemic oxidative stress in Parkinson's disease. *J Neurol Sci* In press.

23. Reilly MP, Delanty N, Roy L, Rokach J, Callaghan PO, Crean P, Lawson JA, and FitzGerald GA. Increased formation of the isoprostanes IPF$_2$-I and 8-epi-prostag-landin F$_2$ in acute coronary angioplasty. Evidence for oxidant stress during coro-nary reperfusion in humans. *Circulation* 1997; 96: 3314–3320.

5 Wine, Alcohol, and a Healthy Heart

Dipak K. Das and Nilanjana Maulik
University of Connecticut School of Medicine,
Farmington, Connecticut

CONTENTS

5.1 OVERVIEW

Both epidemiologic and experimental evidence exist to indicate that mild-to-moderate alcohol consumption is associated with a reduced incidence of mortality and morbidity from coronary heart disease. The consumption of wine, particularly red wine, imparts a greater benefit in the prevention of coronary heart disease than the consumption of other alcoholic beverages. The cardioprotective effects of red wine have been attributed to several polyphenolic antioxidants including resveratrol and proanthocyanidins. The mechanism of cardioprotection afforded by alcohol consumption remains highly speculative. It appears that moderate alcohol consumption induces a significant amount of oxidative stress to the heart, which then induces the expression of several cardioprotective oxidative stress-inducible proteins including heat shock proteins (HSPs) and antioxidants. Feeding rats with red wine extract or its polyphenolic antioxidants, as well as alcohol, results in the improvement of postischemic ventricular function. Both wine and alcohol trigger a signal transduction cascade by reducing proapoptotic transcription factors and genes, such as JNK-1 and c-Jun, thereby potentiating an antideath signal. Several HSPs and antioxidant proteins are induced in the heart after three weeks of alcohol consumption, resulting in the reduction of myocardial infarct size and cardiomyocyte apoptosis. It appears, therefore, that both wine and alcohol reduce myocardial ischemic–reperfusion injury, although the mechanisms of cardioprotection are different in the two cases.

5.2 INTRODUCTION

Epidemiologic studies suggest that mild-to-moderate alcohol consumption is associated with a reduced incidence of mortality and morbidity from coronary heart disease.[1-6] Proposed mechanisms for such cardioprotective effects have included, among others, an increase in high-density lipoprotein (HDL) cholesterol,[7] reduction or inhibition of platelet aggregation,[8] reduction in clotting factor concentrations,[9] reduction in thromboxane synthesis,[10] increase in vasodilatory prostacyclin synthesis,[11] inhibition of low-density lipoprotein (LDL) oxidation,[12] and increase in free radical scavenging.[13]

During the past 20 years, studies in different ethnic groups, starting from an American cohort to the recently performed analysis in the MONICA project, have provided evidence for decreased morbidity and mortality from coronary heart disease in those who have one to three drinks a day when compared to total abstainers.[14] Although it is comprehensible that antioxidants such as flavonoids and polyphenols found in wines may be responsible for the cardioprotective effects of wine consumption, the mechanism of cardioprotection afforded by alcohol remains obscure. Most of the studies have focused on alcohol-induced increase in HDL cholesterol,[15] which has been found by a number of investigators to be cardioprotective.[16] However, several other factors including improved platelet function[17] and fibrinolytic activity[18] are likely to play a role in alcohol-mediated cardioprotection.

Similar to alcohol, mild-to-moderate consumption of red wine is also associated with a reduced incidence of mortality and morbidity from coronary heart disease. This finding gave rise to what is now popularly termed the *French paradox*.[19] Red wine is rich in several typical polyphenols including resveratrol (3,5,4-trihydroxystilbene), a naturally occurring antioxidant phytoalexin, and proanthocyanidins. These polyphenols, which are mostly present in the seeds and skins of grapes, possess many biological functions including protection against atherosclerosis, lipid peroxidation, and cell death.[20] Both proanthocyanidins and resveratrol possess potent cardioprotective properties.[21,22] Grape seed proanthocyanidins have recently found to potentiate antideath signaling events.[23]

White wine contains many of the polyphenolic antioxidants that are also present in red wine, but it contains relatively small amounts of proanthocyanidins and resveratrol. White wines are rich in certain other polyphenols, but whether these polyphenols contribute to cardioprotection is not known. The goal of this chapter is to discuss the possible mechanisms by which alcohol and wine provide cardioprotection.

5.3 RED WINE AND CARDIOPROTECTION

Recently, a number of studies have been devoted to understanding the cause of the so-called French paradox, which refers to the anomaly that in several parts of France and other Mediterranean countries morbidity and mortality due to coronary heart diseases in absolute terms, as well as in comparison with the other causes of mortality, is significantly lower than that in other developed countries, despite the high consumption of fat and saturated fatty acids in these populations.[24,25] The cause of this cardioprotective effect is believed to be, among others, regular consumption of wine. Wines, especially red wines, contain about 1800 to 3000 mg/l of polyphenolic compounds. Many polyphenolic compounds are potent antioxidants, capable of scavenging free radicals and inhibiting lipid peroxidation both *in vitro* and *in vivo*. It has been shown that flavonoids as well as nonflavonoids present in wines inhibit the oxidation of LDL, eicosanoid synthesis, and platelet aggregation, as well as promote nitric oxide synthesis.[26,27]

In a recent study, our laboratory demonstrated that an alcohol-free red wine extract could protect the heart from the detrimental effects of ischemia/reperfusion injury, as evidenced by improved postischemic ventricular function and reduced myocardial infarction.[21] Red wine also reduced oxidative stress in the heart as indicated by decreased malonaldehyde (MDA) formation; it also reduced myocardial infarct size and cardiomyocyte apoptosis (Figure 5.1).

The nuclear binding of NFκB was increased after ischemia and reperfusion. The binding activity was increased slightly with red wine, compared to the control.[22] In contrast, both resveratrol and wine dramatically increased the binding activity of NFκB. Thirty minutes of ischemia followed by 2 h of reperfusion significantly enhanced the amount of the protein levels of p38 MAPK, JNK1, and c-Jun (Figure 5.2). The ischemia/reperfusion-mediated enhancement of c-Jun and JNK-1 protein levels, but not that of p38 MAPK, were significantly reduced by red wine or by its polyphenolic components proanthocyanidin and resveratrol.

5.4 WHITE WINE AND CARDIOPROTECTION

White wine, unlike red wine, is not rich in proanthocyanidins and resveratrol, but it contains many other polyphenolic antioxidants; whether these polyphenols contribute to cardioprotection is not known. Compared to red wine, white wine contains minimal amounts of proanthocyanidins and resveratrol. To study if white wines, similar to red wine, can also protect the heart from ischemia/reperfusion injury, ethanol-free extracts of three different white wines from the Tuscany regions of Italy (WW #1, WW #2, and WW #3) (100 mg/100 g of body weight) were given orally to Sprague-Dawley rats (200-g body weight) for 3 weeks.[28] Control rats were given only water for the same period of time. After 3 weeks, the rats were anesthetized, sacrificed, and the hearts excised for the preparation of isolated working rat hearts. All hearts were subjected to a 30-min global ischemia followed by 2 h of reperfusion. The results demonstrated that among

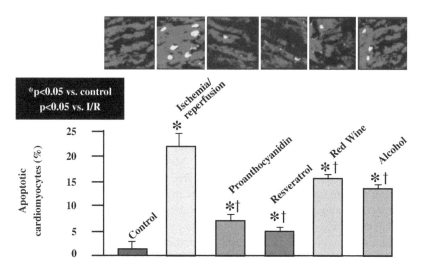

FIGURE 5.1 Effects of wine, its polyphenolic components, proanthocyanidin and resveratrol, and alcohol on cardiomyocyte apoptosis. Rats were fed orally with red wine extract, proanthocyanidin, resveratrol, or alcohol up to a period of 3 weeks. At the end of 3 weeks, rats were anesthetized and sacrificed, and isolated perfused working hearts were subjected to 30-min ischemia followed by 2 h of reperfusion. The apoptotic cardiomyocytes were detected by Tunnel staining in conjunction with a specific antibody against a-myosin heavy chain to specifically stain the cardiomyocytes. Representative apoptotic cells were detected by laser scanning microscopy (top). The results (average of at least 6/group) expressed in a bar graph are shown. $*p < .05$ vs. I/R or control.

the three different white wines, only WW #2 showed cardioprotection, as evidenced by improved postischemic ventricular recovery compared to control (results are for 100 mg/kg) (AF: 24.8 ± 2 vs. 11.7 ± 1.1 ml/min at 30R (30-min reperfusion), 28.3 ± 1.8 vs. 17.1 ± 1.5 ml/min at 60R, and 22.9 ± 1.4 vs. 16.3 ± 1.3 at 120R; LVDP: 60 ± 7 vs. 34 ± 5 mm Hg at 30R, 68 ± 8 vs. 36 ± 3 mm Hg at 60R, and 49 ± 6 vs. 23 ± 5 mm Hg at 120R; LV dp/dt_{max}; 2012 ± 88 vs. 1650 ± 116 mm Hg/sec at 30R, 2194 ± 113 vs. 1759 ± 104 mm Hg/sec at 60R, and 2000 ± 123 vs. 1568 ± 65 mm Hg/sec at 120R; LVSP: 95 ± 6 vs. 77 ± 5 mm Hg at 30R, 100 ± 7 vs. 76 ± 4 mm Hg at 60R, 84 ± 5 vs. 67 ± 4 mm Hg at 120R; LVEP: 35 ± 2 vs. 43 ± 3 mm Hg at 30R; 32 ± 3 vs. 40 ± 4 mm Hg at 60R, and 35 ± 2 vs. 44 ± 4 mm Hg at120R). The amount of MDA production in white-wine-fed rat hearts were lower than that found in control hearts, indicating reduced formation of the reactive oxygen species (ROS). *In vitro* studies using the chemiluminescence technique revealed that these white wines scavenged both superoxide anions and hydroxyl radicals (OH·). The results demonstrated that only WW #2 provided cardioprotection, as evidenced by the improved postischemic contractile recovery and reduced myocardial infarct size. The cardioprotective effect of this white wine may be attributed, at least in part, to its ability to function as an *in vivo* antioxidant. The phenolic composition of the white wines was

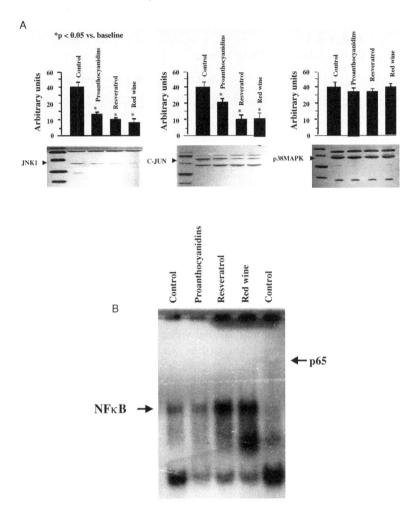

FIGURE 5.2 Effects of wine, its polyphenolic components, proanthocyanidin and resveratrol, and red wine on the induction of the expression of p38 MAPK, JNK-1, and c-Jun protein content (A) and DNA binding of NFκB (B) of the heart. Rats were fed orally with red wine extract, proanthocyanidin, resveratrol, or alcohol up to a period of 3 weeks. At the end of 3 weeks, rats were anesthetized and sacrificed, and isolated perfused working hearts were subjected to 30-min ischemia followed by 2 h of reperfusion. P38 MAPK, JNK-1, and c-Jun were measured by Western blot analysis and DNA binding of NFκB was determined by EMSA. Results are representative of at least six hearts per group.

determined by HPLC. As shown in Table 5.1, tyrosol content of WW #1 and WW #2 were identical. WW #3 contained about a third of the tyrosol compared to that present in WW #1 and WW #2. WW #2 was unique in the sense that it was the only white wine that contained significant amounts of caffeic acid, catechin, and resveratrol compared to other groups of white wines. The results

TABLE 5.1
Composition of White Wines

	WW #1	WW #2	WW #3
Tyrosol	28.1 ± 5.4	30.4 ± 1.9	10.6 ± 1.2
Caftaric acid (as caffeic acid)	<0.1	13.6 ± 0.9	<0.1
Catechin	17.4 ± 2.8	55.4 ± 2.4	<0.5
t-Resveratrol	<1.0	1.04	<1.0
Quercetin	<0.5	<0.5	<0.5

Note: The white wine extracts were dissolved in a mixture of water and ethanol (95:5, v/v). The clear solutions were separated with HPLC and detected at 280 nm. A Hyersil-2-column with 3 µ material was used. The compounds were identified and quantitated against authentic standards.

of this study demonstrated that white wine can also protect the heart from the detrimental effects of ischemia/reperfusion injury, as evidenced by improved postischemic ventricular function and reduced myocardial infarction. Among the three different white wines used in the study, only one group, WW #2, was found to be cardioprotective. In contrast, all the white wines reduced the amount of oxidative stress in the heart, as indicated by decreased MDA formation. These wines were also found to be highly effective in directly scavenging superoxide and OH· (Figure 5.3). The results indicate that reduction of oxidative stress may not be the only criteria that needs to be met for the wines to be cardioprotective. The phenolic composition of the WW #2 was completely different from that found in the other two groups of wines. Although it was not shown, the presence of significant amounts of caffeic acid, catechin, and resveratrol in the WW #2 group could play a role in its ability to protect the myocardium from ischemic–reperfusion injury.

The polyphenol fraction of wine includes phenolic acids (*p*-coumaric, cinnamic, caffeic, gentisic, ferulic, and vanillic acids), trihydroxy stilbenes (resveratrol and polydatin), and the flavonoids (catechin, epicatechin, and quercetin).[29] The free-radical-scavenging property of various flavonoids has been observed in relation to hydroxyl and superoxide as well as peroxyl and alkoxyl radicals.[30] Flavonoids also possess the ability to chelate iron ions, which are known to catalyze many free-radical-generating processes. This property probably also contributes to their antioxidant effectiveness. Antioxidants have long been known to protect against the damaging effects of free-radical-mediated tissue injury especially ischemia/reperfusion injury of the heart and other organs.[31] Substantial evidence exists to support the notion that ischemia and reperfusion generates oxygen free radicals, which contribute to the pathogenesis of ischemic–reperfusion injury.[32] The presence of ROS was confirmed directly by estimating free radical formation and indirectly by assessing lipid peroxidation and DNA breakdown products.[31,32] Virtually, all the biomolecules including lipids, proteins, and DNA molecules are

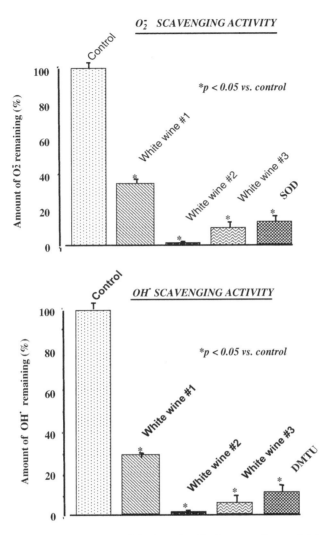

FIGURE 5.3 Superoxide-anion- and hydroxyl-radical-scavenging activities of white wines. Superoxide anions and hydroxyl radicals were generated chemically, and their scavenging was determined with white wines. Scavenging activities were compared against SOD for superoxide anion and DMTU for hydroxyl radicals.

potential targets for the OH radical attack. Thus, the free-radical-scavenging ability of WW #2 may, at least in part, be responsible for its cardioprotective properties. However, it should be noted that all the white wines effectively scavenged OH·, although the antioxidant action was found to be the highest for WW #2. This would suggest that there is some uniqueness in the WW #2. Indeed, WW #2 contained significant amount of resveratrol, caffeic acid, and catechin, compared to those present in the other white wines. Resveratrol, a stilbene polyphenol, has recently been found to protect the heart from ischemic–reperfusion injury by its ability to reduce both necrosis and apoptosis.[33] Proanthocyanidins were shown to be cardioprotective by their ability to reduce ventricular arrhythmias and cardiomyocyte apoptosis.[20–23,34] Catechin is also cardioprotective and is mostly found in green tea.[35] The role of caffeic acid in cardioprotection is not known. It is also possible that there is an as-yet unknown compound present in WW #2, which contributes to its cardioprotective properties.

5.5 ALCOHOL AND CARDIOPROTECTION

As mentioned earlier, mild-to-moderate amounts of alcohol consumption, independent of any antioxidant action, can be beneficial for the heart. In a recent study, hearts from rats were arrested by global ischemia for 30 min followed by 2 h of reperfusion. Control hearts subjected to ischemia and reperfusion had an infarct size of $34.9 \pm 3.3\%$ compared to the infarct size that was almost zero when a heart was subjected to 2.5 h perfusion without ischemia/reperfusion. The percentage of infarct size was significantly reduced with alcohol ($23.2 \pm 1.8\%$). The number of apoptotic cells was significantly higher (21.5%) in the ischemic/reperfused myocardium compared to the baseline levels (Figure 5.1). The alcohol-fed rat hearts displayed reduced numbers of apoptotic cardiomyocytes ($10.2 \pm 1.0\%$) compared to control hearts.

The production of MDA is an indicator for lipid peroxidation and development of oxidative stress. MDA content was increased within hours of alcohol feeding. As shown in Figure 5.4, the MDA content of the heart increased significantly after 1 d of alcohol feeding. This value remained the same for up to 2 d and then dropped slightly. MDA content came down significantly after 7 d of alcohol feeding. This level was maintained for up to 21 d. The results, thus, indicate that alcohol causes development of oxidative stress in the heart; however, such oxidative stress was only transient and progressively came down to the baseline levels.

Alcohol caused the induction of the expression of several cardioprotective proteins including HSP 70, HO-1, and MnSOD, as shown in Figure 5.5. Western blot analysis revealed increased expression of these proteins in the alcohol-fed rat hearts compared to the control hearts. HSP 70 was increased 10-fold, HO-1 by 32-fold, and MnSOD by 15-fold. It should be noted that induction of these cardioprotective and antioxidant proteins parallels the reduction in oxidative stress.

5.6 MECHANISMS OF CARDIOPROTECTION WITH WINE VS. ALCOHOL

It appears that the mechanisms of cardioprotection afforded by wine and alcohol are different from each other. Whereas polyphenolic antioxidants including resveratrol and proanthocyanidin are responsible for the cardioprotection provided

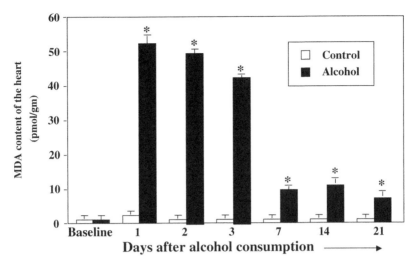

FIGURE 5.4 Effects of alcohol on the MDA content of the heart. The rats were fed alcohol orally up to a period of 3 weeks. At the end of 3 weeks, rats were anesthetized and sacrificed, and isolated perfused working hearts were subjected to 30-min ischemia followed by 2 h of reperfusion. The MDA content was measured by HPLC as described in text. The results are expressed as means ± SEM of at least six hearts per group. $*p < .05$ vs. control.

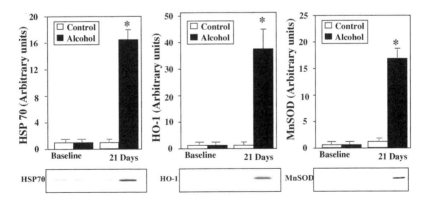

FIGURE 5.5 Effects of alcohol on the induction of the expression of HSP 70, HO-1, and MnSOD protein content of the heart. Rats were fed alcohol orally up to a period of three weeks. At the end of 3 weeks, rats were anesthetized and sacrificed, and isolated perfused working hearts were subjected to 30-min ischemia followed by 2 h of reperfusion. The HSP 70, HO-1, and MnSOD were measured by Western blot analysis as described in text. Results are representative of at least six hearts per group. The results are expressed as means ± SEM of at least six hearts per group. $*p < .05$ vs. control.

by wine, alcohol protects the heart by adapting this organ to oxidative stress. Alcohol consumption appears to induce oxidative stress, which is subsequently translated into oxidative-stress-inducible proteins. Indeed, alcohol feeding resulted in the induction of the expression of cardioprotective antioxidants and heat shock proteins (HSPs).

A large amount of evidence exists in the literature indicating the cardioprotective abilities of HSPs including HSP 70.[36–39] Currie and colleagues were probably the first group to demonstrate enhanced postischemic ventricular recovery of hearts subjected to heat stress.[36] The results of many recent studies including our own support this observation by Currie and further indicate that HSP 70 can also be induced by other stresses, such as oxidative stress, as shown in the present study.[40,41] Another related study showed an induction of HSP 70 in conjunction with rendering the cells resistant to H_2O_2 by ethyl alcohol.[42] In the present study, the expression of HSP 70 was associated with improved postischemic ventricular function and decreased infarct size and cardiomyocyte apoptosis.

HO-1 is another cardioprotective protein that is induced by a variety of stresses including oxidative stress. HO-1 induction has been found to be associated with cellular protection against injury caused by ROS.[43] ROS produced in the ischemic–reperfused myocardium has been shown to induce HO-1 gene expression.[44] Another recent study showed that H_{mox-1} plays a crucial role in ischemic–reperfusion injury not only by functioning as an intracellular antioxidant but also by inducing its own expression under stressful conditions such as preconditioning.[45] It seems likely that alcohol probably functions in an identical manner, resulting in an induction of HO-1 expression.

A significant number of studies show the cardioprotective role of MnSOD. For example, overexpression of MnSOD has been found to protect myocardial ischemia/reperfusion injury in transgenic mice.[46] Increased expression of MnSOD in hearts transfected with MnSOD showed significant improvement in tolerance against ischemia/reperfusion injury.[47] The activation of MnSOD by free radicals produced during hyperthermia was found to be important for early-phase and late-phase cardioprotection against ischemia/reperfusion injury in rats. In another related study, the same investigators found that heat-shock-induced MnSOD enhanced the tolerance of cardiac myocytes to hypoxia–reoxygenation injury.[48] Preconditioning of the heart by cyclic episodes of ischemia and reperfusion has been found to induce MnSOD, suggesting its importance in the myocardial defense mechanism.[49] MnSOD could protect mitochondrial complex I against adriamycin-induced cardiomyopathy in transgenic mice.[50] Another recent study also showed the induction of MnSOD, supporting this previous observation.[51] The induction of the expression of MnSOD observed in response to chronic low-dose alcohol consumption appears to occur also due to an adaptive response to alcohol-induced oxidative stress.

The results of the present study, however, do not preclude other mechanisms for the cardioprotective properties of low-dose alcohol consumption. Alcoholic vasodilation, decreases in platelet aggregability, changes in prostacyclin/thromboxane ratios, and increased fibrinolytic activities can be considered as additional benefits conferred by mild-to-moderate alcohol consumption. The results of this study indicate that the cardioprotective effects of alcohol are mediated by the ability of the heart to adapt to alcohol-induced oxidative stress. The induction of the expression of the cardioprotective proteins such as HSP 70, HO-1, and MnSOD may be at least partially responsible for cardioprotection.

5.7 SUMMARY AND CONCLUSION

From the discussion in this chapter, it should be clear that the mechanism of the cardioprotection afforded by red wine is likely to be mediated by its polyphenolic components, resveratrol and proanthocyanidin. As shown in a recent study, wine, as well as resveratrol and proanthocyanidin, decreased the myocardial infarct size and reduced apoptotic cell death to the same extent.[33] Antioxidants have long been known to protect against the damaging effects of free-radical-mediated tissue injury, especially ischemia/reperfusion injury of the heart and other organs.[52] Substantial evidence exists to support the notion that ischemia and reperfusion generate oxygen free radicals, which contribute to the pathogenesis of ischemic–reperfusion injury.[53,54] The presence of ROS was confirmed directly by estimating free radical formation and indirectly by assessing lipid peroxidation and DNA breakdown products.[55] Among the oxygen free radicals, superoxide anion (O_2) is the least harmful, whereas OH· is the most detrimental to cells. In spite of their relatively low oxidizing ability compared to OH· radicals, in biological systems, organic peroxyl radicals could be extremely damaging to the tissue.[56] Tissues such as the heart are protected from the detrimental actions of peroxyl radicals by the presence of naturally occurring antioxidants, such as bilirubin and biliverdin, as well as plasma antioxidants.[57] Ascorbic acid and vitamin E are the other potent peroxyl radical traps in biological systems.[58] Generally, lipid-soluble antioxidants can scavenge chain-carrying lipid peroxyl radicals, thereby preventing propagation of lipid peroxidation after its initiation. Recent study in our laboratory has demonstrated that not only are red wine extract, proanthocyanidin, and resveratrol potent scavengers of peroxyl radicals, but they also reduce the extent of lipid peroxidation in the ischemic–reperfused myocardium.[20–23] These findings seem to be important because these peroxyl radicals are formed *in vivo* in membranes and lipoproteins as intermediate products of lipid peroxidation.

Wine, as opposed to the other sources of polyphenols and antioxidants, is unique in a number of ways. First, resveratrol and a few other polyphenols are virtually absent from commonly consumed fruits and vegetables and, as such, red wine may constitute the only source of these substances in the diet. The only other resveratrol source in the human diet is peanuts. Second, the various procedures involved in the production of red wine further enrich its polyphenol content. In addition, the increased alcohol content as a result of the fermentation process allows for enrichment of total polyphenol content and also for better solubilization of the polyphenols, resulting in a greater bioavailability than in other foodstuffs.[34] With all these factors, red wine might possibly be the richest effective source of natural polyphenol antioxidants. Alcoholic vasodilation, decreases in platelet aggregability, changes in prostacyclin/thromboxane ratios, and increased fibrinolytic activities are to be considered as additional benefits provided by mild-to-moderate alcohol consumption.

In contrast, regular consumption of low doses of alcohol is associated with an induction of the expression of several cardioprotective proteins including HSPs and antioxidant proteins. The upregulation of these proteins may, at least in part, explain the cardioprotective property of alcohol. Alcohol initially increased the amount of MDA content in the heart; however, the MDA content was reduced to the baseline level after 7 d. This could be due to an increased expression of several cardioprotective proteins including HSP 70, HO-1, and MnSOD, suggesting that, at least in part, these proteins may be responsible for the cardioprotective effects of alcohol.

Alcohol consumption appears to induce oxidative stress, which is subsequently translated into oxidative-stress-inducible proteins. These results are consistent with previous reports, which demonstrated an increase in circulating catecholamine levels after alcohol consumption.[59,60] Increased catecholamines can explain the enhancement of oxidative stress in the heart after alcohol consumption. It appears that after 7 d of repeated low-dose alcohol consumption, an adaptive modification of the oxidative stress occurs, resulting in an induction of the expression of oxidative-stress-inducible proteins.

ACKNOWLEDGMENTS

This study was partially supported by NIH HL 34360, HL 22559, and HL 33889.

REFERENCES

1. St. Leger AS, Cochrane AL, and Moore F. Factors associated with cardiac mortality in developed countries with particular reference to the consumption of wine. *Lancet* 1979; 1: 1017–1020.
2. Hertog MGL, Feskens EJM, and Kromhout D. Antioxidant flavonols and coronary heart disease risk. *Lancet* 1997; 349: 699.
3. Klatsky AL, Armstrong MA, and Friedman GD. Risk of Cardiovascular mortality in alcohol drinkers, ex-drinkers and non-drinkers. *Am J Cardiol* 1990; 66: 1237–1242.
4. Goldberg DM, Hahn SE, and Parks JG. Beyond alcohol: beverage consumption and cardiovascular mortality. *Clin Chim Acta* 1995; 237: 155–187.
5. Kannel WB and Ellison RC. Alcohol and coronary heart disease: The evidence for a protective effect. *Clin Chim Acta* 1996; 246: 59–76.
6. Criqui MH. Alcohol and coronary heart disease: consistent relationship and public health implications. *Clin Chim Acta* 1996; 246: 51–57.

7. Fuhrman B, Lavy A, and Aviram M. Consumption of red wine with meals reduces the susceptibility of human plasma and low-density lipoprotein to lipid peroxidation. *Am J Clin Nutr* 1995; 61: 549–554.

8. Galli C, Colli S, and Gianfranceshi G. Acute effects of ethanol, caffeine, or both on platelet aggregation, thromboxane formation and plasma free fatty acids in normal subjects. *Drug Nutr Interact* 1984; 3: 6107.

9. Ridker PM, Vaughan DE, Stampfer MJ, Glynn RJ, and Hennekens CH. Association of moderate alcohol consumption and plasma concentration of tissue-type plasminogen activator. *JAMA* 1994; 272: 929–933.

10. Mikhailidis DP, Jeremy JY, Barradas MA, Green N, and Dandona P. Effect of ethanol on vascular prostacyclin synthesis, platelet aggregation and platelet thromboxane release. *Br Med J* 1983; 287: 1495–1498.

11. Landolfi R and Steiner M. Ethanol raises prostacyclin in vivo and in vitro. *Blood* 1984; 64: 679–682.

12. Frankel EN, Kanner J and German JB. Inhibition of human low-density lipoprotein by phenolic substances in red wine. *Lancet* 1993; 341: 454–457.

13. Saija A, Marzullo D, Scalese M, Bonina F, Laiza M, and Castelli F. Flavonoids as antioxidant agents: importance of their interaction with biomembranes. *Free Radic Biol Med* 1995; 19: 481–486.

14. Criqui MH and Ringel BL. Does diet or alcohol explain the French Paradox? *Lancet* 1994; 344: 1719–1723.

15. Valimaki M, Nikkila EA, Taskinen A, and Ylikahri R. Rapid decrease in high density lipoprotein subfractions and post heparin plasma lipase activity after cessation of chronic alcohol intake. *Atherosclerosis* 1986; 59: 147–153.

16. Das DK. Cardioprotection with high-density lipoproteins. Fact or fiction? *Circ Res* 2003; 92: 258–260.

17. Rubin R and Rand ML. Alcohol and platelet function. *Alcohol Clin Exp Res* 1994; 18: 105–110.

18. Abou-Agag LH, Tabengwa EM, Tresnak JA, Wheeler CG, Taylor KB, and Booyse FM. Ethanol-induced increased surface-localized fibrinolytic activity in cultured human endothelial cells: kinetic analysis. *Alcohol Clin Exp Res* 2001; 25: 351–361.

19. Renaud S and De Lorgeril M. Wine, alcohol, platelets and the French Paradox for coronary heart disease. *Lancet* 1992; 339:1523–1526.

20. Sato M, Maulik G, Ray PS, Bagchi D, and Das DK. Cardioprotective effects of grape seed proanthocyanidin against ischemic reperfusion injury. *J Mol Cell Cardiol* 1999; 31: 1289–1297.

21. Sato M, Ray PS, Maulik G, Maulik N, Engelman RM, Bertelli AAE, and Das DK. Myocardial protection with red wine extract. *J Cardiovasc Pharmacol* 2000; 35: 263–268.

22. Sato M, Bagchi D, Tosaki A, and Das DK. Grape seed proanthocyanidin reduces cardiomyocyte apoptosis by inhibiting ischemia/reperfusion-induced activation of JNK-1 and c-Jun. *Free Radic Biol Med* (in press).

23. Ray PS, Maulik G, Cordis GA, Bertelli AAE, Bertelli A, and Das DK. The red wine antioxidant resveratrol protects isolated rat hearts from ischemia reperfusion injury. *Free Rad Biol Med* 1999; 27: 160–169.

24. Rimm EB, Giovannucci EL, Willett WC, Colditz GA, Archerio A, Rosner B, and Stampfer MJ. Prospective study of alcohol consumption and risk of coronary disease in men. *Lancet* 1991; 338: 464–486.

25. Gaziano JM, Buring JE, Breslow JL, Goldhaber SZ, Rosner B, VanDenburgh M, and Willett W, Hennekens CH. Moderate alcohol intake, increased levels of high-density lipoprotein and its sub-fractions and decreased risk of myocardial infarction. *N Engl J Med* 1993; 329: 1829–1834.

26. Mikhailidis DP, Jeremy JY, Barradas MA et al. Effect of ethanol on vascular prostacyclin synthesis, platelet aggregation and platelet thromboxane release. *Br Med J* 1983; 287: 1495–1498.

27. Saija A, Scalese M, Laiza M et al. Flavonoids as antioxidant agents: importance of their interaction with biomembranes. *Free Radic Biol Med* 1995; 19: 481–486.

28. Cui J, Tosaki A, Cordis GA, Bertelli AA, Bertelli A, Maulik N, and Das DK. Cardioprotective abilities of white wine. *Ann NY Acad Sci.*

29. Soleas GJ, Diamandis EP, and Goldberg DM. Wine as a biological fluid: history, production and role in disease prevention. *J Clin Lab Anal* 1997; 11: 287–313.

30. Rice-Evans CA, Miller N, Bolwell PG, Bramley PM, and Pridham JB. The relative antioxidant activities of plant-derived polyphenolic flavonoids. *Free Rad Res* 1995; 22: 375–383.

31. Das DK and Maulik N. Evaluation of antioxidant effectiveness in ischemia reperfusion tissue injury. *Methods Enzymol* 1994; 233: 601–610.

32. Tosaki A, Bagchi D, Hellegouarch D, Pali T, Cordis GA, and Das DK: Comparisons of ESR and HPLC methods for the detection of hydroxyl radicals in ischemic/reperfused hearts. A relationship between the genesis of oxygen-free radicals and reperfusion-induced arrhythmias. *Biochem Pharmacol* 1993; 45: 961–969.

33. Hattori R, Otani H, Maulik N, and Das DK. Pharmacological preconditioning with resveratrol: role of nitric oxide. *Am J Physiol* 2002; 282: H1988–H1995.

34. Pataki T, Bak I, Kovacs P, Bagchi D, Das DK, and Tosaki A. Grape seed proanthocyanidins reduced ischemia/reperfusion-induced injury in isolated rat hearts. *Am J Clin Nutr* (in press).

35. Hertog MG, Feskens EJ, Hollman PC, and Katan MB. Dietary antioxidant flavonoids and risk of coronary heart disease: the Zutphen Elderly study. *Lancet* 1993; 342: 1007–1011.

36. Currie RW. Effects of ischemia and reperfusion on the synthesis of stress-induced (heat shock) proteins in isolated and perfused rat hearts. *J Mol Cell Cardiol* 1987; 19: 795–808.

37. Das DK, Engelman RM, and Kimura Y. Molecular adaptation of cellular defences following preconditioning of the heart by repeated ischemia. *Cardiovasc Res* 1993; 27: 578–584.

38. Hutter MM, Sievers RE, Barbosa V, and Wolfe CL. Heat shock protein induction in rat hearts. A direct correlation between the amount of heat-shock protein induced and the degree of myocardial protection. *Circulation* 1994; 89: 355–360.

39. Knowlton AA, Brecher P, and Apstein CS. Rapid expression of heat shock protein in the rabbit after brief cardiac ischemia. *J Clin Invest* 1991; 87: 139–147.

40. Liu X, Engelman RM, Moraru II, Rousou JA, Flack JE, Deaton DW, Maulik N, and Das DK. Heat shock: a new approach for myocardial preservation in cardiac surgery. *Circulation* 1992; 86 (Suppl. 2): 358–363.

41. Maulik N, Liu X, Cordis GA, Engelman RM, and Das DK. The reduction of myocardial ischemia reperfusion injury by amphetamine is linked with its ability to induce heat shock. *Mol Cell Biochem* 1994; 137: 17–24.

42. Su CY, Chong KY, Owen OE, Dillman WH, Chang C, and Lai CC. Constitutive and inducible hsp70 are involved in oxidative resistance evoked by heat shock or ethanol. *J Mol Cell Cardiol* 1998; 30: 587–598.

43. Nath KA, Balla G, and Vercellotti G. Induction of heme oxygenase is a rapid, protective response in rhabdomyolysis in the rat. *J Clin Invest* 1992; 90: 267–270.

44. Sharma HS and Das DK. Induction of the heme oxygenase gene expression during the reperfusion of ischemic rat myocardium. *J Mol Cell Cardiol* 1996; 28: 1261–1270.

45. Yoshida T, Maulik N, Ho YS, Alam J, and Das DK. Hmox-1 constitutes an adaptive response to effect antioxidant cardioprotection. *Circulation* 2001; 103: 1695–1701.

46. Chen Z, Siu B, Ho YS, Vincent R, Chua CC, Hamdy RC, and Chua BH. Over-expression of MnSOD protects against myocardial ischemia/reperfusion injury in transgenic mice. *J Mol Cell Cardiol* 1998; 30: 2281–2289.

47. Sawa Y, Kadoba K, Suzuki K, Bai HZ, Kaneda Y, Shirakura R, and Matsuda H. Efficient gene transfer method into the whole heart through the coronary artery with hemagglutinating virus of Japan liposome. *J Thoracic Cardiovasc Surg* 1997; 113: 512–518.

48. Yamashita N, Hoshida S, Nishida M, Igarashi J, Tanigichi N, Tada M, Kuzuya T, and Hori M. Heat shock-induced manganese superoxide dismutase enhances the tolerance of cardiac myocytes to hypoxia-reoxygenation injury. *J Mol Cell Cardiol* 1997; 29: 1805–1813.

49. Das DK, Engelman RM, and Kimura Y. Molecular adaptation of cellular defenses following preconditioning of the heart by repeated ischemia. *Cardiovasc Res* 1993; 27: 578–584.

50. Yen HC, Oberley TD, Gairola CG, Szweda LI, and St. Clair DK. Manganese superoxide dismutase protects mitochondrial complex I against adriamycin-induced cardiomyopathy in transgenic mice. *Arch Biochem Biophys* 1999; 362: 59–66.

51. Yamashita N, Hoshita S, Taniguchi N, Kuzuya T, and Hori M. A "second window of protection" occurs 24h after ischemic preconditioning in the rat heart. *J Mol Cell Cardiol* 1998; 30: 1181–1189.

52. Das DK and Maulik N. Protection against free radical injury in the heart and cardiac performance. In: *Exercise and Oxygen Toxicity.* Sen CK, Packer L, and Hanninen O. Eds. Elsevier Science. Amsterdam. 1995. pp. 359–388.

53. Arroyo CM, Kramer JH, Dickens BF, and Weglicki WB. Identification of free radicals in myocardial ischemia/reperfusion by spin trapping with nitrone DMPO. *FEBS Lett* 1987; 221: 101–104.

54. Tosaki A, Bagchi D, Hellegouarch D, Pali T, Cordis GA, and Das DK: Comparisons of ESR and HPLC methods for the detection of hydroxyl radicals in ischemic/reperfused hearts. A relationship between the genesis of oxygen-free radicals and reperfusion-induced arrhythmias. *Biochem Pharmacol* 1993; 45:961–969.

55. Cordis GA, Maulik G, Bagchi D, Riedel W, and Das DK. Detection of oxidative DNA damage to ischemic reperfused rat hearts by 8-hydroxydeoxyguanosine formation. *Mol Cell Cardiol* 1998; 30: 1939–1944.

56. Chance B, Sies H, and Boveris A. Hydroperoxide metabolism in mammalian organs. *Physiol Res* 59: 527-540, 1979.

57. Das DK, Maulik N, Moraru II. Gene expression in acute myocardial stress. Induction by hypoxia, ischemia, reperfusion, hyperthermia and oxidative stress. *J Mol Cell Cardiol* 1995; 27: 181–193.

58. Stocker R and Peterhans E. Synergistic interaction between vitamin E and the bile pigments bilirubin and biliverdin. *Biochim Biophys Acta* 1989; 1002: 238–243.

59. Fantidis P, Del Cerro MJ, Martinez I, Rubio G, Ruiz Villaespesa A, Gamallo C, Leon G, and Santodomingo J. Ethanol intake, plasma catecholamine levels, and ST-segment changes without myocardial injury in rats with short-term ethanol consumption. *J Electrocardiol* 1995; 28: 307–312.

60. Patel VA and Pohorecky LA. Acute and chronic ethanol treatment on beta-endorphin and catecholamine levels. *Alcohol* 1989; 6: 59–63.

6 Diabetes Treatment by Alpha-Lipoic Acid: Modulation of Glucose Uptake and Adipocyte Differentiation

An-Sik Chung and Kyung-Joo Cho
Korea Advanced Institute of Science and Technology,
Daejeon, South Korea

Hadi Moini and Lester Packer
University of Southern California, Los Angeles, California

CONTENTS

6.1   INTRODUCTION

α-Lipoic acid (1,2-dithiolane-3-pentanoic acid) and its reduced form, dihydro-lipoic acid (DHLA), have been considered as universal antioxidants that function both in membrane and aqueous phases (Figure 6.1). Recent studies have demon-strated that α-lipoic acid facilitates glucose transport and utilization in fully differentiated adipocytes as well as in animal models of diabetes.[1–3] We have reported that the oxidized isoforms of α-lipoic acid stimulate glucose uptake into 3T3-L1 adipocytes by changing the intracellular redox status toward an oxidizing condition,[4] and it is hypothesized that α-lipoic acid may alter the thiol reactivity of redox-sensitive components of the insulin-signaling pathway based on its oxidant property. Moreover, although most insulin sensitizers, including thiazoli-donediones, are reported to induce adipocyte differentiation,[5] the regulation of adipogenesis by α-lipoic acid has not been elucidated before.

In this chapter, the action of α-lipoic acid on the thiol groups of the insulin receptor (IR) and protein tyrosine phosphatase 1B (PTP1B) in 3T3-L1 adipocytes will be reviewed. In addition, the mechanism by which α-lipoic acid regulates differentiation of preadipocytes to mature adipocytes will be discussed.

6.2   LIPOIC ACID AND GLUCOSE UPTAKE

α-Lipoic acid, a disulfide derivative of octanoic acid, functions as the prosthetic group of enzymes such as α-keto-acid dehydrogenases and pyruvate dehydroge-nase complex that catalyze various redox-based reactions.[6] It is known that α-lipoic acid has the ability to alter the redox status of cells and interact with thiols and other antioxidants.[7] In clinical trials, α-lipoic acid improves glucose metabolism in diabetic patients[8] and is found beneficial in the treatment of diabetic neuropathy.[9] α-Lipoic acid treatment also increases insulin sensitivity in patients

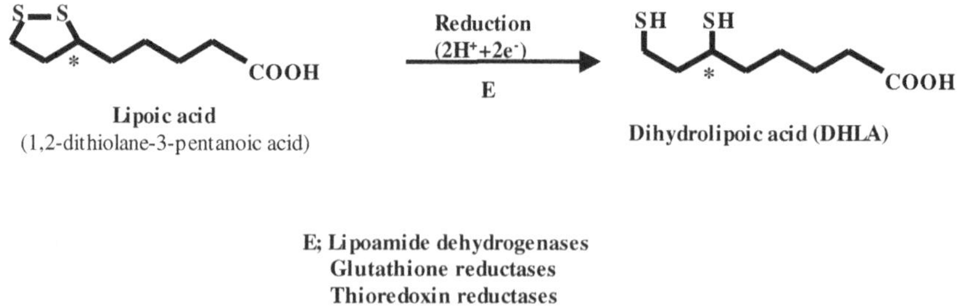

Lipoic acid
(1,2-dithiolane-3-pentanoic acid)

Reduction
(2H⁺+2e⁻)
E

Dihydrolipoic acid (DHLA)

E; Lipoamide dehydrogenases
Glutathione reductases
Thioredoxin reductases

FIGURE 6.1   Structure of α-lipoic acid and dihydrolipoic acid.

with type 2 diabetes[8,10,11] and enhances glucose transport into skeletal muscle isolated from both obese and lean Zucker rats.[12] In a comprehensive series of studies, Klip and colleagues have demonstrated that α-lipoic acid increases tyrosine phosphorylation and/or the activity of several components of the insulin-signaling pathway, including IR, insulin receptor substrate-1 (IRS-1), type-I phosphatidylinositol 3-kinase (PI3-K), protein kinase B/Akt1, and p38-mitogen-activated protein kinase (p38K).[13,14] Based on these findings, it has been proposed that α-lipoic acid stimulates glucose uptake into fat cells by activating the insulin signaling pathway, which enhances the translocation and intrinsic activity of glucose transporter 4 (Glut4).[14]

Accumulating evidence suggests that oxidant signals might be integrally involved in the regulation of the insulin-signaling pathway. Treatment of IR-transfected Chinese hamster ovary cells with antioxidants such as N-acetyl-L-cysteine (NAC) or butylated hydroxyanisole inhibits insulin responsiveness, whereas partial inhibition of glutathione metabolism, which intracellularly induces a mild oxidative stress condition, stimulates IR tyrosine phosphorylation when measured *in vitro*.[15] A similar increase in IR kinase activity is observed following cell treatment with hydrogen peroxide (H_2O_2).[16,17] Moreover, oxidation of critical cysteine residues in the IR β subunit results in the increase of its intrinsic tyrosine kinase activity, whereas low concentration of dithiothreitol inactivates the IR kinase, supporting the importance of oxidation of critical thiol groups in activation of the insulin-signaling pathway.[15] Agents such as H_2O_2, Cu^{2+}, and diamide that readily oxidize sulfhydryls to disulfides markedly enhance 3-O-methylglucose uptake in brown fat cells.[1] IR has also been demonstrated to couple, via $G\alpha_{i2}$, to the NADPH-dependent H_2O_2-generating system, which upon insulin stimulation produces H_2O_2 in 3T3-L1 adipocytes.[2] Insulin-dependent H_2O_2 production is associated with a decreased protein tyrosine phosphatase (PTP) activity,[18] which is also found essential for the activation of PI3-K.[3] These findings suggest that redox signals are involved in the regulation of both the early tyrosine phosphorylation cascade and downstream insulin-signaling events.

6.3 DECREASES IN THIOL REACTIVITY OF THE IR AND PTP1B BY α-LIPOIC ACID IN 3T3-L1 ADIPOCYTES

6.3.1 EFFECT OF α-LIPOIC ACID ON GLUCOSE UPTAKE, CELL REDOX STATUS, AND ACTIVATION OF IR

α-Lipoic acid increases uptake of glucose into 3T3-L1 adipocytes. Although α-lipoic acid (2.5 mM) is reported to increase insulin-stimulated glucose uptake in L6 myotubes,[12] no additive effect of α-lipoic acid (500 μM) with 100 nM of insulin is observed in 3T3-L1 adipocytes. The lack of additivity and the wortmannin sensitivity of α-lipoic-acid-stimulated glucose uptake indicate that α-lipoic acid, at least in part, exploits the same signaling pathway as does insulin in order to increase glucose uptake into 3T3-L1 adipocytes. In undifferentiated

3T3-L1 cells, the number of IRs, the potency of postreceptor-signaling events, and the magnitude of the biological responses to insulin are known to be dramatically lower.[19–21] However, α-lipoic acid stimulation of glucose uptake in undifferentiated 3T3-L1 cells is far less than that observed in fully differentiated adipocytes; nevertheless, it is significant and sensitive to inhibition by wortmannin, suggesting that α-lipoic acid might more efficiently activate existing IR and/or postreceptor-signaling events in undifferentiated cells.

The evidence now indicates that redox state of IR is involved in activation of the IR.[15,22] Reduced cysteine residues present in IR subunit, it is suggested, prevent spontaneous autophosphorylation of IR, whereas their oxidative modification, it is proposed, activates both tyrosine kinase domains.[15] α-Lipoic acid as well as insulin and H_2O_2 decrease the level of reduced cysteine residues in IR subunit by increasing intracellular oxidant level. Recently, α-lipoic acid has also been demonstrated to inhibit glycogen synthesis via its oxidative activity and the uncoupling of mitochondria in rat soleus muscle.[23] α-Lipoic acid increases activity of the redox-sensitive transcription factor NF-κB both in 3T3-L1 adipocytes and in preadipocytes, suggesting its pro-oxidant effect. Long-term (24 to 48 h) treatment with α-lipoic acid is known to increase GSH levels in a variety of cells, including 3T3-L1 adipocytes.[4,24] However, short-term (0 to 2 h) treatment of 3T3-L1 adipocytes with α-lipoic acid does not affect cellular GSH levels. α-Lipoic acid is also negligibly reduced to DHLA at this period of time (less than 1-μM DHLA was detected after a 30-min incubation of cells with 250-μM α-lipoic acid). These findings indicate that in 3T3-L1 adipocytes, α-lipoic acid shifts intracellular redox status toward an oxidizing condition, which may lead to oxidation of IR thiol groups and its activation. Moreover, α-lipoic acid facilitates IR autophosphorylation *in vitro*, indicating direct interaction of oxidized α-lipoic acid with the thiols of the IR.[4] When compared to α-lipoic acid or H_2O_2, insulin is more potent in stimulating IR tyrosine phosphorylation but is less effective in inducing oxidation of IR thiols, suggesting that the oxidation of IR thiol groups may not be the only determining factor for IR tyrosine phosphorylation.

6.3.2 ROLE OF NAPDH OXIDASE IN α-LIPOIC-ACID-STIMULATED GLUCOSE UPTAKE

The biochemical properties of plasma membrane NADPH-dependent oxidase in fat cells have been partially characterized. Similar to the phagocyte enzyme, the adipocyte oxidase displays a severalfold preference for NADPH over NADH, and requires a flavin nucleotide for optimal activity. In contrast to the phagocyte system, diphenylene iodonium (DPI) has no effect on basal or insulin-stimulated rates of NADPH-dependent H_2O_2 generation.[25] In contrast, DPI has recently been reported to inhibit insulin-stimulated H_2O_2 generation, IR autophosphorylation, and glucose uptake into 3T3-L1 adipocytes.[18] DPI effectively inhibits both α-lipoic-acid- or insulin-stimulated glucose uptake and IR autophosphorylation, whereas it marginally affects α-lipoic-acid- or insulin-stimulated oxidant production. The underlying reason for the observed discrepancies related to the effect of DPI on insulin-stimulated oxidant production is not known. Nevertheless, the fact that DPI inhibits IR activation without attenuating the oxidant production suggests that other unidentified signaling components may function in the interface of NADPH oxidase and IR. A thiol-reactive membrane-associated protein termed *molecule X* has recently been reported to covalently cross-link human IR β subunit, although its function is not known.[26]

6.3.3 INHIBITION OF CELLULAR PTP ACTIVITY BY α-LIPOIC ACID

PTPs have in common a conserved 230-amino-acid domain containing a cysteine residue that can catalyze the hydrolysis of protein phosphotyrosine residues by the formation of a cysteinyl-phosphate intermediate.[27] Considerable evidence indicates that reactive oxygen species (ROS), including H_2O_2, oxidize and inactivate PTPs *in vivo*.[28] Insulin is also shown to inhibit cellular PTP activity and the specific activity of PTP1B by increasing H_2O_2 production in 3T3-L1 adipocytes.[18] Consistently, pretreatment of 3T3-L1 adipocytes with insulin, α-lipoic acid, or H_2O_2 inhibits total PTP activity, whereas NAC, a thiol antioxidant, increases total PTP activity. Treatment of the cells with α-lipoic acid or H_2O_2, also decreases the level of thiol-biotinylated PTP1B, a tyrosine phosphatase that is strongly implicated in the downregulation of insulin signaling *in vivo*.[29] These findings suggest that α-lipoic acid might modify the catalytic cysteine residue in PTPs, resulting in their inactivation and thereby enhancing the early tyrosine phosphorylation cascade of the insulin-signaling pathway.

6.4 REGULATION OF ADIPOCYTE DIFFERENTIATION

Obesity is closely correlated with the prevalence of diabetes and cardiovascular disease. Plasma levels of leptin, tumor necrosis factor (TNF)-α and nonesterified fatty acid are elevated in obesity and substantially contribute to the development of insulin resistance.[30] Obesity is caused not only by hypertrophy of adipose tissue but also by adipose tissue hyperplasia, which triggers the transformation of preadipocytes into adipocytes.[31]

The program of adipocyte differentiation is a complex process that involves coordinated expression of specific genes and proteins associated with each stage of differentiation. This process is regulated by several signaling pathways.[32] Insulin, the major anabolic hormone, promotes *in vivo* accumulation of adipose tissue.[33] Structurally unrelated inhibitors of PI3-K, LY294002 and wortmannin, are shown to block adipocyte differentiation in a time- and dose-dependent fashion,[34] suggesting that the IR/Akt-signaling pathway is important in transducing the proadipogenic effects of insulin. In contrast, mitogen-activated protein kinases (MAPKs) such as extracellular signal-regulated kinase (ERK) and c-Jun N-terminal kinase (JNK) suppress the process of adipocyte differentiation.[35,36] TNF-α is known to exert its antiadipogenic effects, at least in part, through activation of the ERK pathway.[35] However, p38K is shown to promote adipocyte differentiation.[37]

The signals that regulate adipogenesis either promote or block the cascade of transcription factors that coordinate the differentiation process. CCAAT element-binding protein (C/EBP) β and δ and sterol response element-binding protein 1 (ADD1/SREBP1) are active during the early stages of the differentiation process and induce the expression and/or activity of the peroxisome proliferator-activated-receptor gamma (PPARγ), a pivotal coordinator of adipocyte differentiation. Activated PPARγ induces exit from the cell cycle and, in cooperation with C/EBPα, stimulates the expression of many metabolic genes such as Glut-4, lipoprotein lipase (LPL),[38] and adipocyte-specific fatty-acid-binding protein (aP2),[39] thus constituting a functional lipogenic adipocyte. JNK and ERK suppress this process by phosphorylating and thereby attenuating the transcriptional activity of PPAR.[35,36] Besides these integral members of the adipogenesis program, other transcription factors such as AP-1[40] and CREB[41] are known to promote adipogenesis, whereas nuclear factor-κB (NF-κB) suppresses adipocyte differentiation.[42] Therefore, the activity and/or expression of these transcription factors are attractive pharmacological targets for modulating adipocyte tissue formation and deposition.

6.5 α-LIPOIC ACID INHIBITION OF ADIPOCYTE DIFFERENTIATION VIA MAPK PATHWAYS

6.5.1 MAPK-Signaling Pathways Mediate Actions of α-Lipoic Acid on Adipogenesis

Several lines of evidence indicate that proadipogenic transcription factors such as PPARγ and members of C/EBP family can be negatively regulated by MAPKs. Epidermal growth factor, platelet-derived growth factor, lipoxygenase-1 metabolites, and prostaglandin $F_{2\alpha}$ are shown to phosphorylate and attenuate transcriptional activity of PPARγ by activating MAPK-signaling pathways.[43–45] Similarly, α-lipoic-acid treatment of preadipocytes inhibits insulin-induced or hormonal-cocktail-induced transcriptional activity of PPARγ and C/EBPα, which is accompanied by strong activation of ERK and JNK. Furthermore, inhibitors of ERK or JNK activity partly abolish the inhibitory effect of α-lipoic acid on insulin-induced or hormonal-cocktail-induced adipogenesis. On the other hand, α-lipoic acid hardly stimulates phosphorylation of IR or IRS-1 both in preadipocytes and adipocytes at the early stage of differentiation. In particular, upon α-lipoic-acid treatment, a transient Akt phosphorylation is detected in preadipocytes, although it is not detectable in adipocytes at the early stage of differentiation. In contrast, insulin strongly activates IR and IRS-1, and induces long-lasting Akt activation in preadipocytes and adipocytes at the early stage of differentiation. Taken together, these findings exclude possible involvement of Akt activation in α-lipoic-acid-induced inhibition of adipogenesis and demonstrate that LA downregulates PPARγ and C/EBPα through activation of MAPK-signaling pathways.

6.5.2 Modulation of Auxiliary Transcription Factors in Adipogenesis by α-Lipoic Acid

Transcriptional activities of AP-1 (activator protein-1) and CREB (cAMP-responsive element binding protein) are increased in fully differentiated 3T3-L1 adipocytes, as well as after a 2-h treatment with the hormonal cocktail in 3T3-L1 preadipocytes. AP-1 is involved in transcriptional regulation of aP2 (adipocyte-specific fatty acid binding protein) and LPL genes.[46,47] CREB appears to stimulate transcription of several adipocyte-specific genes such as aP2, fatty acid synthetase, and phosphoenolpyruvate carboxykinase.[41] α-Lipoic acid, however, strongly downregulates AP-1 and CREB activities, whereas it upregulates NF-κB activity in preadipocytes. Many antiadipogenic factors, such as proinflammatory cytokines,[48] TNFα,[15] and endrin,[18] are also known to upregulate NF-κB activity, whereas proadipogenic factors such as troglitazone display an opposite effect in 3T3-L1 cells.[24] Considering that AP-1,[49] CREB,[50] and NF-κB[51] mediate major downstream effects of MAPK-signaling pathways, our findings suggest that activation of the MAPK-signaling pathways by α-lipoic acid leads to differential regulation of these transcription factors, which eventually results in decreased expression of the adipocyte-specific genes, and consequently suppresses adipogenesis.

6.5.3 Mediation of α-Lipoic-Acid Actions on Cell Cycle and Clonal Expansion by MAPK-Signaling Pathways

In the course of adipogenesis, one of the first events that occur following hormonal induction is reentry of growth-arrested preadipocytes into the cell cycle. α-Lipoic acid is demonstrated to inhibit the process of clonal expansion when induced by insulin or the hormonal cocktail, indicating that insulin and α-lipoic acid regulate cell cycle progression in opposite ways. This differential effect seems to be due to the potency and/or kinetics of the activation of MAPK- and IR/Akt-signaling pathways. Both insulin and α-lipoic acid activate MAPK-signaling pathways in preadipocytes. However, insulin, but not α-lipoic acid, also strongly activates the IR/Akt-signaling pathway. This observation indicates that progression in the cell cycle and clonal expansion may require activation of both MAPK- and IR/Akt-signaling pathways. On the other hand, insulin-induced MAPK activation is transient, whereas that of α-lipoic acid lasts longer, indicating that duration of MAPK activation might be another important factor in determining the fate of a cell in the cell cycle. Indeed, transient activation of MAPK has been considered as a contributor to cell cycle progression, whereas its prolonged activation can result in cell cycle arrest via induction of p21[Cip1/Waf1] expression and inhibition of cyclin-dependent kinase activity.[52,53] It should be emphasized that JNK is known to activate p53, which triggers activation of several proteins involved in cell cycle arrest, such as p21[Cip1/Waf1].[54] This evidence supports the notion that activation of MAPKs mediates the inhibitory effect of α-lipoic acid on the clonal expansion process by suppressing the expression of the immediate early genes.

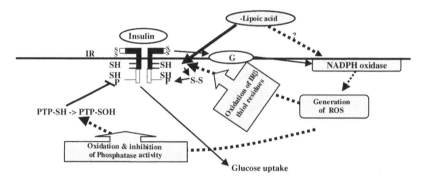

FIGURE 6.2 Proposed mechanism of α-lipoic-acid-induced glucose uptake. α-Lipoic acid may activate adipocyte plasma-membrane-associated NADPH oxidase and increase localized production of H_2O_2, which oxidizes IR thiol groups. Alternatively, α-lipoic acid may directly interact with IR subunit and oxidize its thiol groups. Oxidation of IR thiol groups facilitates its tyrosine autophosphorylation, which eventually results in increased glucose uptake into cells. Intracellularly generated oxidants also inhibit protein tyrosine phosphatases such as PTP1B by oxidation of their catalytic cysteine residues contributing to the activation of the insulin-signaling pathway.

6.6 CONCLUDING REMARKS

The antioxidant activity of α-lipoic acid is considered to contribute to its beneficial effects in the treatment of diabetes. However, our findings now demonstrate that oxidants produced by α-lipoic acid are involved in activation of IR and in the inactivation of PTP, leading to elevated glucose uptake into 3T3-L1 adipocytes (Figure 6.2). These findings indicate that the pro-oxidant effects of α-lipoic acid in tissues such as adipose tissue with a relatively low capacity to reduce α-lipoic acid may contribute to the glucose-lowering effects of α-lipoic acid.

Although α-lipoic acid increases glucose uptake into muscle or adipose tissue by activating the IR/Akt-signaling pathway, the underlying mechanism for regulation of adipogenesis by α-lipoic acid appears to be different. Our findings strongly suggest that α-lipoic acid regulates adipogenesis mainly through activation of MAPKs such as ERK and JNK independent of activation of IR/Akt-signaling pathway (Figure 6.3). Several PPAR agonists, such as thiazolidinediones, have been recommended for the treatment of diabetes as they improve insulin sensitivity and glucose uptake.[55] Treatment with current PPAR agonists, however, leads to increased adiposity and body-weight gain in rodents,[5] which subsequently contribute to the enhanced insulin resistance. In our study, α-lipoic acid at lower concentrations (100 μ*M*) promotes adipogenesis, whereas at higher concentrations (250 and 500 μ*M*), it is inhibitory. Importantly, α-lipoic acid inhibits adipogenesis induced by insulin or troglitazone, indicating that cotreatment with α-lipoic acid may be beneficial in preventing obesity induced by PPAR agonists by maintaining optimal adipogenesis.

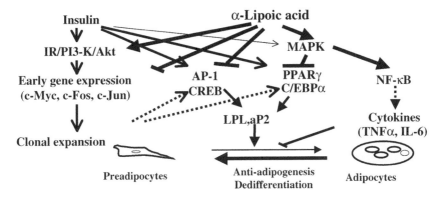

FIGURE 6.3 Regulation of adipocyte differentiation by α-lipoic acid. α-Lipoic acid activates both the IR/PI3-K/Akt pathway and the MAPK pathway. Activation of MAPK pathway mainly leads to the inhibition of proadipogenic transcription factors, such as PPARγ, C/EBPα, AP-1, and CREB, and the activation of antiadipogenic factors such as NF-κB. The inhibition of proadipogenic transcription factors decreases transcription of several proteins including LPL and aP2, which are two major markers of adipogenesis. In contrast, activation of NF-κB may enhance cytokine release and inhibit adipogenesis. α-Lipoic acid may also prevent the adipogenic process by regulating the expression of immediate early genes, which are involved in the process of clonal expansion.

In conclusion, α-lipoic acid may be beneficial in type 2 diabetes by means of two major mechanisms (Figure 6.4). α-Lipoic acid may oxidize IR thiol groups and thereby increase glucose uptake into adipocytes. Moreover, α-lipoic acid may inhibit obesity-induced insulin resistance by decreasing adipocyte formation and inducing process of adipocyte dedifferentiation. Further studies are warranted to evaluate the importance of alteration of intracellular redox status and adipogenesis by α-lipoic acid in controlling glucose metabolism in animal models of diabetes as well as in diabetic patients.

ACKNOWLEDGMENTS

This study was supported by Molecular and Cellular Biodiscovery Research Group from the Ministry of Science and Technology, Republic of Korea. The authors thank Hee-Kyung Shon and Hae-Eun Moon for their technical assistance.

ABBREVIATIONS

aP2: adipocyte-specific fatty acid binding protein
AP-1: activator protein-1
C/EBP: CCAAT element binding protein
CREB: cAMP-responsive element binding protein
DHLA: dihydrolipoic acid

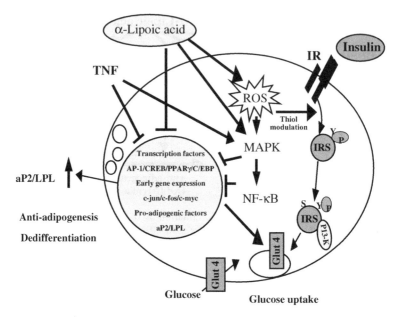

FIGURE 6.4 Mechanisms by which α-lipoic acid may exert beneficial effects in type 2 diabetes. α-Lipoic acid enhances glucose uptake into adipocytes by activating IR. Moreover, α-lipoic acid inhibits adipogenesis and induces dedifferentiation by modulating several pro- or antiadipogenic transcription factors through regulation of the MAPK-signaling pathway.

DPI: diphenylene iodonium
ERK: extracellular signal-regulated kinase
GSH: glutathione
IR: insulin receptor
IRS: insulin receptor substrate
JNK: c-Jun NH (2)-terminal kinase
LA: α-lipoic acid
LPL: lipoprotein lipase
MAPK: mitogen-activated protein kinase
NAC: N-acetyl-L-cysteine
NF-κB: nuclear factor-κB
PTP: protein tyrosine phosphatase
PI3-K: phosphoinositol-3-kinase
p38K: p38 mitogen-activated protein kinase
PPAR: peroxisome-proliferator-activated receptor
PPRE: PPAR-responsive element
Rb: retinoblastoma
TNF: tumor necrosis factor

REFERENCES

1. Czech MP, Fain JN. Cu ++ -dependent thiol stimulation of glucose metabolism in white fat cells. *J Biol Chem* 1972; 247(19): 6218–6223.
2. Krieger-Brauer HI, Medda PK, Kather H. Insulin-induced activation of NADPH-dependent H2O2 generation in human adipocyte plasma membranes is mediated by Galphai2. *J Biol Chem* 1997; 272(15): 10135–10143.
3. Mahadev K, Wu X, Zilbering A, Zhu L, Lawrence JT, Goldstein BJ. Hydrogen peroxide generated during cellular insulin stimulation is integral to activation of the distal insulin signaling cascade in 3T3-L1 adipocytes. *J Biol Chem 2001*; 276(52): 48662–48669.
4. Moini H, Tirosh O, Park YC, Cho KJ, Packer L. R-α-lipoic acid action on cell redox status, the insulin receptor, and glucose uptake in 3T3-L1 adipocytes. *Arch Biochem Biophys* 2002; 397(2): 384–391.
5. Hallakou S, Doare L, Foufelle F, Kergoat M, Guerre-Millo M, Berthault MF, Dugail I, Morin J, Auwerx J, Ferre P. Pioglitazone induces *in vivo* adipocyte differentiation in the obese Zucker fa/fa rat. *Diabetes* 1997; 46(9): 1393–1399.
6. Berg A, de Kok A. 2-Oxo acid dehydrogenase multienzyme complexes. The central role of the lipoyl domain. *Biol Chem.* 1997; 378(7): 617–634.
7. Packer L. Biothiols: Monothiols and dithiols, protein thiols, and thiyl radicals. *Methods in Enzymology.* Vol. 251. New York: Academic Press, 1995.
8. Jacob S, Henriksen EJ, Schiemann AL, Simon I, Clancy DE, Tritschler HJ, Jung WI, Augustin HJ, Dietze GJ. Enhancement of glucose disposal in patients with type 2 diabetes by alpha-lipoic acid. *Arzneimittelforschung* 1995; 45(8): 872–874.
9. Ziegler D, Hanefeld M, Ruhnau KJ, Hasche H, Lobisch M, Schutte K, Kerum G, Malessa R. Treatment of symptomatic diabetic polyneuropathy with the antioxidant alpha-lipoic acid: a 7-month multicenter randomized controlled trial (ALADIN III Study). ALADIN III Study Group. Alpha-Lipoic Acid in Diabetic Neuropathy. *Diabetes Care* 1999; 22(8): 1296–1301.
10. Jacob S, Henriksen EJ, Schiemann AL, Simon I, Clancy DE, Tritschler HJ, Jung WI, Augustin HJ, Dietze GJ. Improvement of insulin-stimulated glucose-disposal in type 2 diabetes after repeated parenteral administration of thioctic acid. *Exp Clin Endocrinol Diabetes* 1996; 104(3): 284–288.
11. Jacob S, Ruus P, Hermann R, Tritschler HJ, Maerker E, Renn W, Augustin HJ, Dietze GJ, Rett K. Oral administration of RAC-alpha-lipoic acid modulates insulin sensitivity in patients with type 2 diabetes mellitus: a placebo-controlled pilot trial. *Free Radic Biol Med* 1999; 27(3–4): 309–314.
12. Henriksen EJ, Jacob S, Streeper RS, Fogt DL, Hokama JY, Tritschler HJ. Stimulation by alpha-lipoic acid of glucose transport activity in skeletal muscle of lean and obese Zucker rats. *Life Sci* 1997; 61(8): 805–812.
13. Yaworsky K, Somwar R, Ramlal T, Tritschler HJ, Klip A, Yaworsky K. Engagement of the insulin-sensitive pathway in the stimulation of glucose transport by alpha-lipoic acid in 3T3-L1 adipocytes. *Diabetologia* 2000; 43(3): 294–303.
14. Konrad D, Somwar R, Sweeney G, Yaworsky K, Hayashi M, Ramlal T, Klip A. The antihyperglycemic drug alpha-lipoic acid stimulates glucose uptake via both GLUT4 translocation and GLUT4 activation: potential role of p38 mitogen-activated protein kinase in GLUT4 activation. *Diabetes* 2001; 50(6): 1464–1471.

15. Schmid E, Hotz-Wagenblatt A, Hacj V, Droge W. Phosphorylation of the insulin receptor kinase by phosphocreatine in combination with hydrogen peroxide: the structural basis of redox priming. *FASEB J* 1999; 13(12): 1491–1500.

16. Hayes GR, Lockwood DH. Role of insulin receptor phosphorylation in the insulinomimetic effects of hydrogen peroxide. *Proc Natl Acad Sci USA* 1987; 84(22): 8115–8119.

17. Koshio O, Akanuma Y, Kasuga M. Hydrogen peroxide stimulates tyrosine phosphorylation of the insulin receptor and its tyrosine kinase activity in intact cells. *Biochem J* 1988; 250(1): 95–101.

18. Mahadev K, Zilbering A, Zhu L, Goldstein BJ. Insulin-stimulated hydrogen peroxide reversibly inhibits protein-tyrosine phosphatase 1b *in vivo* and enhances the early insulin action cascade. *J Biol Chem* 2001; 276(24): 21938–21942.

19. Karlsson FA, Grunfeld C, Kahn CR, Roth J. Regulation of insulin receptors and insulin responsiveness in 3T3-L1 fatty fibroblasts. *Endocrinology* 1979; 104(5): 1383–1392.

20. Rubin CS, Rubin CS, Hirsch A, Fung C, Rosen OM. Development of hormone receptors and hormonal responsiveness *in vitro*. Insulin receptors and insulin sensitivity in the preadipocyte and adipocyte forms of 3T3-L1 cells. *J Biol Chem* 1978; 253(20): 7570–7578.

21. Shimizu M, Torti F, Roth RA. Characterization of the insulin and insulin-like growth factor receptors and responsitivity of a fibroblast/adipocyte cell line before and after differentiation. *Biochem Biophys Res Commun* 1986; 137(1): 552–558.

22. Marin-Hincapie M, Garofalo RS. Drosophila insulin receptor: lectin-binding properties and a role for oxidation-reduction of receptor thiols in activation. *Endocrinology* 1995; 136(6): 2357–2366.

23. Dicter N, Madar Z, Tirosh O. Alpha-lipoic acid inhibits glycogen synthesis in rat soleus muscle via its oxidative activity and the uncoupling of mitochondria. *J Nutr* 2002;132(10): 3001–3006.

24. Sen CK, Roy S, Han D, Packer L. Regulation of cellular thiols in human lymphocytes by alpha-lipoic acid: a flow cytometric analysis. *Free Radic Biol Med* 1997; 22(7): 1241–1257.

25. Krieger-Brauer HI, Kather H. Human fat cells possess a plasma membrane-bound H2O2-generating system that is activated by insulin via a mechanism bypassing the receptor kinase. *J Clin Invest* 1992; 89(3): 1006–1013.

26. Kirsch M, De Groot H. NAD(P)H, a directly operating antioxidant? *FASEB J* 2001; 15(9): 1569–1574.

27. Denu JM, Dixon JE. Protein tyrosine phosphatases: mechanisms of catalysis and regulation. *Curr Opin Chem Biol* 1998; 2(5): 633–641.

28. Herrlich P, Bohmer FD. Redox regulation of signal transduction in mammalian cells. *Biochem Pharmacol* 2000; 59(1): 35–41.

29. Elchebly M, Payette P, Michaliszyn E. Increased insulin sensitivity and obesity resistance in mice lacking the protein tyrosine phosphatase-1B gene. *Science* 1999; 283: 1544–1548.

30. Leong KS, Wilding JP. Obesity and diabetes: *Baillieres Best Pract Res Clin Endocrinol Metab* 1999; 13(2): 221–237.

31. Caro JF, Dohm LG, Pories WJ, Sinha MK. Cellular alterations in liver, skeletal muscle, and adipose tissue responsible for insulin resistance in obesity and type II diabetes. *Diabetes Metab Rev* 1989; 5(8): 665–689.

32. Torti FM, Torti SV, Larrick JW, Ringold GM. Modulation of adipocyte differentiation by tumor necrosis factor and transforming growth factor beta. *J Cell Biol* 1989; 108(3): 1105–1113.

33. Zhang B, Berger J, Zhou G, Elbrecht A, Biswas S, White-Carrington S, Szalkowski D, Moller DE, Zhang B. Insulin- and mitogen-activated protein kinase-mediated phosphorylation and activation of peroxisome proliferator-activated receptor gamma. *J Biol Chem* 1996; 271(50): 31771–31774.

34. Xia X, Serrero G. Inhibition of adipose differentiation by phosphatidylinositol 3-kinase inhibitors. *J Cell Physiol* 1999; 178(1): 9–16.

35. Font de Mora J, Porras A, Ahn N, Santos E. Mitogen-activated protein kinase activation is not necessary for, but antagonizes, 3T3-L1 adipocytic differentiation. *Mol Cell Biol* 1997; 17(10): 6068–6075.

36. Camp HS, Tafuri SR, Leff T. c-Jun N-terminal kinase phosphorylates peroxisome proliferator-activated receptor-gamma1 and negatively regulates its transcriptional activity. *Endocrinology* 1999; 140(1): 392–397.

37. Engelman JA, Lisanti MP, Scherer PE. Specific inhibitors of p38 mitogen-activated protein kinase block 3T3-L1 adipogenesis. *J Biol Chem* 1998; 273(48): 32111–32120.

38. Schoonjans K, Peinado-Onsurbe J, Lefebvre AM, Heyman RA, Briggs M, Deeb S, Staels B, Auwerx J. PPARα and PPARγ activators direct a distinct tissue-specific transcriptional response via a PPRE in the lipoprotein lipase gene. *EMBO J* 1996; 15(19): 5336–5348.

39. Tontonoz P, Hu E, Graves RA, Budavari AI, Spiegelman BM. mPPAR gamma 2: tissue-specific regulator of an adipocyte enhancer. *Genes Dev* 1994; 8(10): 1224–1234.

40. Yang VW, Christy RJ, Cook JS, Kelly TJ, Lane MD. Mechanism of regulation of the 422(aP2) gene by cAMP during preadipocyte differentiation. *Proc Natl Acad Sci USA* 1989; 86(10): 3629–3633.

41. Reusch JE, Colton LA, Klemm DJ. CREB activation induces adipogenesis in 3T3-L1 cells. *Mol Cell Biol* 2000; 20(3): 1008–1020.

42. Ruan H, Hacohen N, Golub TR, Van Parijs L, Lodish HF. Tumor necrosis factor-alpha suppresses adipocyte-specific genes and activates expression of preadipocyte genes in 3T3-L1 adipocytes: nuclear factor-κB activation by TNF-alpha is obligatory. *Diabetes* 2002; 51(5): 1319–1336.

43. Camp HS, Tafuri SR. Regulation of peroxisome proliferator-activated receptor gamma activity by mitogen-activated protein kinase. *J Biol Chem* 1997; 272(16): 10811–10816.

44. Hsi LC, Wilson L, Nixon J, Eling TE. 15-Lipoxygenase-1 metabolites downregulate peroxisome proliferator-activated receptor gamma via the MAPK signaling pathway. *J Biol Chem* 2001; 276(37): 34545–34552.

45. Reginato MJ, Krakow SL, Bailey ST, Lazar MA. Prostaglandins promote and block adipogenesis through opposing effects on peroxisome proliferator-activated receptor gamma. *J Biol Chem* 1998; 273(4): 1855–1858.

46. Distel RJ, Ro HS, Rosen BS, Groves DL, Spiegelman BM. Nucleoprotein complexes that regulate gene expression in adipocyte differentiation: direct participation of c-fos. *Cell* 1987; 49(6): 835–844.

47. Homma H, Kurachi H, Nishio Y, Takeda T, Yamamoto T, Adachi K, Morishige K, Ohmichi M, Matsuzawa Y, Murata Y. Estrogen suppresses transcription of lipoprotein lipase gene. Existence of a unique estrogen response element on the lipoprotein lipase promoter. *J Biol Chem* 2000; 275(15): 11404–11411.

48. Renard P, Raes M. The proinflammatory transcription factor NFκB: a potential target for novel therapeutical strategies. *Cell Biol Toxicol* 1999; 15(6): 341–344.

49. Czech MP, Lawrence JCJ, Lynn WS. Evidence for the involvement of sulfhydryl oxidation in the regulation of fat cell hexose transport by insulin. *Proc Natl Acad Sci USA* 1974; 71(10): 4173–4177.

50. Wilden PA, Pessin JE. Differential sensitivity of the insulin-receptor kinase to thiol and oxidizing agents in the absence and presence of insulin. *Biochem J* 1987; 245(2): 325–331.

51. Clark S, Konstantopoulos N. Sulphydryl agents modulate insulin- and epidermal growth factor (EGF)-receptor kinase via reaction with intracellular receptor domains: differential effects on basal versus activated receptors. *Biochem J* 1993; 292 (Pt. 1): 217–223.

52. Chen JJ, Kosower NS, Petryshyn R, London IM. The effects of N-ethylmaleimide on the phosphorylation and aggregation of insulin receptors in the isolated plasma membranes of 3T3-F442A adipocytes. *J Biol Chem*, 1986; 261(2): 902–908.

53. Jhun BH, Hah JS, Jung CY. Phenylarsine oxide causes an insulin-dependent, GLUT4-specific degradation in rat adipocytes. *J Biol Chem*, 1991; 266(33): 22260–22265.

54. Yang J, Clark AE, Harrison R, Kozka IJ, Holman GD. Trafficking of glucose transporters in 3T3-L1 cells. Inhibition of trafficking by phenylarsine oxide implicates a slow dissociation of transporters from trafficking proteins. *Biochem J* 1992; 281(Pt. 3): 809–817.

55. Tirosh O, Sen CK, Roy S, Packer L. Cellular and mitochondrial changes in glutamate-induced HT4 neuronal cell death. *Neuroscience* 2000; 97(3): 531–541.

7 The Mechanism of Immune Suppression by GIF and Cysteinylated MIF at Cys-60

Yasuyuki Ishii
Research Center for Allergy and Immunology,
Yokohama, Japan

Aoi Son
Kyoto University, Kyoto, Japan

Norihiko Kondo
Research Center for Allergy and Immunology,
Yokohama, Japan

Junko Sakakura-Nishiyama
Redox Bioscience, Kyoto, Japan

Yoshiyuki Matsuo and Junji Yodoi
Kyoto University, Kyoto, Japan

CONTENTS

7.1 INTRODUCTION

Thioredoxin (TRX) superfamily proteins that contain a conserved redox active site -Cys-X-X-Cys- include the proinflammatory cytokine, macrophage-migration-inhibiting factor (MIF), and the immune regulatory cytokine, glycosylation-inhibiting factor (GIF). Although they share an identical gene structure, each performs a unique biological activity. We have previously described GIF, a 13-kDa cytokine, as a product of suppressor T (Ts) cells [1,2] and a subunit of antigen-specific Ts cell factor [3,4]. Repeated injections of partially purified GIF into BDF1 mice resulted in suppression of both IgE and IgG antibody responses to ovalbumin [5]. After molecular cloning of this cytokine, however, we realized that the sequence of the coding region of human GIF cDNA [6] was identical to the sequence of human MIF cDNA [7], except for one base. In the human genome, Paralkar and Wistow [8] identified only one functional MIF-like gene whose predicted transcript sequence agreed exactly with that of human GIF cDNA, indicating that the one base difference between GIF and MIF cDNA is due to an error in sequencing. The nucleotide sequence of mouse MIF cDNA, described by Bernhagen et al. [9], is identical to the sequence of mouse GIF cDNA [6]. Thus, it appears that GIF and MIF share an identical gene in both species.

Another unexpected finding was that the GIF or MIF gene is expressed in essentially all the murine and human cell lines examined [6]. Many of these cell lines secreted the 13-kDa peptide that reacted with polyclonal antibodies against recombinant human (rh) GIF; however, only the 13-kDa peptide secreted from suppressor T (Ts) hybridomas demonstrated GIF bioactivity [10]. *Escherichia coli*–derived rhGIF/MIF also lacked GIF bioactivity. It was also found that both the murine Ts hybridomas and the stable transfectant of hGIF cDNA in the Ts hybridoma secreted bioactive GIF and contained a substantial quantity of inactive GIF in the cytosol, and that the amino acid sequence of the cytosolic inactive GIF peptide was identical to that of the bioactive homologue in culture supernatants [11]. These findings collectively suggested to us the possibility that bioactive GIF is generated by posttranslational modifications of the inactive GIF peptide in Ts cells and that heterogeneity of GIF bioactivity is due to conformational transition of the same peptide [10]. In our present study, we attempted to identify the biochemical nature of the posttranslational modification of GIF and MIF in Ts cells.

7.2 DISCUSSION

GIF in the culture supernatants of the cedar pollen-specific Ts hybridoma 31E9 cells was affinity-purified by using anti-GIF Ab-coupled HiTrap column. Mass-spectrometric analysis of the purified GIF revealed four molecular weight (m.w.) species of 12,346, 12,429, 12,467, and 12,551, respectively (Figure 7.1). The m.w. of the smallest species was identical to the theoretical value calculated from the amino acid sequence of GIF, which lacked the first methionine in the deduced sequence, and to the m.w. of cytosolic GIF [13]. Thus, the remaining three species with higher m.w. appear to be the molecules posttranslationally modified in 31E9

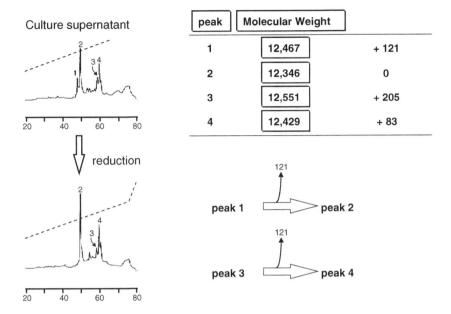

FIGURE 7.1 Analysis of the GIF species in the culture supernatant of 31E9 cells.

cells. As reported in a previous paper [13], inactivation of bioactive GIF from the 31E9 cells by treatment with 1 m*M* DTT was accompanied by conversion of the 12,467 and 12,551 m.w. species to the 12,346 and 12,429 m.w. species, respectively (Figure 7.1). The results indicate that posttranslational modifications consist of two steps: covalent binding of a chemical group of m.w. 83 and the binding of a group of m.w. 121, with the latter susceptible to reducing reagents.

We have also examined the physicochemical properties of GIF protein secreted from BUC cells, which were employed as the fusion partner for establishing the Ts hybridoma 31E9 cells. Affinity purification of the GIF protein followed by mass-spectrometric analysis showed that the GIF protein secreted from BUC cells was homogeneous and represented the 12,346 m.w. species (data not shown).

To determine the posttranslational modifications, both the purified GIF from the 31E9 cells, described earlier, and rhGIF were digested with API and AspN and the digests were fractionated by reverse-phase chromatography. Comparisons between the peptide maps of rhGIF and the 31E9-derived GIF revealed that two peptides (peptide A and peptide B) present in the digest of the latter preparation do not exist in the digest of rhGIF (data not shown). The molecular weights of peptide A and peptide B, determined by mass-spectrometry, were 2469.0 and 2719.5, respectively. To determine which amino acids in peptide A and peptide B were modified, the peptides were digested with carboxypeptidase Y and the fragments isolated using reverse-phase chromatography were analyzed by mass-spectrometry. The results indicated that a chemical group of 119.2 Da must be

peak	Molecular Weight		activity
1	12,467	+ 121	active
2	12,346	0	inactive
3	12,551	+ 205	active
4	12,429	+ 83	inactive?

FIGURE 7.2 Biological activity of native GIF species.

covalently bound to one of the four amino acids from Cys-60 through His-63. Previous experiments have shown that a chemical group of 120 Da dissociated from the GIF sequence on treatment with a reducing reagent [13], suggesting that the group was bound to the peptide through a disulfide bond. Among the four amino acid residues, Cys-60 was the only amino acid that could form a disulfide bond. Thus, the results collectively indicate that the modification would be cysteinylation or attachment of a second cysteine residue to Cys-60 via a disulfide bond, which increases the m.w. of the peptide by 120.

Similar analysis of the digest of peptide B showed that the m.w. of one fragment, Leu-79 to Asp-93, was larger than the theoretical value by 80.0 Da, whereas the m.w. of another fragment, Leu-79 to Ile-90, was identical to that of the theoretical value (Figure 7.2). The results suggested that a chemical group of 80.0 Da is associated with one of the amino acid residues of Ser-91, Phe-92, and Asp-93. An increase in the m.w. of 80.0 Da is consistent with phosphorylation. Among the three amino acid residues, only Ser-91 has the potential to be phosphorylated. Thus, we concluded that one of the chemical modifications of GIF in the Ts cells would be phosphorylation of Ser-91.

The affinity-purified GIF from the culture supernatant of the 31E9 cells was fractionated by gel filtration through a Superdex 75 column. As shown in Figure 7.3, GIF could be fractionated into three peaks. Mass-spectrometric analysis of each peak showed that peak 1 represented cysteinylated GIF (m.w. 12,467), whereas peak 2 and peak 3 were mixtures of phosphorylated GIF (m.w. 12,429)

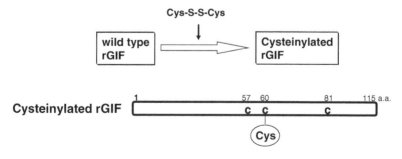

FIGURE 7.3 Preferential cysteinylation at Cys-60 of recombinant GIF/MIF protein.

and unmodified GIF (m.w. 12,346). Approximately 30% of the total GIF in a pool of peak 2 and peak 3 was phosphorylated GIF, and the reminder was unmodified GIF. Because the concentration of the 12,551 m.w. species in the original GIF preparation was too low (lower than that observed in Figure 7.1), this species could not be isolated in this fractionation. Experiments were carried out to determine the immunosuppressive activities of posttranslationally modified GIF species. Determination of the immunosuppressive activities of the peak 1 fraction and that of the pool of peak 2 and peak 3 on the *in vitro* IgE antibody response clearly showed that cysteinylated GIF (peak 1) suppressed the response, whereas the mixture of phosphorylated GIF and unmodified species failed to do so even though the latter fraction contained twice as much GIF protein compared to the peak 1 fraction (Figure 7.2).

In order to confirm the critical role of cysteinylation at Cys-60 for the generation of immunosuppressive activity, attempts were made to prepare a cysteinylated derivative of rhGIF. The wild-type rhGIF was treated with cystine, and the product was fractionated on a CM-5PW column. To confirm that Cys-60 in rhGIF was selectively cysteinylated, the cysteinylated rhGIF was digested with API and AspN and the digest was subjected to peptide mapping. The profile of the peptides was essentially the same as that of 31E9-derived GIF, except that peptide B was not detectable. Digestion of peptide A from the cysteinylated rhGIF with carboxypeptidase Y followed by mass-spectrometric analysis of the fragments confirmed that Cys-60 in the sequence was cysteinylated. Thus, we determined that the immunosuppressive activity of the cysteinylated rhGIF was comparable to that of the cysteinylated GIF isolated from the culture supernatant of the 31E9 cells.

7.3 BINDING CAPACITY OF CYSTEINYLATED GIF DERIVATIVES TO TARGET CELLS

Previous experiments have shown that bioactive GIF from Ts cells and bioactive derivatives of rhGIF bind to the receptors on Th hybridomas, activated T and B cells with high affinity, whereas inactive cytosolic GIF and *E. coli*–derived wild-type rhGIF failed to do so [20,21]. High-affinity binding capacity of GIF molecules for the target cells was generated by replacement of Cys-57 with Ala and of Asn-106 with Ser, or by binding of 5-thio-2-nitrobenzoic acid group to Cys-60 in the C57A molecules. Equilibrium dissociation constant (K_d) of the specific binding between the high-affinity receptors on the Th hybridoma 12H5 cells and bioactive rhGIF derivatives, such as C57A/N106S and C57A-DTNB, was in the 10 to 100 pM range [20].

Because the bioactive GIF species in the culture supernatant of 31E9 cells appears to be cysteinylated GIF, we determined the ability of the 31E9-derived cysteinylated GIF and cysteinylated rhGIF to inhibit the binding of ^{125}I-labeled C57A/N106S to the 12H5 cells. The results of the experiments are shown in Figure 7.4. As expected, wild-type rhGIF failed to inhibit the binding of radiolabeled C57A/N106S even when present with a 100-fold excess, but both the

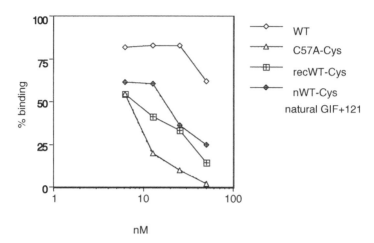

nM

FIGURE 7.4 Inhibition of the binding of ^{125}I-labeled C57A/N106S to 12H5 cells by cysteinylated GIF proteins. Radiolabeled C57A/N106S was mixed with various concentrations of rhGIF (), C57A/N106S (), nGIF-Cys (), or rhGIF-Cys () and the mixtures were incubated with 1×10^6 cells. The final concentration of ^{125}I-labeled C57A/N106S was 1 nM. The ordinate represents the ratio between specifically bound radioactivity in the presence of unlabeled GIF and that in the absence of GIF.

31E9-derived cysteinylated GIF and cysteinylated rhGIF inhibited the binding in a dose-dependent manner. It should be noted that the ability of the cysteinylated rhGIF to inhibit the binding of ^{125}I-labeled C57A/N106S was comparable to that of the homologous ligand. Conversely, C57A/N106S inhibited the specific binding of ^{125}I-labeled cysteinylated rhGIF to 12H5 cells as effectively as unlabeled cysteinylated rhGIF (data not shown).

The present results indicate that bioactivity of GIF is generated by the binding of cysteine to Cys-60. The x-ray crystal structure of rhGIF suggested high conformational flexibility in the α-helices and adjacent loop region in the rhGIF [22] (Figure 7.5). We wondered whether cysteinylation of Cys-60 might have induced conformational changes in these regions. To test this possibility, we prepared polyclonal antibodies against the peptide region, Ala-58 to Arg 74 (HG3), which forms the β4 strand, loop, and amino terminal one third of the α2 helix in rhGIF. SDS-PAGE of rhGIF followed by immunoblotting with the specifically purified anti-HG3 clearly showed that the antibodies were bound to the 13-kDa GIF under these experimental conditions (Figure 7.6). However, fractionation of the wild-type rhGIF and cysteinylated rhGIF on the anti-HG3-coupled immunosorbent showed that rhGIF failed to be retained in the column. In contrast, cysteinylated rhGIF bound to the column and was recovered by elution at acid pH (Figure 7.6). The results suggested that the epitopes recognized by the antibodies are hidden in the wild-type GIF molecules but exposed by the modifications that result in the bioactive GIF derivatives.

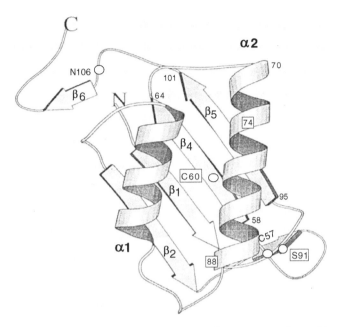

FIGURE 7.5 Ribbon diagram of GIF monomer. Numbers represent the position of amino acid residues.

Because the peptide employed for the preparation of the antibodies covers the amino acid sequence for the β4 strand and α2 helix in rhGIF (Figure 7.6), attempts were made to further map the epitope recognized by the antibodies. Thus, cysteinylated rhGIF was fractionated on the anti-HG3-coupled column in the presence of a synthetic peptide, HG3a, which covers the β4 strand and loop region, or in the presence of peptide HG3b, which covers the loop region and α2 helix in the original rhGIF molecules. The results, shown in Figure 7.6, indicated that HG3a did not inhibit the binding of the cysteinylated rhGIF to the immunosorbent, whereas the peptide HG3b completely inhibited the binding. These findings do not exclude the possibility that the antibody preparation contains antibodies against HG3a but simply indicate that such antibodies, if they exist, do not bind either rhGIF or the cysteinylated derivative. In any event, binding of the antibodies specific for HG3b to cysteinylated rhGIF, but not to the wild-type rhGIF, indicates that certain critical amino acid residues in the region of the HG3b peptide are exposed to the solvent face of the cysteinylated rhGIF and are capable of binding the anti-HG3 antibodies, but the epitope is hidden in the wild-type rhGIF. These results strongly suggest that cysteinylation of Cys-60 causes a conformational change in the α2 helix region (Figure 7.5). Taken together, cysteinylation of Cys-60 in the GIF sequence occurs in Ts cells, and this modification is responsible for the generation of bioactivity.

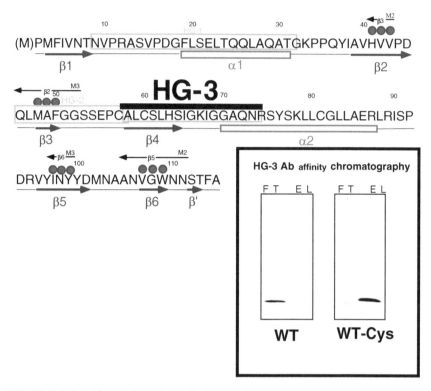

FIGURE 7.6 Specific binding of cysteinylated rhGIF derivatives to anti-HG3. Immuno-blotting of rhGIF with anti-HG3 antibodies. rhGIF (1) or cysteinylated rhGIF (2) were fractionated on anti-HG3 Abs-coupled HiTrap, and the flow-through (FT) and acid-eluate (EL) fractions were analyzed by immunoblotting with anti-GIF. (C) A 10 μg aliquot of cysteinylated rhGIF was mixed with 1 mg/ml of either HG3a or HG3b peptide and the mixture was fractionated on anti-HG3-coupled column. The fractions were analyzed by immunoblotting with anti-GIF. Amino acid sequences of HG3a and HG3b peptides are shown above the immunoblot.

   To clarify the mechanism of the binding of cysteinylated GIF but not MIF to the activated lymphocytes, attempts were made to biochemically identify the GIF receptor on murine B cell line using a cross-linking reagent, which forms uncleavable covalent linkages under reducing conditions. The treated cells were solubilized in lysis buffer and a portion of the fraction was analyzed by 2-D gel electrophoresis followed by immunoblotting with anti-GIF polyclonal antibodies. As shown in Figure 7.7, a single spot of 75 to 80 kDa with pI 5 to 6 was obtained when the cells were incubated with cysteinylated GIF prior to treatment with the cross-linking reagent, but not wild-type GIF. Further research on biochemical analysis of this candidate protein of the GIF receptor is ongoing.

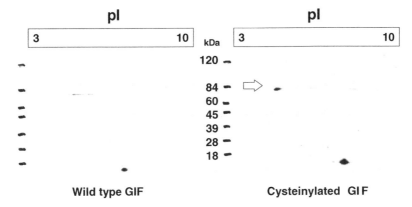

FIGURE 7.7 2-D gel electrophoresis analysis of the microsomal fractions from cells treated with wild-type rGIF (in left panel) or cysteinylated GIF (in right panel). The complexes of rGIFs with binding targets were identified by Western blot analysis using anti-GIF antibodies.

REFERENCES

1. Ishizaka, K. (1984) *Annu Rev Immunol* 2, 159–182.
2. Jardieu, P., Uede, T., and Ishizaka, K. (1984) *J Immunol* 133, 3266–3273.
3. Nakano, T., Ishii, Y., and Ishizaka, K. (1996) *J Immunol* 156, 1728–1734.
4. Ishii, Y., Nakano, T., and Ishizaka, K. (1996) *J Immunol* 156, 1735–1742.
5. Akasaki, M., Jardieu, P., and Ishizaka, K. (1986) *J Immunol* 136, 3172–3179.
6. Mikayama, T., Nakano, T., Gomi, H., Nakagawa, Y., Liu, Y.-C., Sato, M., Iwamatsu, A., Ishii, Y., Weiser, W.Y., and Ishizaka, K. (1993) *Proc Natl Acad Sci USA* 90, 10056–10060.
7. Weiser, W.Y., Temple, P.A., Witek-Giannotti, J.S., Remold, H.G., Clark, S.C., and David, J.R. (1989) *Proc Natl Acad Sci USA* 86, 7522–7526.
8. Paralkar, V. and Wistow, G. (1994) *Genomics* 19, 48–51.
9. Bernhagen, J., Mitchell, R. A., Calandra, T., Voelter, W., Cerami, A., and Bucala, R. (1994) *Biochemistry* 33, 14144–14155.
10. Liu, Y.-C., Nakano, T., Elly, C., and Ishizaka, K. (1994) *Proc Natl Acad Sci USA* 91, 11227–11231.
11. Nakano, T., Liu, Y.-C., Mikayama, T., Watarai, H., Taniguchi, M., and Ishizaka, K. (1995) *Proc Natl Acad Sci USA* 92, 9196–9200.
12. Nakano, T., Watarai, H., Liu, Y.-C., Oyama, Y., Mikayama, T., and Ishizaka, K. (1997) *Proc Natl Acad Sci USA* 94, 202–207.
13. Tomura, T., Watarai, H., Honma, N., Sato, M., Iwamatsu, A., Kato, Y., Kuroki, R., Nakano, T., Mikayama, T., and Ishizaka, K. (1999) *J Immunol* 162, 195–202.
14. Huff, T. F., and Ishizaka, K. (1984) *Proc Natl Acad Sci USA* 81, 1514–1515.
15. Iwata, M., Adachi, M., and Ishizaka, K. (1988) *J Immunol* 140, 2534–2542.
16. Sato, T., Sasahara, T., Nakamura, Y., Osaki, T., Hasegawa, T., Tadakuma, T., Arata, Y., Kumagai, Y., Katsuki, M., and Habu, S. (1994) *Eur J Immunol* 24, 1512–1516.

17. Sette, A., Buss, S., Colon, S., Miles, C., and Grey, H. M. (1988) *J Immunol* 141, 45–48.

18. Liu, F. T., Bohn, J. W., Ferry, E. L., Yamamoto, H., Molinaro, C. A., Sherman, L. A., Klinman, N. R., and Katz, D. H. (1980) *J Immunol* 124, 2728–2737.

19. Hirano, T., Miyajima, H., Kitagawa, H., Watanabe, N., Azuma, M., Taniguchi, O., Hashimoto, H., Hirose, S., Yagita, H., Furusawa, S., Ovary, Z., and Okumura, K. (1988) *Int Arch Allergy Appl Immunol* 85, 47–

20. Sugie, K., Nakano, T., Tomura, T., Takakura, K., Mikayama, T., and Ishizaka, K. (1997) *Proc Natl Acad Sci USA* 94, 5278–5283.

21. Sugie, K., Tomura, T., Takakura, K., Kawano, T., Taniguchi, M., and Ishizaka, K. (1999) *Int Immunol* 11, 1149–1156.

22. Kato, Y., Muto, T., Tomura, T., Tsumura, H., Watarai, H., Mikayama, T., Ishizaka, K., and Kuroki, R. (1996) *Proc Natl Acad Sci USA* 93, 3007–3010.

23. Hwang, C., Sinskey, A.J., and Lodish, H. F. (1992) *Science* 257, 1496–1502.

24. Yu, Y.Y.L., Myers, N.B., Hilbert, C.M., Harris, M.R., Balendiran, G.K., and Hansen, T.H. (1999) *Int Immunol* 11, 1897–1905.

25. Wagner, B.L., Bauer, A., Schutz, G., and Montminy, M. (2000) *J Biol Chem* 275, 8263–8266.

26. Calandra, T., Bernhagen, J., Mitchell, R. A., and Bucala, R. (1994) *J Exp Med* 179, 1895–1902.

8 Role of Free Radicals in Cellular Damage and Its Modification by Antioxidants: Implications for Cancer Therapy

K.P. Mishra
Bhabha Atomic Research Centre, Mumbai, India

CONTENTS

8.1 OVERVIEW

Studies in our laboratory have shown that cellular damage, especially that involving membranes, plays an important role in the modification of cellular radiosensitivity. We have investigated the free radical mechanism of radiation oxidative damage to model and cellular membrane, employing fluorescence and electron spin resonance techniques. Results have shown that γ-irradiation of liposomal membrane increased their bilayer rigidity as measured by fluorescence polarization of 1,6-diphenyl hexatriene (DPH) and enhanced lipid peroxidative damage, which was, however, significantly prevented by antioxidants. Moreover, radiation oxidative damage caused radiation-dose-dependent hemolysis in human red cells, which was, however, prevented by catechin, an antioxidant from plant sources. Membrane protein oxidation was found inhibited, as studied by sodium dodecyl sulfate/polyacrylamide gel electrophoresis (SDS-PAGE). Radiation-induced increase in intracellular reactive oxygen species (ROS) in thymocytes was monitored by 2,7-dichlorodihydrofluorescein diacetate (DCH-FDA), which was markedly inhibited in the presence of antioxidants such as ascorbic acid and tocopherol. Results suggest that ROS-induced membrane peroxidative damage contributed to radiation induction apoptosis. Studies on human cervical cancer cells have shown that radiation and ellagic acid affected ROS generation synergistically. It is concluded that radiation-generated free-radical-mediated membrane injury was involved in the mechanism of apoptotic death; this is of relevance to cancer radiotherapy.

8.2 INTRODUCTION

In recent years, considerable interest has been generated among researchers in identifying the molecular targets of radiation and in understanding the mechanism of its action in cell killing. However, the recent advancement in techniques in molecular radiobiology have revealed that radiation produces significant effects on cell membrane and cytoplasm elements/organelles, apart from DNA, leading to alterations in cell signaling, gene expression, and regulation of cell cycle. Radiation-mediated damage is known to occur at different levels of biological organization, involving free radicals generated either directly or through indirect pathways of cells. Moreover, involvement of radicals and the associated pathways in various clinical disorders has been investigated [5]. The major biologic effects of radiation manifest in various end points such as genetic alterations, carcinogenesis, and cell killing. However, there is a significant gap in knowledge about events between radiation exposure and the resultant biologic manifestations, which need to be explored in the future. Evidence exists to show that the cellular membrane is a sensitive target of radiation action, and in recent years researchers have shown keen interest in investigating the involvement of radiation-induced membrane oxidative damage in the loss of cell function and cell killing [1–4]. Damage to critical biomolecules such as DNA, proteins, and lipids of membrane by radiation-generated ROS initiates biochemical reactions resulting in cell death [4,5]. Radiation-mediated damage of model/cellular membrane has been known to result in structural and functional alterations [6–8], which have been implicated in oxidative-stress-induced apoptosis [2,3]. Apoptosis is a significant biological process known to involve pathogenesis, cancer, and organogenesis [9,10]. Therefore, understanding the molecular mechanism and the factors modifying apoptosis [11,12] may have significant implications for improvements in cancer radiotherapy [13]. In the present investigation, the role of free radicals in radiation-induced changes in model and cellular membrane was studied with a view to modifying the radiation response in normal and cancer cells by antioxidants.

8.3 MATERIALS AND METHODS

Egg yolk lecithin (EYL), dipalmitoyl phosphatidyl choline (DPPC), DPH, and thiobarbituric acid (TBA) were purchased from Sigma Chem Co., St. Louis (MO), and RPMI-1640 and fetal calf serum (FCS) were obtained from ICN Chem. Co. 2′-7′-dichlorofluorescein diacetate was procured from Molecular Probes Co., α-tocopherol acetate from Fluka Chem., and ascorbic acid from S.D. Fine Chemicals, Mumbai, India.

Unilamellar liposomes were prepared either from EYL or DPPC by the thin-film hydration method through sonication, as mentioned elsewhere [7]. The radiation-induced membrane oxidative damage was studied by measuring of malondialdehyde (MDA) spectrophotometrically [14]. The change in membrane fluidity in control, as well as in irradiated liposomes, was determined by DPH polarization [7].

Blood samples were collected from volunteers. The suspension of erythrocytes was prepared according to the method described by Hanahan and Ekholm [15]. The washed cells were suspended in isotonic (310 mOsM) tris buffer, pH 7.6. RBC cells were treated with catechin; this was followed by measurement of hemolysis. SDS-polyacrylamide (7.5%) gel electrophoresis was carried out according to the method of Laemmli [16]. Catechin was added to the irradiated ghost membrane immediately after irradiation. Equal amounts (66 µg) of control and irradiated membrane protein were loaded on gel. Gels were fixed and stained with Coomassie R-250.

Thymus was dissected out from Swiss female mice (4 to 5 weeks) according to the instructions from the BARC Animal Ethics Committee (BAEC). The thymocyte cell suspension was prepared in ice-cold RPMI-1640 [11]. Cell suspensions (1×10^7 cells/ml) were irradiated in a culture medium supplemented with FCS by ^{60}Co γ-rays (dose rate: 0.77 Gy/min) followed by incubation at 37°C in a 5% CO_2 environment. The intracellular ROS in the thymocytes after radiation exposure was estimated using DCH-FDA as a fluorescence probe [17]. The percentage apoptosis in the thymocytes was determined by the annexin-V-fluos protocol (Boehringer Mannheim, GmbH, Germany) provided with the kit.

Human cervical cancer cell line, HeLa cells, were procured from the National Centre for Cell Sciences, Pune, India. These cells were cultured and maintained using Dulbecco's Modified Eagle Medium (DMEM) supplemented with 10% FCS and antibiotics (penicillin 100 U ml^{-1} and streptomycin 250 µg ml^1). Cancer cells were treated with ellagic acid (Sigma Chem Co., St. Louis [MO]) or radiation in culture condition, or both, followed by harvesting with trypsinization.

8.4 RESULTS AND DISCUSSION

8.4.1 EFFECT OF RADIATION ON LIPOSOMAL MEMBRANE

Research activities have been pursued in our laboratory to investigate radiation-induced molecular damage in model membranes such as liposomes and human erythrocytes to quantify the associated structural and chemical changes. Results from fluorescence anisotropy measurements in unilamellar EYL and DPPC liposomes showed a significant increase in fluorescence polarization in irradiated liposomes, suggesting a decrease in membrane fluidity by radiation exposure [7,18]. However, the radiation-induced changes were found to be lower in saturated phospholipids (DPPC) than in unsaturated (EYL) lipids. The molecular mechanism of radiation-induced membrane damage in the presence of natural antioxidants was investigated in the liposomal membrane. It has been found that membrane-specific antioxidants such as α-tocopherol are more effective in protecting the DPPC membrane than are antioxidants (ascorbic acid) localized in the aqueous compartment [18]. Furthermore, it was also observed that the membrane oxidative damage measured in terms of MDA formation was decreased significantly in irradiated liposomes (unpublished result), suggesting involvement of the free radical mechanism in membrane oxidative damage induced by radiation.

8.4.2 Modification of Radio-Oxidative Membrane Damage in Erythrocytes by Catechin

An enhancement in hemolysis was observed in irradiated erythrocytes when compared to unirradiated controls monitored spectrophotometrically. However, the membrane damage, expressed in terms of percentage hemolysis, was substantially inhibited when erythrocytes were treated with catechin, an antioxidant from an Indian medicinal tree (Table 8.1) independent of treatment before or after radiation. Catechin was further shown to protect against oxidative damage in membrane proteins owing to radiation in erythrocytes by the SDS-PAGE method (our unpublished result).

TABLE 8.1
Effect of Catechin on Radiation-Induced Hemolysis

Treatment Group	Hemolysis (% ± SEM)[a]
Radiation (20 Gy)	40 ± 2
Catechin before radiation	20 ± 0.8
Catechin after radiation	22 ± 0.5

Note: A 5% (v/v) cell suspension was irradiated to a dose of 20 Gy at the rate of 10 Gy/min. at ambient temperature. Catechin of increasing concentration (0 to 30 mM) was added treated before radiation or immediately after irradiation. Following 2-h irradiation in isotonic PBS, cells were resuspended in hypotonic (0.45% NaCl) PBS. The hemolysis was measured at 540 nM after 1-h incubation in hypotonic PBS, which was the ratio of absorbance of a particular sample to that of a completely hemolyzed sample in distilled water multiplied by 100.

[a]SEM = Standard error of the mean.

8.4.3   OXIDATIVE DAMAGE IN THYMOCYTES AND CANCER CELLS BY RADIATION AND ELLAGIC ACID

An increase in generation of intracellular ROS was observed in irradiated thymocytes, which could be sensitively detected by using DCH-FDA as a fluorescence probe. The presence of natural antioxidants such as ascorbic acid and eugenol resulted in significant inhibition of intracellular ROS generation in thymocytes [19]. Moreover, cellular oxidative stress induced by radiation has been found to be linked to the mechanism of apoptotic death in thymocytes, measured by the externalization of phosphatidyl serine (PS) in membrane and by the propidium iodide (PI) method. Furthermore, treatment of cervical cancer cells with ellagic acid resulted in significant increase in ROS, which was dependent on applied concentrations of ellagic acid (Figure 8.1). It was further observed that treatment of cancer cells with ellagic acid in combination with radiation resulted in significant increase in ROS generation.

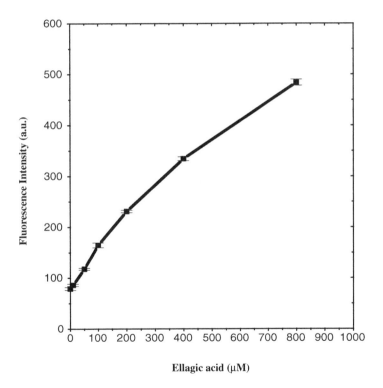

FIGURE 8.1 Effect of ellagic acid on generation of intracellular ROS in HeLa cells. Cultured cancer cells harvested by trypsinization were labeled with DCH-FDA, followed by treatment with ellagic acid of increasing concentrations at 37°C. The change in fluorescence intensity was monitored with a steady-state fluorimeter (LS 50B, Perkin Elmer, USA).

8.5 CONCLUSIONS

The results suggest that radiation-induced structural changes occur in liposomal and cellular membrane after radiation, which seem to be linked to the mechanism of cellular oxidative damage and the resultant apoptotic death.

REFERENCES

1. Halliwell B, Gutteridge JMC. *Free Radical in Biology and Medicine*. Oxford University Press: Oxford, 1989.
2. Ramakrishnan N, McClain DE, Catravas GN. Membrane as a sensitive target in thymocyte apoptosis. *Intl J Radiat Biol* 1993; 63: 693–701.
3. Ojeda F, Diehl HA, Folch, H. Radiation induced membrane changes and programmed cell death: possible interrelationships. *Scann Micros* 1994; 3: 645–651.
4. Stark G. The effect of ionizing radiation on lipid membranes. *Biochim Biophys Acta* 1991; 1071: 103–122.
5. Pandey BN, Mishra KP. Radiation biology for cancer radiotherapy. In: *advances in radiation biology and peace, UP J Zool Society,* Muzaffarnagar, 1999; Suppl. 2: 45–51.
6. Kolling A, Maldonado C, Ojeda F, Diehl HA. Membrane fluidity of microsomal and thymocyte membranes after X-ray and UV irradiation. *Radiat Environ Biophys* 1994; 33: 303–313.
7. Pandey BN, Mishra KP. Radiation induced oxidative damage modification by cholesterol in liposomal membrane. *Radiat Phys Chem* 1999; 54: 481–489.
8. Pandey BN, Mishra KP. Fluorescence and ESR studies on membrane oxidative damage by gamma radiation. *Appl Magn Res* 2000; 18: 483–492.
9. Kerr JFR, Wyllei AH, Currie AR. Apoptosis: a basic biological phenomenon with wide ranging implications in tissue kinetics. *Br J Can* 1972; 26: 239–257.
10. Bowen ID, Bowen SM. *Programmed Cell Death in Tumors and Tissues*. Cambridge University Press: Cambridge, 1990.
11. Pandey BN, Mishra KP. Role of membrane peroxidative damage in cellular apoptotic response to ionizing radiation. *Res Bull PU* 1999; 49: 175–182.
12. Pandey BN, Mishra KP. Oxidative membrane damage and its involvement in radiation induced apoptotic cell death. *Iran J Radiat Res* 2003; 1: 17–22.
13. Thulsidas S, Mishra KP. Radiation oxidative membrane damage and apoptosis with relevance to cancer therapy. *Natl Acad Sci Lett* 2002; 25: 327–338.
14. Konings AWT, Drijver EB. Radiation effects on membrane. 1. Vitamin E deficiency and lipid peroxidation. *Radiat Res* 1979; 80: 494–501.
15. DJ Hanahan, JE Ekholm. The preparation of red cell ghosts (membranes). *Methods Enzymol* 1974; 31: 168–172.
16. UK Laemmli. Cleavage of structural proteins during the assembly of the head of bacteriophage. *Nature* 1970; 227: 680–685.
17. Pandey BN, Mishra KP. *In vitro* studies on radiation induced membrane oxidative damage in apoptotic death of mouse thymocytes. *Intl J Low Radiat* 2003; 1: 113–119.
18. Marathe DL, Mishra KP. Radiation induced changes in permeability in unilamellar phospholipid liposomes. *Radiat Res* 2002; 157: 685–92.
19. Pandey BN, Mishra KP. Modification of thymocytes membrane radio-oxidative damage and apoptosis by eugenol. *J Environ Pathol Toxicol Oncol* (in press).

9 Free Radicals, Cigarette Smoke, and Health Scavenging Effects of Pycnogenol® on Free Radicals Generated from Cigarette Smoke and Its Detoxification Effects

Baolu Zhao
Institute of Biophysics, Beijing, China

CONTENTS

9.1 SMOKING AND HEALTH

It is well known that smoking is harmful for human health as is marked on any pack of cigarettes: "Smoking damages your health" (China); "Tobacco causes serious damage to your health" (U.S.); "Cigarette smoke contains carbon monoxide" (555); "Smoking causes cancer, heart disease, and emphysema and may complicate pregnancy" (Canada). It has been found by epidemiological studies that smoking causes high mortality by initiating heart diseases and cancer. The incidence of heart diseases is proportional to the number of cigarettes smoked, and about 30% of lung cancers are caused by cigarette smoking [1,2]. The excess risk was largely confined to long-term and heavy smokers, with significant two-fold excess risk among men who smoked 40 or more packs per year. Among women, an 80% increase in risk was observed in both current and former smokers but dose-response trends were less consistent than among men [3,4]. The noxious pro-oxidant effects of smoking extend beyond the epicardial arteries to the coronary microcirculation and affect the regulation of myocardial blood flow, causing carotid-media thickness [5,6]. Cigarette smoking is also associated with significantly higher rates of both low birth weight and preterm and other diseases [7,8]. It had been considered in the past that nicotine in tobacco smoke is the most poisonous material. However, recent studies showed that the free radicals in smoke are closely related to the diseases caused by cigarette smoking [4–8]. The free radicals generated by cigarette smoke not only cause lung injury, lung cancer, and heart diseases, but also pollute the environment. Studies showed that some of the polycyclic aromatic hydrocarbon free radicals are direct chemical carcinogens.

Numerous epidemiological studies have reported a highly significant negative association between cigarette smoking and neurodegenerative disorders, especially in Parkinson's disease (PD) [9,10] and Alzheimer's disease (AD) [11,12]. Nicotine is a predominant component of cigarette smoke and is currently being used in pilot clinical studies for the treatment of AD [13]. $\alpha 4\beta 2$ nicotinic receptor activation plays an important role in neuroprotection against $A\beta$ cytotoxicity [14]. Garrido et al. [15] reported that nicotine could exert potent neuroprotective effects by inhibiting arachidonic-acid-induced apoptotic cascades (caspases-3 activation and cytochrome c release) of spinal cord neurons. We found that nicotine could scavenge oxygen free radicals [16] and attenuate β-amyloid-peptide-induced neurotoxicity, and free-radical and calcium accumulation in hippocampal neuronal cultures [17].

9.2 FREE RADICALS PRODUCED BY CIGARETTE SMOKE

There are two kinds of free radicals, the first existing in the condensate and the second in the gas phase of cigarette smoke. The free radicals in condensate are stable free radicals from polycyclic aromatic hydrocarbons and quinones, which can be partly retained by cigarette filters. The free radicals in the gas phase are very reactive, and it is extremely difficult to inactivate these radicals in a filter. First, the cigarette generates NO and NO_2 free radicals; these then react with the alkene compounds generated in the burning process to produce alkyl, alkoxyl, and alkylperoxyl free radicals [1,2].

Free radicals are generated by cigarette smoking through a dynamic process as follows [18]:

$$Carbohydrogen + O_2 \rightarrow R \text{ (alkyl)}$$

$$Proteins + O_2 \rightarrow NO$$

$$2NO + O_2 \rightarrow 2 NO_2$$

$$NO_2 + R \rightarrow R.\text{(carbon center radicals)}$$

$$R. + O_2 \rightarrow ROO\cdot$$

$$ROO\cdot + NO \rightarrow RO\cdot + NO$$

9.3 DETECTION OF FREE RADICALS IN CIGARETTE SMOKE

Free radicals generated in the gas phase of cigarette smoke were detected by the ESR spin-trapping technique. The cigarette was "smoked" by a smoke machine with continuous flow of 400 ml/min or 35 ml/2 sec at a 1-min interval. The collected cigarette tar was filtrated by three Cambrage papers and, thereafter, the smoke was passed through 2 ml of spin-trapping solution. The ESR spectra were measured with a ESR spectrometer at room temperature. ESR conditions: X-band, microwave power 20 mW, 100 kHz modulation with amplitude 1G.

The ESR spectra of free radicals generated in the gas phase of cigarette smoking and trapped by PBN in CCl_4 and in benzene are shown in Figure 9.1a and Figure 9.1b, respectively. The ESR spectrum parameters are $a_N = 13.7$ G, $a^{\alpha}_H = 2.1$ G; $a^{\beta}_N = 14.1$G, $a^{\gamma}_H = 3.3$ G. The free radicals are recognized mainly as alkoxyl and alkyl radicals [1,2]. The signal intensity of the ESR is proportional to the relative concentration of free radicals. Results are shown in Table 9.1.

The free radicals generated in the gas phase of cigarette smoking are not only alkoxyl (RO·), alkyl (R·), and alkylperoxide (ROO·) radicals. However, other free radicals were not detected, because PBN was not sensitive enough to detect these. The generation of free radicals from cigarette smoke is a kinetic process as mentioned earlier [18].

9.4 CIGARETTE SMOKING: PATHOLOGICAL CHANGES

9.4.1 DENATURE OF α1-MACROGLOBULIN PROTEINASE INHIBITOR

The α1-macroglobulin proteinase inhibitor releases from liver to plasma to inhibit proteinases; if not inhibited, they will hydrolyze the catalage in lungs. The catalage is an important protein for respiratory function in the lung tissue. Patients

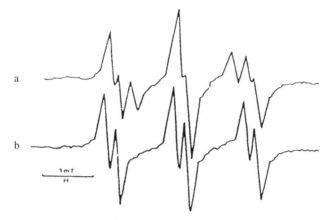

FIGURE 9.1 ESR spectra of free radicals generated in cigarette smoke and trapped by PBN (0.1 M) in (a) CCl_4 and (b) benzene, respectively. ESR conditions are described in methods.

TABLE 9.1
Scavenging Effects of Pycnogenol on Free Radicals in the Gas Phase Generated by Cigarette Smoke (Relative Amount)

Concentrations	Free Radicals	Scavenging Effects	Number
Control	8.62 ± 2.28	0	100
0.15 mg	6.67 ± 2.22[a]	22.6%	95
0.3 mg	6.57 ± 2.27[a]	27.6%	100
0.5 mg	6.11 ± 3.01[a]	29.1%	98
1.0 mg	6.90 ± 2.02[a]	20.0%	100

[a] Compared to control $p < 0.05$.

with a congenital defect in $\alpha1$-macroglobulin proteinase inhibitor usually have pulmonary emphysema and lose vital capacity. Results of Epidemiological studies show that even when the $\alpha1$-macroglobulin proteinase inhibitor is normal, cigarette smoke often causes pulmonary emphysema. This may be due to the inactivity of $\alpha1$-macroglobulin proteinase inhibitor, caused by free radicals generated in cigarette smoke, which thus cannot inhibit the proteinase released by neutrophil. The activity of $\alpha1$-macroglobulin proteinase inhibitor in the reperfusion from the smoker's lung is usually too low to inhibit the protenase released from neutrophils in the lung. This may be an important reason for the damage to lung tissue caused by cigarette smoke. Free radicals generated by cigarette smoke, including nitric oxide, nitric dioxide, oxygen radicals, and nitrate, can inactivate α1-macroglobulin proteinase inhibitor [18].

9.4.2 LIPID PEROXIDATION INJURY

Free radicals attack lung cell membrane and cause lipid peroxidation. Lipid peroxidation causes aggregation of macrophages in the lungs, generation of respiratory burst, and production of more oxygen free radicals. Both the free radicals cause more serious damage to lung and other tissues, possibly leading to cancer, heart disease, etc. Free radicals in the gas phase such as NO, NO_2, alkoxyl, and alkylperoxide free radicals can cause lipid peroxidation of lung cell membranes. The free radicals generated by lipid peroxidation of lung cell membranes can be detected by ESR spin trapping technique, and lipid peroxidation can be detected by the TBA method. It was found that lipid peroxidation increased with the duration of smoking. The study on lung cells showed that not only could the free radicals in gas phase change fluidity of the lung cell membrane, but they damaged the structure and conformation of membrane proteins, too. Antioxidant green tea polyphenols could protect the cell membrane and protein against the damage [20–28].

9.5 SCAVENGING EFFECTS OF PYCNOGENOL ON FREE RADICALS

Recent studies have shown that pycnogenol isolated from the bark of the French maritime pine (Horphag Research, Geneva) has very strong antioxidant and scavenging effects on free radicals [29–32]. We studied the scavenging effects of pycnogenol on free radicals in the gas phase of cigarette smoke and its detoxification effects [33].

9.5.1 SCAVENGING EFFECT OF PYCNOGENOL ON FREE RADICALS GENERATED IN SMOKE

The relative concentration of free radicals generated in cigarette smoke and the scavenging effects of pycnogenol in the filter are shown in Table 9.1. It has been found that filters with lower concentration of pycnogenol (0.1 to 0.5 mg) had higher scavenging effects in a dose-dependent manner (23% at 0.1 mg, 27% at 0.3 mg, and 29% at 0.5 mg, all values being significantly different from control, $p < 0.01$). At a high concentration (1 mg), the scavenging effects do not increase but, in fact, decrease (20%). This may be due to the phenomenon in which pycnogenol at high concentrations blocks the filter channels, decreasing the chances of contact between free radicals in the gas phase of smoke and the filter surface.

9.5.2 ANTIMUTATION EFFECT OF PYCNOGENOL AGAINST SMOKING

9.5.2.1 Inhibition of Mutagenicity of Cigarette Smoke by Pycnogenol in Ames Test

Smoke from cigarettes with (0.15 and 5 mg) and without pycnogenol was filtered by three Cambrage papers and then passed through cold acetone. After evaporation of acetone, the condensed materials were dissolved in DMSO in different concentrations for the Ames test.

The Ames test, widely used to test the mutagenic effects of materials, was performed according to the literature with some modifications [34,35]. One dish was used for solvent control and three dishes for positive control, e.g., 2.5 µg NaN_3, 10.0 µg 2-AF, and 0.2 µg 9-fluorenone/CD (culture dish). The other dishes contained 50, 150, and 500 µg condensed materials/CD from control and test cigarettes. Three dishes were provided for each dose, and all tests were repeated twice. Results were evaluated according to the Ames-test criteria. Only results that showed a reverse mutation colony two times higher than that of the control in a dose-dependent manner with reproducibility were accepted as indicating positive mutagenic effects.

Results showed no mutagenic effect of condensates from cigarette smoke on TA100 (with or without S9) and T98 without S9. A mutagenic effect could be detected only in the dishes containing the condensate in the case of T98 with S9 (Table 9.2).

The reverse mutation colony numbers of TA98 + S9 caused by 50, 150, and 500 µg/CD condensate from control cigarette were 1.2, 1.6, and 2.8 times higher than those from solvent alone, in a dose-dependent manner, and the experiments

---

TABLE 9.2
Mutagenic Effects of Condensates of Cigarette Smoke on TA100 and TA98 Colonies

Concentration (µg/CD)	TA98 +S9	TA98 S9	TA100 +S9	TA100 S9
Solvent control	24.17 ± 3.54	29.5 ± 3.45	173.67 ± 15.0	203.5 ± 20.0
Positive Control				
NaN_3(2.5mg/CD)				1389 ± 190
2-AF(10mg/CD)	1741 ± 119		1516 ± 200	
9-Fluorene(0.2mg/CD)		2121 ± 154		
Condensate				
50 µg/CD	29.2 ± 3.76[a]	30.7 ± 5.35	180.2 ± 21.27	221.3 ± 25.13
150 µg/CD	38.7 ± 5.47[a]	33.7 ± 4.80	182.3 ± 19.14	219.3 ± 26.32
500 µg/CD	67.5 ± 8.11[a]	36.1 ± 6.18	249.3 ± 32.62	209.5 ± 40.25

[a] Compared to control $p < .01$.

TABLE 9.3
**The Mutagenic Effects of the Condensate from Gas of Cigarette Smoke
and the Inhibition of Pycnogenol on TA98+S9 Colony**

Case	Concentration (μg/CD)	Reverse Mutation (Number/CD)	Inhibition Rate (%)
Control cigarette	50	29.2 ± 3.76	
	150	38.7 ± 5.47	
	500	67.2 ± 8.11	
Cigarette (0.15 mg Pycnogenol)	50	28.2 ± 4.67	3.5
	150	34.6 ± 9.54[a]	11.9
	500	49.6 ± 11.41[b]	35.4
Cigarette (0.5 mg Pycnogenol)	50	26.5 ± 2.90[a]	10.1
	150	31.8 ± 6.10[b]	21.6
	500	48.4 ± 8.60[b]	39.5

[a] Compared to control $p < .05$.

[b] Compared to control $p < .01$.

were reproducible. Thus, the condensate from control cigarettes had a mutagenic effect.

The mutagenic effects of condensates and the inhibition by pycnogenol are shown in Table 9.3. It has been found that condensates from smoke passed through pycnogenol-containing filters still have mutagenic effects but to a much smaller extent than smoke from control cigarettes. The inhibitory effects of 0.15 mg pycnogenol in the filter are 3.5%, 11.9%, and 35.4%, respectively. For 0.5 mg, reduction of 10.1%, 21.6%, and 39.5% was observed for doses of 50, 150, and 500 μg/CD condensate, respectively. The inhibition rates of pycnogenol in the filter are clearly dose dependent.

9.5.2.2 Acute Toxicity of Cigarette Smoke for Mice and the Inhibition by Pycnogenol

Cigarette smoke was let into a cabin with ten mice until all were dead. The number of cigarettes needed and the time were recorded. The dead mice were inspected for histopathological changes. Forty mice were randomly divided into four groups. The mice in group 1 were treated by control cigarette smoke; the mice in groups 2 and 3 were treated by cigarette smoke containing 0.15 mg and 0.5 mg pycnogenol in the filter, respectively; mice in group 4, a control group, were not treated by cigarette smoke.

The survival times and lung coefficients of the mice in different cigarette smoking treatments are shown in Table 9.4. The inhalation of cigarette smoke caused the death of mice due to acute toxicity. Adding pycnogenol in small amounts of 0.15 and 0.5 mg in the cigarette filter could inhibit toxicity of the

TABLE 9.4
The Survival Times and Lung Coefficients of the Mice in Different Cigarette Smoking Treatments

Case	Number	BW(g)	Cigarette (Number)	Survival Time (min)	Lung Coefficient	Inhibition Rate (%)
Control	10	29.7 ± 1.9	8	12.2 ± 1.0	0.81 ± 0.10	0
P (0.15 mg)[a]	10	27.0 ± 1.5	12	18.0 ± 0.3[c]	0.68 ± 0.07[d]	47.5
P (0.5 mg)[b]	10	28.5 ± 2.4	14	20.8 ± 2.1[c]	0.67 ± 0.08[d]	70.5
Normal	10	30.0 ± 2.1			0.59 ± 0.05[d]	

[a] P (0.15 mg) = cigarette smoke containing (0.15 mg) pycnogenol in the filter.
[b] P (0.5 mg) = cigarette smoke containing (0.5 mg) pycnogenol in the filter.
[c] The survival time is longer than that of control $p < .01$.
[d] Compared to control $p < .05$.

cigarette smoke to about 47% and 70%, respectively. The significantly prolonged survival times indicate that pycnogenol in the cigarette filter can protect the mice to some extent against the acute toxicity of cigarette smoke.

In the control group, there was obvious congestion, as well as even hemorrhage in lung tissue, in 80% of the mice. There was vasodilation and congestion in the small blood vessels of the kidney, and there was also slight vasodilation and congestion in the central veins of the liver, but there was no abnormal change in the heart and spleen. However, these pathogenic changes appeared only in 40% of mice in the group treated with cigarette smoke containing 0.15 mg pycnogenol in the filter; virtually no such change appeared in mice treated with cigarette smoke containing 0.5 mg pycnogenol in the filter. These results underline the fact that pycnogenol in cigarette filters does protect the mice from acute toxicity caused by cigarette smoking.

9.5.2.3 *In vivo* Mutagenicity and Toxicity of Chronic Smoking

The reduction of mutagenicity in the Ames test and of acute toxicity in mice prompted us to demonstrate the prevention of mutagenicity by pycnogenol filters *in vivo*. We investigated the incidence of micronuclei in polychromatic erythrocytes as a measure of mutagenicity and the ratio of change of the polychromatic erythrocytes to that of normochromatic erythrocytes as an indicator of toxicity in rats exposed to cigarette smoke for 75 d. The incidence of micronuclei in polychromatic erythrocytes increased by more than twofold when mice were chronically exposed to cigarette smoke, in accordance with a mutagenic activity test of cigarette smoke (Table 9.5). When cigarette filters contained 0.12 mg pycnogenol, the incidence of micronuclei was inhibited by 24% as compared to the rats that inhaled smoke from cigarettes with normal filters. The presence of 0.4 mg pycnogenol in filters was accompanied by a considerable improvement in micronuclei formation, which was inhibited by 48%. The incidence of micronuclei

TABLE 9.5
Chronic Toxicity of Cigarette Smoke in Rats

Cigarette Filter	PCE/NCE	PCE with Micronuclei (%)	Micronuclei Inhibition (%)
None (control	1.02	4.2 ± 1.4	—
Standard filter	0.88	10.0 ± 0.9	0
Filter with 0.12 mg Pycnogenol®	0.88	7.6 ± 2.5	24
Filter with 0.4 mg Pycnogenol®	1.01	5.2 ± 2.3	48

Note The number of micronuclei per 1000 polychromatic erythrocytes (PCE) was determined as a measure of *in vivo* mutagenicity. The ratio of PCE to normochromatic erythrocytes (PCE/NCE) is given as a measure of cytotoxicity. Mean values for 10 rats per group are given.

was statistically insignificant in rats not exposed to smoke. Hence, there was a significant reduction in mutagenicity of cigarette smoke due to the presence of 0.4 mg pycnogenol in cigarette filters.

Additionally, the ratio of polychromatic erythrocytes to normochromatic erythrocytes should normally be about 1 in healthy animals, whereas a lower ratio is known to indicate cytotoxicity insults. Accordingly, the polychromatic erythrocytes to normochromatic erythrocytes ratio was decreased in the rats exposed to cigarette smoke for 75 d. The presence of 0.12 mg pycnogenol in filters does not protect against the toxic effect of cigarette smoke; however, 0.4 mg pycnogenol in filters brought the polychromatic erythrocytes to normochromatic erythrocytes ratio back to normal.

The weight of the rodents was followed, and it was found that there was no difference between various groups. Specimens of heart, liver, spleen, kidney, adrenal glands, and lung of each animal from all groups were embedded in paraffin, sectioned, stained with hematoxylin-eosin, and examined under the optical microscope for pathological changes. In lungs of animals exposed to cigarette smoke with normal filters or filters with 0.12 mg pycnogenol, a variety of pathological alterations were discovered, whereas all other organs remained unaffected. We observed an increased thickness of bronchi capillary walls with signs of neutrophil and lymphocyte infiltration. There was low evidence of mucous epithelial cell proliferation or even exfoliation, an increase of inflammatory exudate, and lympho-proliferation in the bronchial lumen. Abscess and low fiber proliferation were found in lung tissue in two rats from the normal filter group and one rat from the group treated with cigarette smoke passed through filters with 0.12 mg pycnogenol. When 0.4-mg pycnogenol filters were used, only one rat showed pathological alterations; five other rats showed minor pathological changes in the lungs (moderate neutrophil and lymphocyte infiltration of alveoli); in four animals, almost normal lung tissue was observed.

REFERENCES

1. Zhao B-L. Cigarette, free radicals and cancer. *Nat J* 1989; 12: 453–460.
2. Zhao B-L. Cigarette, free radicals and heart diseases. *Nat J* 1989; 12: 610–611.
3. Chow WH, Swanson CA, Lissowska J, Groves FD, Sobin LH, Nasierowska-Guttimejer A, Radziewski J, Regula J, Hsing AW, Jagannatha S, Zatonski W, Blot WJ. Risk of stomach cancer in relation to consumption of cigarettes, alcohol, tea and coffee in Warsaw, Poland. *Intern J Cancer* 1999; 81: 871–875.
4. Mizoue T, Tokui N, Nishisaka K, Nishisaka S, Ogimoto I, Ikeda M, Yoshimura T. Prospective study on the relation of cigarette smoking with cancer of the liver and stomach in an endemic region. *Intern J Epidem* 2000; 29: 232–237.
5. Kaufmann PA, Gnecchi-Ruscone T, di Terlizzi M, Schafers KP, Luschers TF, Camici PG. Coronary heart disease in smokers: vitamin C restores coronary microcirculatory function. *Circulation* 2000; 102: 1233–1238.
6. De Waart FG, Smilde TJ, Wollersheim H, Stalenhoef AF, Kok FJ. Smoking characteristics, antioxidant vitamins, and carotid artery wall thickness among life-long smoking. *J Clin Epidem* 2000; 53: 707–714.
7. Moore ML, Zaccaro DJ. Cigarette smoking, low birth weight, and preterm birth in low-income Africa women. *J Perinat* 2000; 20: 176–180.
8. Laskowska-Klita T, Szymborski J, Chelchowska M, Czewinska B, Chazan B. Compensatory antioxidant activity in blood of women whose pregnancy is complicated by cigarette smoking. *Med Wieku Rozmoj* 1999; 3: 485–494.
9. Morens DM, Grandinetti A, Reed D, White LR, Ross GW: Cigarette smoking and protection from Parkinson's disease: false association or etiologic clue? *Neurology* 1995; 45: 1041–1051.
10. Ramon SO, Estefanýa MA, Ricardo, R, Emilio Q, Ines, S, Manuel LL. Studies on the interaction between 1,2,3,4-tetrahydro-β-carboline and cigarette smoke: a potential mechanism of neuroprotection for Parkinson's disease. *Brain Res* 1998; 802: 155–162.
11. Brenner DE, Kukull WA, van Belle G, Bowen JD, McCormick WC, Teri L, Laeson EB. Relationship between cigarette smoking and Alzheimer's disease in a population-based case-control study. *Neurology* 1993; 43: 293–300.
12. Ulrich J, Johannson-Locher G, Seiler WO, Stahelin, HB. Does smoking protect from Alzheimer's disease? Alzheimer-type changes in 301 unselected brains from patients with known smoking history. *Acta Neuropathol* 1997; 945: 450–454.
13. Emilien G, Beyreuther K, Masters CL, Maloteaux JM. Prospects for pharmacological intervention in Alzheimer disease. *Arch Neurol* 2000; 57: 454–459.
14. Kihara T, Shimohama S, Sawada H, Kimura J, Kume T, Kochiyama H, Maeda T, Akaike A. Nicotinic receptor stimulation protects neurons against β-amyloid toxicity. *Ann Neurol* 1997; 42: 159–163.
15. Garrido R, Mattson MP, Heninig B, Toborek M. Nicotine protects against arachidonic-acid-induced caspase activation, cytochrome c release and apoptosis of cultured spinal cord neurons. *J Neurochem* 2001; 76: 1395–403.
16. Liu Q, Tao Y, Zhao B-L. ESR study on scavenging effect of nicotine on free radicals. *Appl Mag Reson* 2003; 24: 105–112.
17. Liu Q, Zhao B-L. Nicotine attenuates β-amyloid peptide induced neurotoxicity, free radical and calcium accumulation in hippocampal neuronal cultures. *Brit J Pharmacol* 2004; 141: 746–754.

18. Church DF, Pryor WA. Free radical chemistry of cigarette smoke and its toxicological implications. *Environ Health Pespect* 1985; 64: 111–130.
19. Zhao B-L, Yan L-J, Hou J-W, Xin W-J. Studies on the free radicals in gas phase of cigarette smoke by spin trapping technique. *Natl Med J China* 1990; 70: 387–391.
20. Yan L-J, Zhao B-L, Li X-J, Xin W-J. Studies on the effect of gas phase of cigarette smoke on the respiratory burst of polymorphonuclear leukocytes using ESR technique. *Acta Sci Circum* 1991; 11: 79–83.
21. Yan L-J, Zhao B-L, Xin W-J. Experimental studies on some aspects of toxicological effect of gas phase cigarette smoke (GPCS). *Res Chem Interm* 1991; 16: 15–28.
22. Yan L-J, Zhao B-L, Xin W-J. Studies on gas phase smoke-induced changes of membrane biophysical characters. *Acta Biophys Sinica* 1991; 7: 305–309.
23. Yang F-J, Zhao B-L, Xin W-J. Toxicological effects of gas phase cigarette smoke on rat lung cells and protection effects of green tea polyphenols in the system. *Environ Chem* 1992; 11: 50–55.
24. Yang F-J, Zhao B-L, Xin W-J: ESR studies on the effects of GPCS-treated liposomes on superoxide generation of rat neutrophils. *Acta Biophys Sinica*, 1992; 8: 659–663.
25. Yang F-J, Zhao B-L, Xin W-J. Studies on toxicological mechanisms of gas-phase cigarette smoke and protective effects of GTP. *Res Chem Interm* 1992; 17: 39–57.
26. Yang F-J, Zhao B-L, Xin W-J. An ESR study on lipid peroxidation of rat liver microsome induced by gas-phase cigarette smoke. *Environ Chem* 1993; 12: 116–120.
27. Yang F-J, Zhao B-L, Xin,W-J. ESR studies on the effect of green tea polyphenols on the production of lipid free radicals in gas phase cigarette smoke treated microsomes. *Acta Biophys Sinica* 1993; 8: 468–471.
28. Zhang S-L, Zhao B-L, Xin W-J. Cytotoxicity of gas phase cigarette smoke and the protection by green tea polyphenols. *China Environ Sci* 1996; 16: 34–39.
29. Noda Y, Anzai K, Mori A, Kohno M, Shinmei M, Packer L. Hydroxyl and superoxide anion radical scavenging activities of natural source antioxidants using the computerized JES-FR 30 ESR spectrometer system. *Biochem Mol Biol Int* 1997; 42: 35–44.
30. Virgili F, Kobuchi H, Packer L. Procyanidins extracted from *Pinus maritima* (Pycnogenol: scavengers of free radical species and modulators of nitrogen monoxide metabolism in actived murine RAW 264.7 macrophages). *Free Radic Biol Med* 1998; 24: 1120–1129.
31. Packer L, Rimbach G, Virgili F. Antioxidant activity and biologic properties of a procyanidin-rich extract from pine (*Pinus maritima*) bark, Pycnogenol. *Free Radic Biol Med* 1999; 27: 704–724.
32. Guo Q, Zhao B-L, Packer L. Electron spin resonance study of free radicals formed from a procyanidin-rich pine (*Pinus maritima*) bark extract, Pycnogenol. *Free Radic Biol Med* 2000; 27: 1308–1312.
33. Zhang D-L, Tao Y, Duan S-J, Rohdwald P, Zhao B-L. Pycnogenol in cigarette filters scavenges free radicals and reduces mutagenicity and toxicity of tobacco smoke in vivo. *Toxicol Indus Health* 2002; 18: 215–224.
34. Mure K, Hayatsu H, Takeruchi T, Takeshita T, Morimoto K. Heavy cigarette smokers show higher mutagenicity in urine. *Mut Res* 1997; 373: 107–111.
35. Zhou R, Li S, Zhou Y, Haug A. Comparison of environmental tobacco smoke concentrations and mutagenicity for several indoor environments. *Mut Res* 2000; 465: 191–200.

Section II

Brain Diseases, and Free Radicals and Their Protection

10 Free Radical Scavengers and Neuroprotection

Midori Hiramatsu
Tohoku University of Community Service and Science,
Yamagata, Japan

CONTENTS

10.1 INTRODUCTION

The Japanese Ministry of Health, Labor, and Welfare recognized the magnitude of the problem of lifestyle-related diseases in 1996 (Figure 10.1). Lifestyle factors, such as alcohol, cigarette, stress, exercise, and food and the related diseases, such as cancer, heart disease, stroke, hypertension, diabetes, kidney disease, and so on, are all associated with free radicals [1,2]. In Japan, people are advised to protect themselves against such diseases, because adequate national funds are not allocated for health insurance. Food, health, and lifestyle are of growing concern in 21st-century Japan.

In brain diseases such as ischemia, Parkinson's disease, Alzheimer's disease, dementia, and posttraumatic epilepsy, free radicals play a major role [3,4]. Senile dementia, especially, is thought to be a lifestyle-related disease. In Japan, the number of very elderly people will increase subsequently in 20 to 30 yr, and the number of those affected by senile dementia would also rise during the same period. The Japanese government's "care insurance" program (a mandatory social insurance) was designed to help those requiring long-term care. The afore-mentioned problem and inadequate funding would pose difficulties for the insurance program.

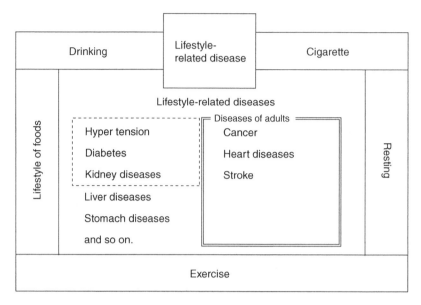

FIGURE 10.1 Concept of lifestyle-related diseases.

Recently, many papers have been written on the relation of free radicals to cellular signaling pathways and neuronal death in brain diseases. That is, transcription factors such as NF-κB, AP-1, p53; apoptosis factors such as Bcl-2, bax; and necrosis are affected by free radicals. We are studying the relationship between free radicals and some neurological diseases using animal models with biomarkers such as lipid peroxidation, oxidized protein and DNA damages (8-hydroxy-2′-deoxyguanosine), and neuron death, treating them with antioxidants or free radical scavengers. In this chapter, we describe the beneficial effects of free radical scavengers as neuron protectors; that is, we study the effects of Japanese herbs, such as sho-saiko-to-go-keishi-ka-shakuyaku-to and toki-shakuyaku-san, flower extract of the safflower flower, and antiepileptic zonisamide on biomarkers and neuron death, using animal models of emotional stress such as senescence-accelerated mice (SAMP8) that have functional recognition disorder [5], cerebral ischemia, and posttraumatic epilepsy [6].

10.2 EMOTIONAL STRESS AND THE JAPANESE HERB, SHO-SAIKO-TO-GO-KEISHI-KA-SHAKUYAKU-TO

A major factor in lifestyle-related diseases is psychological stress. In Japan, we have serious economic problems, and the risk of suicide has increased in recent years due to much emotional-stress-inducing depression. Hence, we examined the relationship between emotional stress and free radicals in the stomach and rat brain.

Sho-saiko-to-go-keishi-ka-shakuyaku-to (Tsumura & Co., Tokyo) is a mix of vacuum-concentrated boiled extract of nine herbs. A 7.5-g sample of sho-saiko-to-go-keishika-shakuyaku-to includes the extracts of 7 g of Bupreuri Radix (*Bupleurum falcatum* L.), 6 g of Paeoniae Radix (*Paeonia albiflora* palla var. trichocarpa Bunge), 4 g of Cinnamomi Cortex (*Cinnamomum cassia* Blume), 2 g of Glycerrhizae Radix (*Glycyrrhiza glabra* L. var. glandulifera Regel et Herder; G. *uralensis* Fisher), 3 g of Ginseng Radix (*Panax ginseng* C.A. Meyer), 3 g of Scutellariae Radix (*Scutearia baicalensis* Georgi), 5 g of Pinelliae tuber (*Pinellia ternate* Breitenbach), 1 g of Zingiberis Rhizoma (*Zingiber officinale* Roscoe), and 4 g of Zizyphi Fructus (*Zyzyphus vulgaris* Lamarck var. inermis Bunge). It is a free radical scavenger against hydroxyl, superoxide, 1,1-diphenyl-2-picrylhydrazyl radical, carbon-centered radicals in rat brain homogenate [7], and tocopheroxyl radicals [8]. We immobilized Wistar male rats with a wire net for 6 h and excised the stomach. Remarkable shrinkage of the stomach and bleeding were observed after immobilization. One-month administration of sho-saiko-to-go-keishi-ka-shakuyaku-to in drinking water almost inhibited shrinkage and bleeding. At the same time, carbon-centered radical level decreased in the homogenate of cerebral cortex and increased in the hippocampus. Mitochondrial superoxide dismutase (SOD) activity was lowered in the homogenate of cerebral cortex, hippocampus, and cerebellum and was elevated in the midbrain. The decreased SOD activity resulted in the scavenging of superoxide, and the increased SOD activity suggested the need of quenching excessive amounts of free radicals generated by strong emotional stress. These data show that excessive emotional stress induces free radical generation in the mitochondria of the brain and stomach and that sho-saiko-to-go-keishi-ka- shakuyaku-to, a free radical scavenger, is a neuroprotector against emotional stress.

Matsumoto et al. [9] showed that psychological stress induced enhancement of brain lipid peroxidation via nitric oxide reaction systems. Lee and Sapolsky [10] observed that major depression involved persistent loss of hippocampus volume, which is suggested to be due to oxygen radical generation, among other factors. Our data and other reports suggest that the free radicals generated in the neurons by emotional stress might induce neuronal damages. For this reason, it is recommended to take free radical scavengers everyday to overcome emotional stress in our present stressful society.

10.3 FREE RADICAL SCAVENGERS AND NEURON DEATH BY ISCHEMIA

Brody has reported that the number of neurons in the frontal lobe cortex and occipital lobe cortex decrease by about 60% and 50%, respectively, at 65 yr of age compared with the number present after birth [11]. This leads to lapse of memory. The mechanism of neuron death was thought to be due to peroxynitrite ion generated by the pathway following excess amounts of glutamate binding to the N-methyl-D-aspartic acid receptor. That is, the influx of calcium ion stimulates nitric oxide synthetase, generating nitric oxide, which reacts with superoxide, generating the formation of peroxynitrite ion. Two kinds of *in vivo* studies show that treatment with sho-saiko-to-go-keishi-ka-shakuyaku-to (an excellent free radical scavenger against superoxide, hydroxyl radical, 1,1-diphenyl-2-picryl-hydrazyl [DPPH] radical, carbon-centered radical generated in rat brain homogenate, and α-tocopheroxyl radical) inhibited neuron death in the hippocampus of rats 4 d after 20 min of ischemia [12]. Baicalein, which is a scavenger of hydroxyl radical, inhibited neuron death in the hippocampus of gerbils for 20 min after 5 min of ischemia [13]. These data showed that treatment with free radical scavengers protect against neuron death induced by ischemia, and these phenomena ascertained the involvement of free radicals in neuron death.

10.4 JAPANESE HERB TOKI-SHAKUYAKU-SAN AND DEMENTIA

Toki-shakuyaku-san (Tsumura & Co., Tokyo) is a mix of vacuum-concentrated boiled extract of six herbs. A 7.5 g sample of toki-shakuyaku-san includes the extracts of 4 g of Paeoniae Radix (*Paeonia lactiflora* Pallas), 4 g of Atractylodis Lanceae rhizome (*Atractylodes lancea* De Candolle or *Atractylodes chinensis* Koidzumi), 4 g of Alismatis rhizome (*Alisma orientale* Juzepczuk), 4 g of hoelen (*Poria cocos* Wolf), 3 g of Cnidii rhizome (*Cnidium officinale* Makino), and 3 g of Angelicae Radix (*Angelica acurtiloba* Kitagawa).

Because toki-shakuyaku-san stimulates the hypothalamus and follicle to secrete the female hormone estrogen, it is used, in Japan, for postmenopausal disorders, irregular menstruation, and infertility. It also increases choline acetyltransferase (CAT) activity and enhances cholinergic-receptor-binding activity [14]. The effects of toki-shakuyaku-san on cholinergic neurons are summarized elsewhere [15]. Thus, it may be effective in the treatment of memory disorders.

Toki-shakuyaku-san is a free radical scavenger against superoxide, hydroxyl, and DPPH radicals. Free radicals oxidize proteins, DNA and membrane, forming the carbonyl compounds, 8-hydroxy-2′-deoxyguanosine (8-OHdG) and malondialdehyde, respectively. In aged rat brain, the level of thiobarbiturate reactive substances (TBARS) was increased, and superoxide dismutase (SOD) activity was decreased in the cortex, hippocampus, and striatum. However, treatment with toki-shakuyaku-san in aged rats lowered TBARS formation and increased SOD activity in these areas of brain [16].

The level of 8-OHdG was higher in aged rat brain than in adult rat brain. Oral administration of toki-shakuyaku-san in drinking water every day for one month decreased the 8-OHdG level in aged rat brain without affecting the same in adult brain. The level of 8-OHdG in the brains of SAMP8 (has functional recognition disorder and has been developed by Takeda et al. [4]) was higher than that of the control (SAMR1) at 3 to 4 months, when aging starts. No difference was found between the levels of 8-OHdG in the brains of SAMR1 and

SAMP8 at 7 to 12 months of age. One-month oral administration of toki-shakuyaku-san in drinking water lowered the 8-OHdG levels in the brain of SAMP8 at 3 to 4 months of age and no effect was found in SAMR1. In the case of the control, SAMR1, one-month administration of toki-shakuyaku-san decreased the level in the brain at 7 to 12 months of age, and did not affect the level at 3 to 4 months of age. In the liver, the level of SAMP8 was higher at 3–4 and 7–12 months of age compared to those of the control SAMR1. However, no change was found in the level of both mouse livers at 3–4 and 7–12 months of age after administration of toki-shakuyaku-san. In conclusion, it was found that toki-shakuyaku-san decreased the level of 8-OHdG in only the aged brain of the rats SAMP8 and SAMR1.

Neuron death is thought to be due to excess amounts of glutamate in the synapse. Toki-shakuyaku-san inhibited death of glutamate-induced C6 glia cells using lactate dehydrogenase method and elevated cell survival ratio with MTT (3-4,5-dimethylthiazol-2-yl- 2,5-diphenyltetrazolium bromide) method. In Japan, we have effective clinical reports on the ability of toki-shakuyaku-san to improve cognitive disorders such as Alzheimer's disease and senile dementia [17,18].

8-OHdG levels were found to be decreased in the brain of those affected by Alzheimer's disease. Furthermore, the level of lipid peroxide was increased and shrinkage due to neuron death was found in the brain of those affected by Alzheimer's disease. Based on these facts and our data, we can infer that toki-shakuyaku-san might be a prophylactic against dementia and other brain diseases associated with free radical damage.

10.5 SAFFLOWER FLOWER IN NEURONAL DEATH AND POSTTRAUMATIC EPILEPSY

The flower of the safflower plant (*Carthamus tinctorius* Linne) is used in the preparation of medicine, dyes, lipstick, and foods. Pharmaceutical effects increase blood circulation [19] and are anti-inflammatory [20], and anticancer [21]. Recently, we found that water extract of the safflower flower has free-radical-scavenging activity against superoxide, hydroxyl, and DPPH radicals. The components responsible for the scavenging activity are yellow pigments of safflower yellow B, safflomin-A, pro-derivative of carthamin, and safflomin C, and red pigment of carthamin [22]. IC_{50} of free-radical-scavenging activity against DPPH radical was 15 μg/ml of safflower yellow, 1.5 μg/ml of carthamin, and 0.3 mg/ml of water extract of dry flower of safflower. As for IC_{50} of antioxidant, water-soluble vitamin C and lipid soluble vitamin E were 1 μg/ml and 30 μg/ml, respectively. Antioxidant activity of carthamin is almost the same as that of vitamin C, and the radical-scavenging activity in safflower yellow is ten times lower than carthamin. The flower of safflower contains lipid-soluble carthamin and water-soluble safflower yellow, suggesting that it might protect membrane functions against free radicals.

Acute and chronic epileptiform discharges and behavioral seizures are induced by a 5-μl injection of ferric chloride solution into the left sensory motor cortex of rats. These effects are thought to reflect neurochemical changes that are initiated by head trauma with subsequent hemorrhaging and the development of posttraumatic epilepsy [5]. Prior studies suggest that the fundamental chemical reactions induced by such injections result in the formation of hydroxyl radicals that alter the double bonds of unsaturated fatty acids of neuronal membranes.

TBARS formation increased 30 min after iron solution was injected into the left cortex, compared to control. Oral administration of the extract of safflower flower in a dose of 1 and 3 g/kg/d in drinking water every day for 1 month decreased the increased TBARS formation in the rat cortex 30 min after an injection of iron solution. The 8-Hydroxy-2'-deoxyguanosine level in the left cortex of rat was elevated 30 min after iron solution injection, and injection of the extract of safflower flower at a dose of 3 g/kg/d into the stomach daily for 1 week inhibited the increase in 8-OHdG formation. Oral administration of red pigment carthamin at a dose of 10 mg/kg/d every day for 1 week retarded the 8-OHdG levels in the cortex of rat. The inhibitory effects of the extract of safflower flower against TBARS and 8-OHdG levels are thought to be because of its free-radical-scavenging activity.

In the case of neuron death, extract of safflower flower in the range of 50 to 400 µg/ml inhibited the glutamate-induced death of C6 glial cells using lactate dehydrogenase method. Furthermore, the extract at a dose of 50 to 400 µg/ml raised survival rate of glutamate-induced death of C6 glial cells.

Based on these facts and our data, we conclude that the extract of the safflower flower effectively fights against aging and lifestyle-related diseases, especially brain diseases associated with free radical injury.

10.6 ANTIEPILEPTIC ZONISAMIDE AND POSTTRAUMATIC EPILEPSY

Zonisamide is being developed by the Dainippon Pharmaceutical Company (Osaka, Japan) as an antiepileptic. It has free-radical-scavenging activity against DPPH radicals. Inhibitory action of zonisamide at a dose of 100 mg/kg on the formation of lipid peroxide and 8-hydroxy-2'-deoxyguanosine can be observed in the ipsilateral cortex of rat 30 min after an injection of ferric chloride solution into the left sensory motor cortex [23,24]. The formation of carbonyl compounds, which are products of protein with free radical reaction, increased in the ipsilateral cortex of rats 30 min after an iron solution injection; however, treatment with zonisamide inhibited the formation of these compounds. The W/S ratio of MAL-6-labelled cortical synaptosomes decreased, showing oxidation of protein 30 min after an injection of iron solution into the left cortex, and treating with zonisamide inhibited this decrease. These effects are attributed to its free-radical-scavenging action.

Epileptiform spikes were observed 3 h after an injection of 15 µl of 0.1-M iron solution into the hippocampus of rats. In the same model, 30 min after injection of iron solution, malondialdehyde level in the ipsilateral hippocampus was elevated, and pretreatment with zonisamide at a dose of 100 mg/kg inhibited this rise. The accumulation of carbamoyl-PROXYL, was found in the ipsilateral hippocampus of rats 30 min after an intraperitoneal injection of carbamoyl-PROXYL. An iron solution was injected into the left hippocampus using an L-band electron-spin-resonance-computed tomography. However, zonisamide inhibited the accumulation. Because carbamoyl-PROXYL does not cross the blood–brain barrier, it was accumulated through the damaged blood–brain barrier induced by iron solution injection. Extracellular glutamate level in the ipsilateral hippocampus was dramatically elevated 24 h after injection of iron solution into the hippocampus, and the elevation recovered to the control level further 24 h after the iron solution injection.

NF-κB immunostaining, TUNEL staining, and Hoechst 33258 staining in the surrounding tissue of the iron solution injection site of the left cortex increased 30 min after injection of iron solution. However, the treatment with zonisamide at a dose of 100 mg/kg inhibited these changes in staining. In the Hoechst 33258 staining no condensation of chromosome was found. These data suggest that necrosis without apoptosis was observed in the neurons of the injection site at the cortex after an iron solution injection, and the phenomena may induce epileptogenesis in rats.

Furthermore, zonisamide is effective on animal models of Parkinson's disease [25] and ischemia in rat [26]. In conclusion, zonisamide is considered to be a neuroprotector against loss of memory, Parkinson's disease, stroke, dementia, and other diseases associated with free radicals, in addition to posttraumatic epilepsy.

10.7 CONCLUSION

Because a blood–brain barrier exists, the question regarding the permeability of free radical scavengers arises. Free radical scavengers that act as protectors against brain diseases associated with free radicals, might be prophylactics against aging and lifestyle-related diseases associated with free radicals. Because of this, we prefer to use iron-induced epileptogenesis in rats as a model.

In brain diseases, the mechanism of stroke, Parkinson's disease, Alzheimer's disease, senile dementia, posttraumatic epilepsy are all partially related to the activity of free radicals. Edalavone (3-,ethyl-1-phenyl-2-pyrazolin-5-on, of which the pharmacological function is only as a free radical scavenger, has been clinically used as the first choice against stroke in Japan [27]. This fact reinforces the role of free radicals in stroke [28]. However, free radicals should be considered even in other diseases of the brain and other organs.

For this reason, in this chapter we described our data in detail, focusing on free radical scavenger. We used free radical scavengers such as sho-saiko-to-go-keishi-ka-shakuyaku-to, bicalein, toki-shakuyaku-san, flower of safflower plant, and zonisamide. They showed satisfactory effects on biomarkers such as lipid peroxidation, 8-hydroxy-2′-deoxyguanosine, and oxidized protein in the brain of rodents and neuron death in culture (Table 10.1). It can be said that these free radical scavengers are neuroprotectors. However, it can also be said that free radical scavengers are general protectors against aging and lifestyle-related diseases.

TABLE 10.1
Free Radical Scavengers and Neuroprotection

Free Radical Scavengers	Neuroprotection Against
Japanese Herbs	
Sho-saiko-to- go-keisha-ka-shakuyaku-to (scavenger against OH, O_2, DPPH, C, and VE radicals)	Free radicals in the stomach and brain generated by emotional stress
	Neuron death due to ischemia in rat
Toki-shakuyakusan (scavenger against OH, O_2 and DPPH radicals)	Lipid peroxidation aged rat brain
	8OHdG level in aged rat brain and SAMP8 brain
	C6 glia cell death
Bicalein (scavenger of OH radicals)	Neuron death due to ischemia in gerbil
Flower of safflower plant (scavenger against OH,O_2 and DPPH radicals)	Lipid peroxidation and 8OHdG level in rat brain due to iron-induced epilepsy
	C6 glia cell death
Zonisamide (scavenger against OH, O_2 and DPPH radicals)	Lipid peroxidation and 8OHdG level in rat brain due to iron-induced epilepsy
Edalavone(28) (scavenger against OH radical)	Stroke

Note: 8-OHdG = 8-Hydroxy-2′-deoxyguanosine; DPPH = 1,1-diphenyl-2-picrylhydrazyl.

REFERENCES

1. Packer, L., Hiramatsu, M., and Yoshikawa, T. Eds., *Antioxidant Food Supplements in Human Health.* Academic Press, 1999.
2. Packer, L. and Colman, C., *The Antioxidant Miracle.* John Wiley & Sons, New York, 1999.
3. Packer, L., Hiramatsu, M., and Yoshikawa, T. Eds., *Free Radicals in Brain Physiology and Disorders.* Academic Press, 1996.
4. Poli, G., Cadenas, E., and Packer, L. Eds., *Free Radicals in Brain Pathophysiology.* Marcel Dekker, New York, 2000.
5. Takeda, T. et al., A new murine model of accelerated senescence mice. *Aging Dev* 17, 183, 1981.
6. Willmore, L.J., Sypert, G.W., and Munson, J.B. Chronic focal epileptiform discharges induced by injection iron into rat and cat cortex, *Science* 200, 1501, 1978.
7. Hiramatsu, M. and Komatsu, M. Mixed Japanese herbs and age-related neuronal functions. In *Antioxidant Food Supplements in Human Health*, Packer, L., Hiramatsu, M., and Yoshikawa, T., Eds., Academic Press, San Diego, 411, 1999
8. Hiramatsu, M., Velasco, R.D., and Packer, L. Vitamin E radical reaction with antioxidants in rat liver membranes, *Free Rad Biol Med* 9, 459, 1990.
9. Matsumoto, K. et al., Psychological stress-induced enhancement of brain lipid peroxidation via nitrite oxide systems and its modulation by anxiolytic and anxiogenic drugs in mice, *Brain Res* 21, 74, 1999.

10. Lee, A.L., Ogl, W.O., and Sapolsky, R.M. Stress and depression: possible links to neuron death in the hippocampus, *Bipolar Disord* 4, 117, 2002.

11. Brody, H. Organization of the cerebral cortex. III. A study of aging in the human cerebral cortex, *J Comp Neurol* 102, 511, 1995.

12. Sugimoto, A. et al., Protective effect of Sho-saiko-to-go-keishi-ka-shakuyaku-to (TJ-960) against cerebral ischemia. In *Recent Advances in the Pharmacology of KAMPO (Japanese Herbal) MEDICINES,* Hosoya, E., and Yamamura, H., Eds., Tokyo: Excerpta Medica, 112, 1988.

13. Hamada, H., Hiramatsu, M., and Mori, A. Free radical scavenging action of bicalein, *Arch Biochem Biophys* 306, 261, 1993.

14. Hiramatsu, M. et al., Herbal Medicine, Toki-shakuyaku-san enhanced cholinergic neuronal functions in the striatum of aged rats, *J Anti-Aging Med* 2, 51, 1999.

15. Hiramatsu, M. and Komatsu, M. Mixed Japanese herbs and age-related neuronal functions. In *Antioxidant Food Supplements in Human Health,* Packer, L., Hiramatsu, M., and Yoshikawa, T. Eds., Academic Press, San Diego, 411,1999.

16. Ueda, Y., Komatsu, M., and Hiramatsu, M. Free radical scavenging activity of the Japanese herbal medicine Toki-Shakauyaku-San (TJ-23) and its effect on superoxide dismutase activity, lipid peroxides, glutamate, and monoamine metabolites in aged rat brain, *Neurochem Res* 21, 909, 1996.

17. Inanaga, K. et al., Effects of Toki-shakuyaku-san in patients with cognitive disorders, *Prog Med* 16, 293, 1996 (in Japanese).

18. Yamamoto, T. and Kondo, K. The treatment of herbs on dementia of the Alzheimer type, *J Trad Med* 6, 454, 1989 (in Japanese).

19. Todoki, K. et al., The increase in peripheral blood flow induced by Carthamus t. and Amica m., *Kanagawashigaku* 18, 64, 1983.

20. Kasahara, Y. Studies on physiologically active natural products in Yamagata prefecture II : Antiinflammatory actions of *Cartamus tinctorius, Yamagata Eikenshoho* 23, 21, 1990.

21. Kasahara, K., Kumaki, K., and Katagiri, S. Carthami Flos extract and its component, stigmasterol, inhibit tumour promotion in mouse skin two-stage carcinogenesis, *Phytotherapy Res* 8, 327, 1994.

22. Obara, H., Onodera, J., and Sato, S. Carthamin, the red pigment of safflower, *Bull Yamagata Univ* 22, 91, 1993.

23. Komatsu, M., Okamura, Y., and Hiramatsu, M. Free radical scavenging activity of zonisamide and its inhibitory effect on lipid peroxide formation in iron-induced epileptogenetic foci of rats, *Neurosciences* 21, 23, 1995.

24. Komatsu, M., Hiramatsu, M., and Willmore, L.J. Zonisamide reduces the increase in 8-hydroxy-2′-deoxyguanosine levels formed during iron-induced epileptogenesis in the brain of rats, *Epilepsia* 41, 1091, 2000.

25. Murata, M. Novel therapeutic effects of the anti-convulsant, zonisamide, on Parkinson's disease, *Curr Pharm Des* 10, 687, 2004.

26. Hayakawa, T. et al., Zonisamide reduces hypoxic-ischemic brain damage in neonatal rats irrespective of its anticonvulsive effect. *Eur J Pharm* 257, 131, 1994.

27. Watanabe, T. et al., Research and development of the free radical scavenger edaravone as a neuroprotectant *Yakugaku Zasshi* 124, 99–111, 2004.

28. Schaller, B. Prospects for the future: the role of free radicals in the treatment of stroke. *Free Radic Biol Med* 2005; 38: 411–425.

11 Oxidative Adaptation in Aging and Alzheimer's Disease: The Roles of Amyloid and Tau

Hyoung-gon Lee and Xiongwei Zhu
Case Western Reserve University, Cleveland, Ohio

Kelly L. Drew
University of Alaska, Fairbanks, Alaska

James A. Joseph
Tufts University, Boston, Massachusetts

Akihiko Nunomura
Asahikawa Medical College, Asahikawa, Japan

Keisuke Hirai
Takeda Chemical Industries, Osaka, Japan

Atsushi Takeda
Tohoku University, Sendai, Japan

George Perry and Mark A. Smith
Case Western Reserve University, Cleveland, Ohio

CONTENTS

11.1 INTRODUCTION

Neurons of the central nervous system are subject to a number of unique conditions that make them particularly vulnerable to oxidative stress and its sequelae that can culminate in cell death [1]. This vulnerability is a likely consequence of the high energy and oxygen consumption rates, the high unsaturated lipid content of the neuronal membrane, the high levels of transition metals, the relative scarcity of antioxidant defense systems as compared to other organs, and the postmitotic nature of neuronal populations [2]. Although there is debate on whether oxidative stress is a cause or a consequence of the disease, evidence showing oxidative stress before any other known change indicates a causative role in the pathogenesis of Alzheimer's disease (AD).

11.2 OXIDATIVE STRESS IN AD

Free radicals produced during oxidative stress are speculated to be pathologically important in AD and other neurodegenerative diseases [3,4]. Over the past decade, posttranslational oxidative modification to virtually all classes of biomacromolecules, including oxidized nucleic acids [5–9], oxidized proteins [10–12], oxidized lipids [13,14], and oxidized carbohydrates [4,15–17], have been described in association with the susceptible neurons of AD.

Analysis of the site of oxidative damage can provide clues to the mechanistic source of such damage. However, in doing so, it is important to compare the differences noted when damage occurs to rapidly turned-over biological macromolecules (active modification) vs. damage to stable structures (accumulative damage), some of which arise by oxidative cross-linking. Obviously, the former represents more recent oxidative modifications, whereas the latter represents the history of oxidative damage. Because the products of lipid peroxidation and glycation often yield cross-linked molecules that are resistant to removal, they can be used as markers for accumulative damage [18]. On the other hand, protein nitration, a non-cross-link-related oxidative modification of protein resulting from either peroxynitrite attack or peroxidative nitration, indicates more recent active modifications. What is remarkable when comparing these different markers is that highly stable accumulative modifications involving cross-linked proteins are predominantly associated with lesions [13,19–21], whereas active modifications are more commonly associated with the cytoplasm of susceptible neuronal populations in AD [22]. These findings point to the cytoplasm, not the lesions, as the source of reactive oxygen species (ROS). Our observation that RNA, a rapidly turned-over cell component, is a target for oxidative stress provided another ideal marker for assessing recent active modifications. 8-Hydroxyguanosine (8OHG), a nucleic acid modification predominantly derived from hydroxyl radical ($\bullet$OH) attack of guanosine, is greatly increased in cytoplasmic RNA in vulnerable neurons [7]. Ultrastructural analysis shows that most 8OHG is in the endoplasmic reticulum, with the majority of mitochondria showing little 8OHG [23]. As the $\bullet$OH can only diffuse nanometer distances and cannot permeate through the plasma membrane, the source of reactive oxygen must be in close physical proximity to the damage. Therefore, 8OHG is likely to form at the site of $\bullet$OH production within the neuronal cytoplasm, a process dependent on redox-active metal-catalyzed reduction of hydrogen peroxide together with cellular reductants such as ascorbate or superoxide [22,24]. Although the cytosolic sites of damage seemingly exclude mitochondria, we suspect a more complex relationship that involves superoxide generated from mitochondria being dismutated to freely diffusible hydrogen peroxide that then interacts with cytosolic redox-active metals and other oxidative stress response elements to cause damage [23,25,26].

11.3 OXIDATIVE STRESS AND AD PATHOLOGY

Although it is clear that oxidative stress is a prominent feature of the AD brain, it is still not clear how oxidative stress is related to the hallmark pathologies, i.e., amyloid-β (Aβ) plaques and neurofibrillary tangles, of the disease. There is increasing evidence that the very earliest neuronal and pathological changes characteristic of AD show oxidative damage [8,9,27,28] indicating that oxidative stress represents a very early contributor to the disease. Indeed, this early role is borne out by clinical management of oxidative stress, which appears to reduce the incidence and severity of AD [29–31]. A systematic examination of the spatiotemporal relationship between the presence of oxidative modification and the hallmark AD lesions during early AD stages suggests that many markers of oxidative damage are present in susceptible neurons in the absence of neurofibrillary pathology [7]. Further, there is marked accumulation of 8OHG and nitrotyrosine in the cytoplasm of cerebral neurons from Down's syndrome cases and elevated glycoxidation in fetal Down's cases [32], which temporally precedes Aβ deposition by decades [8]. A recent time course study in Aβ transgenic mice confirmed these findings because lipid peroxidation was found to be significantly increased preceding any apparent Aβ deposit and increase of Aβ levels [28].

11.3.1 OXIDATIVE STRESS AND AMYLOID-β

Oxidative stress condition is known to increase Aβ production and deposition. For example, *in vitro* studies show that hydrogen peroxide upregulates both secretion of Aβ in the cell medium [33] and levels of Aβ in the cell [34], which can be blocked by antioxidant treatment [34]. Increased Aβ production induced by oxidative stress is likely because of the increased synthesis of amyloid-β protein precursor (AβPP) through the activation of AP-1 transcription factor [35] and also increased generation of Aβ from AβPP through activation of β-secretase

[36]. Given such a role for oxidative stress in Aβ expression, it is not surprising that Aβ production also increases *in vivo* after brain injury [37–39] and in response to ischemic and hypoxic injury and brain trauma [40–42]. Further, in Aβ or ApoE-deficient transgenic mice, antioxidant treatment results in a decrease in overall levels of Aβ [43–46].

Based on the observations that Aβ production and deposition appear to be a response to neuronal injury rather than the mediator of such an injury [40–42], it is likely that the formation of Aβ may provide protection against oxidative stress [2]. In fact, that Aβ levels in the cerebral spinal fluid (CSF) of AD patients are inversely correlated with dementia severity is consistent with such a protective function [47], and other reports showing Aβ as a normal physiological product with protective effects supports an antioxidant role for Aβ [48–51].

11.3.2 OXIDATIVE STRESS AND NEUROFIBRILLARY TANGLES

In our previous study, we showed that oxidative damage is an early event in AD that decreases with disease progression and lesion formation [9]. These findings implicate that neurofibrillary tangles (NFT) are associated with compensatory changes that reduce damage from reactive oxygen [50]. Although it is controversial to think of NFT as having antioxidant properties rather than as mediators of neuronal death, such a function of NFT is suggested by numerous studies [9,50,52]. In fact, both tau and neurofilament proteins, the major components of NFT, appear uniquely adapted to oxidative attack due to their high content of lysine-serine-proline (KSP) domains. Exposure of these domains on the protein surface is affected by extensive phosphorylation of serine residues, resulting in an "oxidative sponge" of surface-modifiable lysine residues [52]. Indeed, oxidative stress and attendant modification of tau by products of oxidative stress including 4-hydroxyl-2-nonenal (HNE) [53] as well as other cytotoxic carbonyls [54], though leading to protein aggregation as NFT, does enable such neurons to survive for decades [55]. Because phosphorylation plays this pivotal role in redox balance, it is perhaps not surprising that oxidative stress, through activation of mitogen-activated protein (MAP) kinase pathways, leads to phosphorylation [56–58]. Indeed, conditions associated with chronic oxidant stress, such as AD, are invariably associated with extensive phosphorylation of cytoskeletal elements [59]. Moreover, other neurological conditions, where phosphorylated tau and neurofilament protein accumulations occur, also show evidence of oxidative adducts, e.g., progressive supranuclear palsy [60] and frontal temporal dementia [61]. Given this protective role of tau phosphorylation, it is not surprising that embryonic neurons that survive treatment with oxidants have more phospho-tau immunoreactivity relative to those that die [62]. It is also shown that paired helical filament (PHF)-like tau phosphorylation occurs during hibernation [63]. Because hibernation is neuroprotective [64], regulating tau phosphorylation is preserved in the adult mammalian brain as a naturally occurring process associated with specific requirements on neuroprotection [63]. Further, the expression of endogenous antioxidants and tau expression and phosphorylation are opposing [53,65], suggesting that the reduced oxidative damage in neurons with tau accumulation may be a part of the antioxidant function of phosphorylated tau.

11.4 CONCLUSION

Although AD is likely associated with multiple etiologies and pathogenic mechanisms, all these share the commonality of oxidative stress. As such, increased oxidative modifications to virtually every class of biological macromolecules are well documented in AD patients, and oxidative damage is the earliest event in AD, which not only well precedes the occurrence of AD pathologies but may also directly and indirectly contribute to the formation of these pathologies. In fact, the formation of AD pathologies may provide protection against oxidative stress, whereas those neurons without pathology appear to rapidly succumb to oxidative insults. In line with this, there is accumulating evidence to support antioxidative stress functions of Aβ and NFT in the pathogenesis of AD. Therefore, current therapeutic strategies based on eliminating Aβ or tau phosphorylation may actually leave the brain open to additional oxidative damage and serve to exacerbate the disease process.

REFERENCES

1. Halliwell B. Reactive oxygen species and the central nervous system. *J Neurochem* 1992; 59: 1609–1623.
2. Joseph J, Shukitt-Hale B, Denisova NA, Martin A, Perry G, Smith MA. Copernicus revisited: amyloid beta in Alzheimer's disease. *Neurobiol Aging* 2001; 22: 131–146.
3. Cross CE, Halliwell B, Borish ET, Pryor WA, Ames BN, Saul RL, McCord JM, Harman D. Oxygen radicals and human disease. *Ann Intern Med* 1987; 107: 526–545.
4. Smith MA, Sayre LM, Monnier VM, Perry G. Radical AGEing in Alzheimer's disease. *Trends Neurosci* 1995; 18: 172–176.
5. Mecocci P, MacGarvey U, Beal MF. Oxidative damage to mitochondrial DNA is increased in Alzheimer's disease. *Ann Neurol* 1994; 36: 747–751.
6. Mecocci P, Beal MF, Cecchetti R, Polidori MC, Cherubini A, Chionne F, Avellini L, Romano G, Senin U. Mitochondrial membrane fluidity and oxidative damage to mitochondrial DNA in aged and AD human brain. *Mol Chem Neuropathol* 1997; 31: 53–64.
7. Nunomura A, Perry G, Pappolla MA, Wade R, Hirai K, Chiba S, Smith MA. RNA oxidation is a prominent feature of vulnerable neurons in Alzheimer's disease. *J Neurosci* 1999; 19: 1959–1964.
8. Nunomura A, Perry G, Pappolla MA, Friedland RP, Hirai K, Chiba S, Smith MA. Neuronal oxidative stress precedes amyloid-beta deposition in Down syndrome. *J Neuropathol Exp Neurol* 2000; 59: 1011–1017.
9. Nunomura A, Perry G, Aliev G, Hirai K, Takeda A, Balraj EK, Jones PK, Ghanbari H, Wataya T, Shimohama S, Chiba S, Atwood CS, Petersen RB, Smith MA. Oxidative damage is the earliest event in Alzheimer disease. *J Neuropathol Exp Neurol* 2001; 60: 759–767.

10. Castegna A, Aksenov M, Thongboonkerd V, Klein JB, Pierce WM, Booze R, Markesbery WR, Butterfield DA. Proteomic identification of oxidatively modified proteins in Alzheimer's disease brain. Part II: dihydropyrimidinase-related protein 2, alpha-enolase and heat shock cognate 71. *J Neurochem* 2002; 82: 1524–1532.

11. Castegna A, Aksenov M, Aksenova M, Thongboonkerd V, Klein JB, Pierce WM, Booze R, Markesbery WR, Butterfield DA. Proteomic identification of oxidatively modified proteins in Alzheimer's disease brain. Part I: creatine kinase BB, glutamine synthase, and ubiquitin carboxy-terminal hydrolase L-1. *Free Radic Biol Med* 2002; 33: 562–571.

12. Castegna A, Thongboonkerd V, Klein JB, Lynn B, Markesbery WR, Butterfield DA. Proteomic identification of nitrated proteins in Alzheimer's disease brain. *J Neurochem* 2003; 85: 1394–1401.

13. Sayre LM, Zelasko DA, Harris PL, Perry G, Salomon RG, Smith MA. 4-Hydroxy-nonenal-derived advanced lipid peroxidation end products are increased in Alzheimer's disease. *J Neurochem* 1997; 68: 2092–2097.

14. Butterfield DA, Drake J, Pocernich C, Castegna A. Evidence of oxidative damage in Alzheimer's disease brain: central role for amyloid beta-peptide. *Trends Mol Med* 2001; 7: 548–554.

15. Smith MA, Taneda S, Richey PL, Miyata S, Yan SD, Stern D, Sayre LM, Monnier VM, Perry G. Advanced Maillard reaction end products are associated with Alzheimer's disease pathology. *Proc Natl Acad Sci USA* 1994; 91: 5710–5714.

16. Vitek MP, Bhattacharya K, Glendening JM, Stopa E, Vlassara H, Bucala R, Manogue K, Cerami A. Advanced glycation end products contribute to amyloidosis in Alzheimer's disease. *Proc Natl Acad Sci USA* 1994; 91: 4766–4770.

17. Castellani RJ, Harris PL, Sayre LM, Fujii J, Taniguchi N, Vitek MP, Founds H, Atwood CS, Perry G, Smith MA. Active glycation in neurofibrillary pathology of Alzheimer's disease: N-(carboxymethyl) lysine and hexitol-lysine. *Free Radic Biol Med* 2001; 31: 175–180.

18. Sayre LM, Smith MA, Perry G. Chemistry and biochemistry of oxidative stress in neurodegenerative disease. *Curr Med Chem* 2001; 8: 721–738.

19. Montine TJ, Amarnath V, Martin ME, Strittmatter WJ, Graham DG. E-4-hydroxy-2-nonenal is cytotoxic and cross-links cytoskeletal proteins in P19 neuroglial cultures. *Am J Pathol* 1996; 148: 89–93.

20. Smith MA, Perry G, Richey PL, Sayre LM, Anderson VE, Beal MF, Kowall N. Oxidative damage in Alzheimer's. *Nature* 1996; 382: 120–121.

21. Smith MA, Sayre LM, Anderson VE, Harris PL, Beal MF, Kowall N, Perry G. Cytochemical demonstration of oxidative damage in Alzheimer's disease by immunochemical enhancement of the carbonyl reaction with 2,4-dinitrophenyl-hydrazine. *J Histochem Cytochem* 1998; 46: 731–735.

22. Smith MA, Harris PLR, Sayre LM, Perry G. Iron accumulation in Alzheimer's disease is a source of redox-generated free radicals. *Proc Natl Acad Sci USA* 1997; 94: 9866–9868.

23. Hirai K, Aliev G, Nunomura A, Fujioka H, Russell RL, Atwood CS, Johnson AB, Kress Y, Vinters HV, Tabaton M, Shimohama S, Cash AD, Siedlak SL, Harris PLR, Jones PK, Petersen RB, Perry G, Smith MA. Mitochondrial abnormalities in Alzheimer's disease. *J Neurosci* 2001; 21: 3017–3023.

24. Sayre LM, Perry G, Harris PLR, Liu Y, Schubert KA, Smith MA. *In situ* oxidative catalysis by neurofibrillary tangles and senile plaques in Alzheimer's disease: a central role for bound transition metals. *J Neurochem* 2000; 74: 270–279.

25. Perry G, Castellani RJ, Hirai K, Smith MA. Reactive oxygen species mediate cellular damage in Alzheimer's disease. *J Alzheimers Dis* 1998; 1: 45–55.

26. Casadesus G, Smith MA, Zhu X, Aliev G, Cash AD, Honda K, Petersen RB, Perry G. Alzheimer's disease: evidence for a central pathogenic role of iron-mediate reactive oxygen species. *J Alzheimers Dis* 2004; 6: 165–169.

27. Perry G, Smith MA. Is oxidative damage central to the pathogenesis of Alzheimer's disease? *Acta Neurol Belg* 1998; 98: 175–179.

28. Pratico D, Uryu K, Leight S, Trojanoswki JQ, Lee VM. Increased lipid peroxidation precedes amyloid plaque formation in an animal model of Alzheimer amyloidosis. *J Neurosci* 2001; 21: 4183–4187.

29. Sano M, Ernesto C, Thomas RG, Klauber MR, Schafer K, Grundman M, Woodbury P, Growdon J, Cotman CW, Pfeiffer E, Schneider LS, Thal LJ. A controlled trial of selegiline, alpha-tocopherol, or both as treatment for Alzheimer's disease: the Alzheimer's disease cooperative study. *N Engl J Med* 1997; 336: 1216–1222.

30. Stewart WF, Kawas C, Corrada M, Metter EJ. Risk of Alzheimer's disease and duration of NSAID use. *Neurology* 1997; 48: 626–632.

31. Zandi PP, Anthony JC, Khachaturian AS, Stone SV, Gustafson D, Tschanz JT, Norton MC, Welsh-Bohmer KA, Breitner JC. Cache County Study Group. Reduced risk of Alzheimer's disease in users of antioxidant vitamin supplements: the Cache County study. *Arch Neurol* 2004; 61: 82–88.

32. Odetti P, Angelini G, Dapino D, Zaccheo D, Garibaldi S, Dagna-Bricarelli F, Piombo G, Perry G, Smith M, Traverso N, Tabaton M. Early glycoxidation damage in brains from Down's syndrome. *Biochem Biophys Res Commun* 1998; 243: 849–851.

33. Olivieri G, Baysang G, Meier F, Muller-Spahn F, Stahelin HB, Brockhaus M, Brack C. *N*-acetyl-L-cysteine protects SHSY5Y neuroblastoma cells from oxidative stress and cell cytotoxicity: effects on beta-amyloid secretion and tau phosphorylation. *J Neurochem* 2001; 76: 224–233.

34. Misonou H, Morishima-Kawashima M, Ihara Y. Oxidative stress induces intracellular accumulation of amyloid beta-protein (Aβ) in human neuroblastoma cells. *Biochemistry* 2000; 39: 6951–6959.

35. Frederikse PH, Garland D, Zigler JS, Jr., Piatigorsky J. Oxidative stress increases production of β-amyloid precursor protein and β-amyloid (Aβ) in mammalian lenses, and Aβ has toxic effects on lens epithelial cells. *J Biol Chem* 1996; 271: 10169–10174.

36. Tamagno E, Bardini P, Obbili A, Vitali A, Borghi R, Zaccheo D, Pronzato MA, Danni O, Smith MA, Perry G, Tabaton M. Oxidative stress increases expression and activity of BACE in NT2 neurons. *Neurobiol Dis* 2002; 10: 279–288.

37. Hall ED, Braughler JM. Role of lipid peroxidation in post-traumatic spinal cord degeneration: a review. *Cent Nerv Syst Trauma* 1986; 3: 281–294.

38. Kitagawa K, Matsumoto M, Oda T, Niinobe M, Hata R, Handa N, Fukunaga R, Isaka Y, Kimura K, Maeda H, et al. Free radical generation during brief period of cerebral ischemia may trigger delayed neuronal death. *Neuroscience* 1990; 35: 551–558.

39. Jenner P, Dexter DT, Sian J, Schapira AH, Marsden CD. Oxidative stress as a cause of nigral cell death in Parkinson's disease and incidental Lewy body disease: the Royal Kings and Queens Parkinson's disease research group. *Ann Neurol* 1992; 32 Suppl: S82–87.

40. Gentleman SM, Graham DI, Roberts GW. Molecular pathology of head trauma: altered beta APP metabolism and the aetiology of Alzheimer's disease. *Prog Brain Res* 1993; 96: 237–246.

41. Roberts GW, Gentleman SM, Lynch A, Murray L, Landon M, Graham DI. Beta amyloid protein deposition in the brain after severe head injury: implications for the pathogenesis of Alzheimer's disease. *J Neurol Neurosurg Psychiatry* 1994; 57: 419–425.

42. Geddes JW, Tekirian TL, Soultanian NS, Ashford JW, Davis DG, Markesbery WR. Comparison of neuropathologic criteria for the diagnosis of Alzheimer's disease. *Neurobiol Aging* 1997; 18: S99–105.

43. Lim GP, Chu T, Yang F, Beech W, Frautschy SA, Cole GM. The curry spice curcumin reduces oxidative damage and amyloid pathology in an Alzheimer transgenic mouse. *J Neurosci* 2001; 21: 8370–8377.

44. Matsubara E, Bryant-Thomas T, Pacheco Quinto J, Henry TL, Poeggeler B, Herbert D, Cruz-Sanchez F, Chyan YJ, Smith MA, Perry G, Shoji M, Abe K, Leone A, Grundke-Ikbal I, Wilson GL, Ghiso J, Williams C, Refolo LM, Pappolla MA, Chain DG, Neria E. Melatonin increases survival and inhibits oxidative and amyloid pathology in a transgenic model of Alzheimer's disease. *J Neurochem* 2003; 85: 1101–1108.

45. Veurink G, Liu D, Taddei K, Perry G, Smith MA, Robertson TA, Hone E, Groth DM, Atwood CS, Martins RN. Reduction of inclusion body pathology in ApoE-deficient mice fed a combination of antioxidants. *Free Radic Biol Med* 2003; 34: 1070–1077.

46. Sung S, Yao Y, Uryu K, Yang H, Lee VM, Trojanowski JQ, Pratico D. Early vitamin E supplementation in young but not aged mice reduces $A\beta$ levels and amyloid deposition in a transgenic model of Alzheimer's disease. *Faseb J* 2004; 18: 323–325.

47. Hock C, Golombowski S, Muller-Spahn F, Naser W, Beyreuther K, Monning U, Schenk D, Vigo-Pelfrey C, Bush AM, Moir R, Tanzi RE, Growdon JH, Nitsch RM. Cerebrospinal fluid levels of amyloid precursor protein and amyloid beta-peptide in Alzheimer's disease and major depression — inverse correlation with dementia severity. *Eur Neurol* 1998; 39: 111–118.

48. Kontush A, Berndt C, Weber W, Akopyan V, Arlt S, Schippling S, Beisiegel U. Amyloid-beta is an antioxidant for lipoproteins in cerebrospinal fluid and plasma. *Free Radic Biol Med* 2001; 30: 119–128.

49. Hou L, Kang I, Marchant RE, Zagorski MG. Methionine 35 oxidation reduces fibril assembly of the amyloid A-β(1-42) peptide of Alzheimer's disease. *J Biol Chem* 2002; 277: 40173–40176.

50. Smith MA, Casadesus G, Joseph JA, Perry G. Amyloid-β and τ serve antioxidant functions in the aging and Alzheimer brain. *Free Radic Biol Med* 2002; 33: 1194–1199.

51. Zou K, Gong JS, Yanagisawa K, Michikawa M. A novel function of monomeric amyloid beta-protein serving as an antioxidant molecule against metal-induced oxidative damage. *J Neurosci* 2002; 22: 4833–4841.

52. Wataya T, Nunomura A, Smith MA, Siedlak SL, Harris PL, Shimohama S, Szweda LI, Kaminski MA, Avila J, Price DL, Cleveland DW, Sayre LM, Perry G. High molecular weight neurofilament proteins are physiological substrates of adduction by the lipid peroxidation product hydroxynonenal. *J Biol Chem* 2002; 277: 4644–4648.

53. Takeda A, Smith MA, Avila J, Nunomura A, Siedlak SL, Zhu X, Perry G, Sayre LM. In Alzheimer's disease, heme oxygenase is coincident with Alz50, an epitope of tau induced by 4-hydroxy-2-nonenal modification. *J Neurochem* 2000; 75: 1234–1241.

54. Calingasan NY, Uchida K, Gibson GE. Protein-bound acrolein: a novel marker of oxidative stress in Alzheimer's disease. *J Neurochem* 1999; 72: 751–756.

55. Morsch R, Simon W, Coleman PD. Neurons may live for decades with neurofibrillary tangles. *J Neuropathol Exp Neurol* 1999; 58: 188–197.

56. Zhu X, Rottkamp CA, Boux H, Takeda A, Perry G, Smith MA. Activation of p38 kinase links tau phosphorylation, oxidative stress, and cell cycle-related events in Alzheimer's disease. *J Neuropathol Exp Neurol* 2000; 59: 880–888.

57. Zhu X, Castellani RJ, Takeda A, Nunomura A, Atwood CS, Perry G, Smith MA. Differential activation of neuronal ERK, JNK/SAPK and p38 in Alzheimer's disease: the "two hit" hypothesis. *Mech Ageing Dev* 2001; 123: 39–46.

58. Zhu X, Sun Z, Lee HG, Siedlak SL, Perry G, Smith MA. Distribution, levels, and activation of MEK1 in Alzheimer's disease. *J Neurochem* 2003; 86: 136–142.

59. Zhu X, Raina AK, Perry G, Smith MA. Alzheimer's disease: the two-hit hypothesis. *Lancet Neurol* 2004; 3: 219–226.

60. Odetti P, Garibaldi S, Norese R, Angelini G, Marinelli L, Valentini S, Menini S, Traverso N, Zaccheo D, Siedlak S, Perry G, Smith MA, Tabaton M. Lipoperoxidation is selectively involved in progressive supranuclear palsy. *J Neuropathol Exp Neurol* 2000; 59: 393–397.

61. Gerst JL, Siedlak SL, Nunomura A, Castellani R, Perry G, Smith MA. Role of oxidative stress in frontotemporal dementia. *Dement Geriatr Cogn Disord* 1999; 10(Suppl. 1): 85–87.

62. Ekinci FJ, Shea TB. Phosphorylation of tau alters its association with the plasma membrane. *Cell Mol Neurobiol* 2000; 20: 497–508.

63. Arendt T, Stieler J, Strijkstra AM, Hut RA, Rudiger J, Van der Zee EA, Harkany T, Holzer M, Hartig W. Reversible paired helical filament-like phosphorylation of tau is an adaptive process associated with neuronal plasticity in hibernating animals. *J Neurosci* 2003; 23: 6972–6981.

64. Zhou F, Zhu X, Castellani RJ, Stimmelmayr R, Perry G, Smith MA, Drew KL. Hibernation, a model of neuroprotection. *Am J Pathol* 2001; 158: 2145–2151.

65. Takeda A, Perry G, Abraham NG, Dwyer BE, Kutty RK, Laitinen JT, Petersen RB, Smith MA. Overexpression of heme oxygenase in neuronal cells, the possible interaction with Tau. *J Biol Chem* 2000; 275: 5395–5399.

12 Coenzyme Q$_{10}$ Stabilizes Mitochondria in Parkinson's Disease

*Manuchair Ebadi, Sawitri Wanpen,
Shaik Shavali, Sushil Sharma, and
Hesham El ReFaey*
University of North Dakota, Grand Forks, North Dakota

CONTENTS

12.1 OVERVIEW

Parkinson's disease (PD) results primarily from the death of dopaminergic neurons in the substantia nigra. The culprits implicated in the neurodegeneration of PD are oxidative stress, mitochondrial dysfunction, misfolding of proteins, and disruption of the ubiquitin–proteasome pathway. Neurotoxin-based models, particularly those involving MPTP, have been important in elucidating the molecular cascade of cell death in dopaminergic neurons.

The main obstacle to developing neuroprotective therapies is a limited understanding of the key molecular events that cause PD. The mitochondrial component coenzyme Q$_{10}$ has been used for many years as a dietary supplement to promote good health because of its ability to trap free radicals, thus preventing lipid peroxidation and DNA damage. Complex I activity in platelet mitochondria is reduced in patients with early untreated PD. The administration of coenzyme Q$_{10}$ increases brain mitochondrial concentrations and exerts neuroprotective effects.

Metallothionein isoforms, which scavenge hydroxyl radicals, superoxide anions, and peroxynitrite, attenuate oxidative stress. In addition, metallothionein, by stimulating the activity of lipoamide dehydrogenase, enhances the formation of ubiquinol, an antioxidant.

12.2 MITOCHONDRIAL DYSFUNCTION IN THE PATHOGENESIS OF PD

PD is the second most common neurodegenerative disorder after Alzheimer's disease, affecting ~1% of the population above the age of 50 yr. There is a worldwide increase in the disease prevalence due to the increasing age of human populations. A definitive neuropathological diagnosis of PD requires proof of the loss of dopaminergic neurons (DA neurons) in the substantia nigra (SN) and related brainstem nuclei, and the presence of Lewy bodies in surviving nerve cells.

The contribution of genetic factors to the pathogenesis of PD is being increasingly recognized. A point mutation that is sufficient to cause a rare autosomal dominant form of the disorder has recently been identified in the α-synuclein (α-Syn) gene on chromosome 4 in the much more common sporadic, or "idiopathic," form of PD, and a defect of complex I of the mitochondrial respiratory chain was confirmed at the biochemical level. A defect, specifically of this factor, has been demonstrated in the parkinsonian SN. These findings, and the observation that the neurotoxin 1-methyl-4-phenyl-1, 2, 3, 6-tetrahydropyridine (MPTP) causes a PD-like syndrome in humans and acts via inhibition of complex I, have triggered research interest in the mitochondrial genetics of PD (see Figure 12.1).

The exact molecular mechanism remains unknown as to how perinuclear and endonuclear translocation and aggregation of α-Syn during oxidative and nitrative stresses cause Lewy body formation and progressive neurodegeneration in aging and PD. In view of the above, we [1–3] studied 1-methyl-4-phenylpyridinium (MPP^+)-mediated induction, translocation, and aggregation of α-Syn in wild-type control ($control_{wt}$) and aging mitochondrial genome knockout (RhO_{mgko}) dopaminergic (SK-N-SH) neurons. α-Syn expression was studied in $control_{wt}$ and aging RhO_{mgko} DA (SK-N-SH) neurons in response to MPP^+ and/or selegiline [1–3]. Basal α-Syn expression was higher in RhO_{mgko} neurons as compared to $control_{wt}$ neurons. α-Syn expression was enhanced in response to overnight treatment with MPP^+. MPP^+ induced perinuclear aggregation of α-Syn in both $control_{wt}$ and aging RhO_{mgko} neurons. In RhO_{mgko} neurons, α-Syn was translocated in the endonuclear region. Pretreatment with selegiline suppressed perinuclear accumulation of α-Syn in $control_{wt}$ neurons and endonuclear accumulation in aging RhO_{mgko} neurons. These observations are interpreted as suggesting that parkinsonian neurotoxins, such as MPP^+ can induce neurodegeneration via α-Syn induction, translocation, and aggregation in DA neurons.

Glutathione deficiency, which causes the accumulation of H_2O_2, leads to mitochondrial damage in the brain. Moreover, coenzyme Q_{10} (CoQ_{10}) attenuates the MPTP-induced loss of striatal DA neurons. Deficiency of striatal glutathione

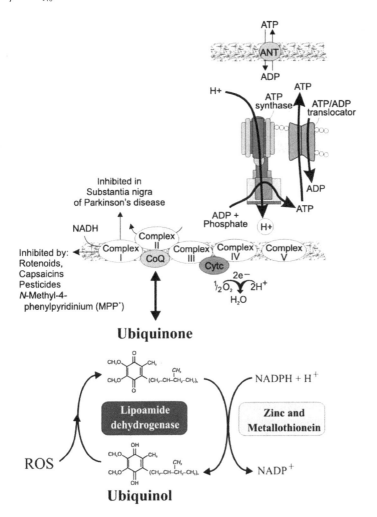

FIGURE 12.1 Oxidative phosphorylation consists of five protein–lipid enzyme complexes located in the mitochondrial inner membrane that contain flavins (FMN, FAD), quinoid compounds (CoQ$_{10}$) and transition metal compounds (iron–sulfur clusters, hemes, protein-bound copper). These enzymes are designated complex I (NADH:Ubiquinone oxidoreductase EC 1.6.5.3), complex II (succinate: ubiquinol oxidoreductase, EC 1.3.5.1), complex III (ubiquinol:ferrocytochrome c oxidoredutase EC 1.10.2.2), complex IV (ferrocytochrome c:oxygen oxidoreductase or cytochrome c oxidase, EC 1.9.3.1), and complex V (ATP synthase, EC 3.6.1.34). A defect in mitochondrial oxidative phosphorylation in terms of a reduction in the activity of NADH CoQ reductase (complex I) has been reported in the striatum of patients with PD. The reduction in the activity of complex I is found in the SN, but not in other areas of the brain, such as the globus pallidus or cerebral cortex. Therefore, the specificity of mitochondrial impairment may play a role in the degeneration of nigrostriatal DA neurons. This view is supported by the fact that MPTP generating MPP$^+$ destroys DA neurons in the SN. Although the serum levels of CoQ$_{10}$ is normal in patients with PD, CoQ$_{10}$ is able to attenuate the MPTP-induced loss in striatal DA neurons. Metallothionein, a zinc-binding protein, not only scavenges free radicals itself but also, by increasing the activity of lipoamide dehydrogenase, increases the concentration of ubiquinol, another antioxidant.

(GSH) in PD fosters oxidative stress and causes apoptosis. γ-Glutamylcysteinylg-lycine assists in maintaining the intracellular reducing environment, protects protein thiol groups from oxidation, and participates as a coenzyme or cofactor in a wide variety of chemical reactions. GSH exerts its antioxidant activity synergistically with both vitamin C and vitamin E. GSH deficiency in PD enhances the susceptibility of the SN to destruction by endogenous or exogenous neurotoxins. Moreover, treatment with lazaroid, which inhibits lipid peroxidation, prevents the death of mesencephalic DA neurons following GSH depletion [4].

The selective vulnerability and loss of certain neurons is a remarkable char-acteristic of age-related degenerative disorders of the brain as exemplified by PD. Glutamate is the major excitatory neurotransmitter in the brain, and excitotoxicity plays a role in PD. Furthermore, growing evidence implicates oxidative stress as a mediator of excitotoxic cell death. Following activation of N-methyl-D-aspartate (NMDA) receptors, the generation of free radicals increases and oxidation damage to lipids occurs; antioxidants prevent the ensuring cell death. Dizocilpine, which blocks NMDA receptors, may provide neuroprotection in PD [5,6].

Damage to the mitochondrial electron transport chain has been suggested to be an important factor in the pathogenesis of a range of neurological disorders, such as PD, Alzheimer's disease, multiple sclerosis, stroke, and amyotrophic lateral sclerosis. There is also a growing body of evidence to implicate excessive or inappropriate generation of nitric oxide (NO) in these disorders. It is now well documented that NO and its toxic metabolite, peroxynitrite (ONOO), can inhibit components of the mitochondrial respiratory chain leading to a cellular energy deficiency state [7].

Oxidative phosphorylation involves five protein–lipid enzyme complexes located in the mitochondrial inner membrane that contain flavins (e.g., FMN, FAD), quinoid compounds (CoQ_{10}), and transition metal compounds (iron–sulfur clusters, hemes, and protein-bound copper). These enzymes are designated *com-plex I* (NADH: ubiquinone oxidoreductase, EC 1.6.5.3), *complex II* (succinate: ubiquinone oxidoreductase, EC 1.3.5.1), *complex III* (ubiquinol: ferrocytochrome c oxidoreductase, EC 1.10.2.2), *complex IV* (ferrocytochrome c: oxygen oxido-reductase or cytochrome c oxidase; EC 1.9.3.1), and *complex V* (ATP synthase, EC 3.6.1.34). A defect in mitochondrial oxidative phosphorylation, in terms of a reduction in the activity of complex I, has been reported in the striatum of patients with PD. The reduction in the activity of complex I is found in the SN but not in other areas of the brain, such as the globus pallidus or cerebral cortex. There-fore, the specificity of mitochondrial impairment may play a role in the degen-eration of nigrostriatal DA neurons.

This view is supported by the fact that MPTP generating MPP+ destroys DA neurons in the SN [8,9]. Lesions produced by the reversible inhibitor of succinate dehydrogenase (complex II), malonate, and the irreversible inhibitor 3-nitropro-pionic acid closely resemble the histologic, neurochemical, and clinical features of Huntington's disease in both rats and nonhuman primates. The interruption of oxidative phosphorylation results in decreased levels of ATP. A consequence is partial neuronal depolarization and secondary activation of voltage-dependent

NMDA receptors, which may result in excitotoxic neuronal cell death (secondary excitotoxicity). The increase in intracellular Ca^{2+} concentration leads to an activation of Ca^{2+}-dependent enzymes, including the constitutive neuronal nitric oxide synthase (cnNOS), which produce NO. NO· may react with the superoxide anion (O$_2$) to form ONOO$^-$. Schulz et al. [10] have shown that systemic administration of 7-nitroindazole (7-NI), a relatively specific inhibitor of neuronal nitric oxide synthase (nNOS) *in vivo,* attenuates lesions produced by striatal malonate injections, or systemic treatment with 3-nitropropionic acid or MPTP. Furthermore, 7-NI attenuates increases in lactate production and hydroxyl radical (·OH) and 3-nitrotyrosine generation *in vivo*, which may be a consequence of ONOO$^-$ formation. These results suggest that nNOS inhibitors may be useful in the treatment of neurologic disease in which excitotoxic mechanisms play a role [10].

In the central nervous system (CNS), NO may play important roles in neurotransmitter release, neurotransmitter reuptake, neurodevelopment, synaptic plasticity, and regulation of gene expression. However, excessive production of NO following a pathologic insult can lead to neurotoxicity. NO plays a role in mediating the neurotoxicity associated with a variety of neurologic disorders, including stroke, PD, and HIV dementia [11,12].

Due to its ability to modulate neurotransmitter release and reuptake, mitochondrial respiration, DNA synthesis, and energy metabolism, it is not surprising that NO can be neurotoxic. Under conditions where NO is abnormally produced, such as when inducible nitric oxide synthase (iNOS) expression is stimulated in the CNS, dysregulation of normal physiologic activities of NO likely contributes to neuronal dysfunction and, subsequently, to neuronal death. However, acute toxicity mediated by NO appears to require the production of O$_2$.

NO by itself is a relatively nontoxic molecule, which in the absence of O$_2$ will not kill cells even at extremely high concentrations. In the presence of O$_2$, however, NO is a potent neurotoxin. The reaction of NO with O$_2$ is the fastest biochemical rate constant currently known, resulting in the formation of the potent oxidant ONOO$^-$, which is a lipid permeable molecule with a wider range of chemical targets than NO. It can oxidize proteins, lipids, RNA, and DNA. Neurotoxicity elicited by ONOO$^-$ formation may have a dual component. Peroxynitrite can potently inhibit mitochondrial proteins. Peroxynitrite inhibits the function of manganese superoxide dismutase (MnSOD), which can lead to increased O$_2$ and ONOO$^-$ formation. Additionally, ONOO$^-$ is an effective inhibitor of enzymes in the mitochondrial respiratory chain, which results in decreased ATP synthesis. Secondly, ONOO$^-$ efficiently modifies and breaks DNA strands and inhibits DNA ligase, which increases the DNA strand breaks. DNA strand breaks activate DNA repair mechanisms. One of the initial proteins activated by DNA damage is the nuclear enzyme poly(ADP-ribose) polymerase (PARP). PARP catalyzes the attachment of ADP-ribose units from NAD to nuclear proteins such as histone and PARP itself. PARP can add hundreds of ADP-ribose units within seconds to minutes of being activated. For every mole of ADP-ribose transferred from NAD, one mole of NAD is consumed, and four free-energy equivalents of ATP are required to regenerate NAD to normal cellular levels. Activation of PARP can result in a rapid drop in energy stores. If this drop is severe and sustained, it can lead to impaired cellular metabolism and, ultimately, to cell death [11].

We [1] developed the multiple fluorescence Comet assay to examine mitochondrial as well as nuclear DNA damage, simultaneously in a single neuron. The multiple fluorescence Comet assay is performed primarily on a single neuron using alkaline gel electrophoresis and digital fluorescence imaging microscopy. In these experiments, we conducted the multiple fluorochrome Comet assay to determine MPP$^+$-induced neurotoxicity in control$_{wt}$ and α-Syn-overexpressed aging RHO$_{mgko}$ DA (SK-N-SH) neurons. Dominance of green fluorescence in control$_{wt}$ neurons suggests that in these neurons, MPP$^+$-induced neuronal damage remains restricted primarily to the mitochondrial region [due to synthesis of the mitochondrial DNA oxidation product, 8-hydroxy-2-deoxyguanosine (8-OH-2dG)]; whereas in α-Syn-overexpressed aging RHO$_{mgko}$ DA (SK-N-SH) neurons, the damage was observed in both mitochondrial and nuclear regions. These observations suggest that α-Syn-overexpressed aging RHO$_{mgko}$ DA (SK-N-SH) neurons are highly susceptible to MPTP-induced neurotoxicity, which could involve both mitochondrial DNA and nuclear DNA. The damage in the nuclear DNA is illustrated with red fluorescence tails due to ethidium bromide staining, and the damage to mitochondrial DNA is represented by green fluorescence tails due to fluorescein isothiocyanate (FITC)-conjugated antibody to the DNA oxidation product, 8-OH-2dG (see Figure 12.2).

It is well known that mitochondrial complex I is downregulated during the progression of neurodegeneration in PD. An experiment was performed to establish the primary involvement of the mitochondrial genome in neuronal repair during aging [1]. In view of these facts, aging RhO$_{mgko}$ neurons were prepared by selective inactivation of the mitochondrial DNA with 5 µg/l of the DNA intercalating agent ethidium bromide for 6 to 8 weeks. The control$_{wt}$ neurons

Coenzyme Q$_{10}$ protects dopaminergic cells by preventing Salsolinol-induced nuclear damage

(Single cell gel-electrophoresis)

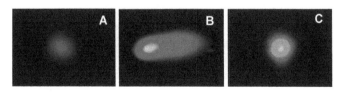

FIGURE 12.2 Neuroprotective effects of CoQ$_{10}$ on nuclear damage induced by SAL. Cells were treated with SAL or CoQ$_{10}$ for 24 h, and subjected to single cell gel-electrophoresis (comet assay) followed by staining with ethidium bromide. The length of the tail indicates the extent of nuclear damage. A: Control; B: SAL (100 µM) C: SAL (100 µM + CoQ$_{10}$ (10 µM). CoQ$_{10}$ reduced nuclear damage induced by SAL.

exhibited structurally intact neuronal morphology with long axons and dendrites; whereas, aging RhO$_{mgko}$ neurons exhibited enhanced granularity, mitochondrial aggregation, elliptical appearance (which was due to stunted neuritogenesis), downregulation of oxidative phosphorylation, and reduced ATP generation. Neuritogenesis and ATP production in aging RhO$_{mgko}$ neurons were regained by transfecting these neurons with the mitochondrial genome encoding complex I activity. The transfection was done using the pEGFP-N1 vector, Qiagen Effectene transfection reagent, and a DNA enhancer as per the manufacturer's recommendations. The transfected neurons were selected using G-418 and enriched by the limiting dilution technique.

There is evidence that NO can displace iron from its binding site on ferritin, an iron-storage protein and, consequently, promote lipid peroxidation. NO can also influence iron metabolism at the posttranscriptional level by interacting with cytosolic aconitase. Cytosolic aconitase has dual functions that are regulated by NO. In the absence of NO, it functions as cytosolic aconitase, but in the presence of NO, it functions as the iron-responsive element binding protein to the iron-responsive element [2,13].

The role of NO in 6-hydroxydopamine-induced parkinsonism has been established [14]. Riobo et al. [15] analyzed the potential reaction between 6-hydroxydopamine and NO. The results showed that NO reacts with the deprotonated form of 6-hydroxydopamine at pH 7 and 37°C with a second-order rate constant of 1.5×10^3 M^1 s^1, as calculated by the rate of NO decay measured with an amperometric sensor. Accordingly, the rate of formation of 6-hydroxydopamine quinone was dependent on NO concentration. The coincubation of NO and 6-hydroxydopamine with either bovine serum albumin or α-Syn led to tyrosine nitration of the protein in a concentration dependent-manner and was sensitive to superoxide dismutase (SOD). These findings suggest the formation of ONOO$^-$ during the redox reactions following the interaction of 6-hydroxydopamine with NO. Hunot et al. [16], using immunohistochemistry and histochemistry, analyzed the production of NO in the mesencephalon of four patients with idiopathic PD and three matched control subjects. Using specific antibodies directed against the inducible isoforms of nitric oxide synthase (NOS), the enzyme involved in the synthesis of nitric oxide, they found evidence to suggest that this isoform was present solely in glial cells displaying the morphological characteristics of activated macrophages. Immunohistochemical analysis performed with antibodies against the neuronal isoforms of NOS, however, revealed perikarya and processes of neurons, but no glial cell staining. The number of NOS-containing cells was investigated by histoenzymology, using the NADPH-diaphorase activity of NOS. Histochemistry revealed (1) a significant increase in NADPH-diaphorase-positive glial cell density in the dopaminergic cell groups characterized by neuronal loss in PD and (2) a neuronal loss in PD that was two-fold greater for pigmented NADPH-diaphorase-negative neurons than for pigmented NADPH-diaphorase-positive neurons. These data suggest a potentially deleterious role for glial cells, producing excessive levels of NO in PD, which may be neurotoxic for a subpopulation of DA neurons, especially those not expressing NADPH-diaphorase activity. However, it cannot be ruled out that the presence of glial cells expressing NOS synthase in the SN of patients with PD represents a consequence of DA neuronal loss [16].

12.3 MPTP DOWNREGULATES DOPAMINE TRANSPORTER

MPP^+, the active metabolite of the neurotoxin MPTP, mediates selective damage to DA neurons and has been widely used to generate a model of PD. However, the mechanisms of the neurotoxic action of MPP^+ are not fully understood. MPP^+ is transported into cells via the dopamine transporter (DAT), where it mediates cellular toxicity [17–19]. DAT is a presynaptic plasma membrane protein responsible for the regulation of extracellular dopamine levels and termination of its action by mediating the reuptake of dopamine [20,21]. Functional impairment of DAT alters many physiological and behavioral processes that are mediated by dopamine. A dysfunction of dopamine transmission could consequentially interrupt motor neural circuits which control movement, as is seen in PD.

Recently, it has been shown that cellular mRNAs encoding DAT and vesicular monoamine transporters are decreased in PD [22,23]. Indeed, an alteration of dopaminergic neurotransmission by the modulation of DAT activity could have an important implication in the cellular events that lead to PD and could be a target for potential therapeutic intervention.

When MPP^+ enters into the cells, it causes the release of dopamine from secretory vesicles and subsequently generates free radicals [24–27]. Inside the cell, MPP^+ disrupts cellular respiration by inhibiting the mitochondrial complex I system [28,29], reducing the level of ATP and, hence, contributing to degeneration of DA neurons.

Several studies have shown that MPP^+ causes a significant decrease in the activity of DAT [30–33]. However, it remained unclear whether the decrease in DAT activity resulted from the selective uptake of MPP^+ through DAT, reduction in DA neurons mediated by the toxin, or changes in the trafficking of the transporter molecules.

Chagkutip et al. [34] have hypothesized that a reduction in cell surface expression of DAT may be an underlying mechanism for the downregulation of DAT function by MPP^+ and the subsequent neurotoxicity. To test this hypothesis, we selected HEK-293 cells, which stably express human DAT in order to monitor, especially, the effects of MPP^+ on DAT function. The data indicated that MPP^+ reduced cell viability, dopamine uptake, and cell surface expression of DAT, thus, directly correlating its neurotoxic effects with alterations in the kinetic parameters of dopamine transporters. In this study [34], we have shown that DAT is required for MPP^+-mediated cytotoxicity in HEK-293 cells stably transfected with human DAT. Furthermore, MPP^+ produced a concentration- and time-dependent reduction in the uptake of [³H]dopamine. We observed a significant decrease in [³H]WIN 35428 binding in intact cells with MPP^+. The saturation analysis of the [³H]WIN 35428 binding obtained from total membrane fractions revealed a decrease in the transporter density (B_{max}) with an increase in the dissociation equilibrium constant (K_d) after

MPP$^+$ treatment. Furthermore, biotinylation assays confirmed that MPP$^+$ reduced both plasma membrane and intracellular DAT immunoreactivity. Taken together, these findings suggest that the reduction in cell surface DAT protein expression in response to MPP$^+$ may be a contributory factor in the downregulation of DAT function and enhanced lysosomal degradation of DAT may signal events leading to cellular toxicity [34].

Several studies have demonstrated that MPP$^+$ elicits a marked decrease in hDAT transport activity, but the mechanisms involved remain largely hypothetical and equivocal. Consistent with earlier studies [23,30–32], we have shown that the downregulation of hDAT observed in response to MPP$^+$ occurred in a concentration- and time-dependent fashion. MPP$^+$ at a concentration of 100 µM was used to investigate the time course of dopamine uptake because this concentration produced the highest reduction in dopamine uptake without affecting the cell culture conditions. Because MPP$^+$ has been shown to impair cell energy metabolism over time [27,32,33,35–38], a 3 h incubation period was chosen to preclude any possible effects of energy deficit and cell death on DAT function and the underlying mechanisms.

We [34] used [^{3}H]WIN 35428 binding assay in intact cells, which was conducted at a low temperature to limit endocytosis of the ligand. A significant reduction in the binding of [^{3}H]WIN 35428 was observed after 3 h incubation with MPP$^+$ at each of the concentration ranges used (1, 10, 100 µM and 1 mM), and this was consistent with a decline in DAT activity. Moreover, a notable decrease in the B_{max} value for [^{3}H]WIN 35428 binding in the total membrane fractions isolated from lysed cells after treatment with 100 µM MPP$^+$ was observed, indicating that the MPP$^+$-induced decrease in DAT is correlated with the reduction in the number of available [^{3}H]WIN 35428 binding sites on the cell surface and also in the cytoplasm.

The cell-impermeable biotinylated sulfo-NHS-biotin reagent and Western blot analysis were used to demonstrate that MPP$^+$ decreased both the biotinylated and nonbiotinylated DAT protein levels; this was consistent with the data obtained from the Scatchard analysis. The data further indicated a smaller percentage change for the intracellular DAT, compared with that of the plasma membrane (72% and 53% of control, respectively). This probably reflects the greater absolute amounts of measurable intracellular DAT, which also includes cell surface membrane DAT. Taken together, these findings suggest that MPP$^+$-mediated downregulation of DAT activity correlates with an overall lowering of DAT protein expression [34].

Another mechanism whereby MPP$^+$ may cause a reduction in DAT activity is phosphorylation. In recent observations, researchers have proposed that DAT activity was regulated by its phosphorylation [39–41]. A consequence of phosphorylation is the possible sequestration of the transporter proteins [42] and, consequently, lower activity. Our data [34], however, provides evidence that MPP$^+$ may reduce the activity of DAT by lowering its level both on the plasma membrane and in the cytoplasm. It is also possible that MPP$^+$ mediates neurotoxcity by increasing the sensitivity of DAT to lysosomal degradation, resulting in a shorter half-life. The data presented here form the basis for further studies toward understanding the neurotoxic processes that occur in PD. The decrease in DAT function could involve (1) DAT phosphorylation, (2) enhanced internalization and degradation of DAT, and (3) reduction in DAT protein expression, or a combination of all three. However, these possibilities need to be explored further.

12.4 1-BENZYL-1,2,3,4-TETRAHYDROISOQUINOLINE (1BnTIQ), AN ENDOGENOUS NEUROTOXIN CAUSING PD, INDUCES DOPAMINERGIC CELL DEATH THROUGH APOPTOSIS

PD is a progressive neurodegenerative disorder characterized by extensive loss of DA neurons in substantia nigra pars compacta, with a substantial decrease in dopamine content, and the appearance of Lewy bodies in surviving neurons. The precise biochemical and molecular mechanisms of dopaminergic cell death in PD remains unclear; however, it is believed to be caused by either environmentally produced or endogenously generated toxins coupled with genetic abnormalities such as complex I deficiency [43]. The neurotoxin MPTP degenerates specifically the nigrostraital DA neurons and induces dopamine deficiency in the striatum causing symptoms similar to PD [44–46].

The active metabolite of MPTP, i.e., MPP+, which is formed in astrocytes, migrates and accumulates specifically in DA neurons by a high-affinity dopamine reuptake through dopamine transporters [34,47]. The neurotoxicity induced by MPP+ is partly related to inhibition of the mitochondrial complex [43–48] and generation of free radicals [49,50] which, eventually, may cause dopaminergic cell death.

It has been speculated that endogenously produced dopamine-derived toxins structurally similar to MPTP or MPP+ may cause PD. Two different groups of endogenous MPTP-like amines have been found in human brains, i.e., tetrahydroisoquinolines (TIQs) and β-carbolines [51].

In addition, two TIQs, specifically, salsolinol (SAL) and 1BnTIQ, have been identified in normal human and mouse brains. Furthermore, the concentrations of these TIQs were reported to be higher in the brain and cerebrospinal fluid (CSF) of PD patients as compared to that in age-matched control subjects [52,53].

Although, TIQs are found in various foods [54,55], they do not cross the blood–brain barrier [56,57], and those found in the brain are believed to be specifically synthesized locally.

In human CSF, 1BnTIQ was first detected as a novel endogenous amine, suggesting that it might be produced by condensation and cyclization of two endogenous compounds, namely, 2-phenylethylamine (PEA) and phenylacetaldehyde, a metabolite of PEA generated by monoamine oxidase type B (MAO-B) [58,59]. The concentration of 1BnTIQ in parkinsonian CSF is three times higher than in control CSF [58]. 1BnTIQ, is known to induce parkinsonism in the monkey and mouse [53,59]. Antkiewicz-Michalukl et al. [59] reported that in rats, 1BnTIQ causes a significant decrease in striatal dopamine level (up to 60%) and increases homovanillic acid (HVA) level (up to 40%), compared to that in untreated control rats.

The biochemical and molecular mechanisms of neurotoxicity of 1BnTIQ-induced neurotoxicity was studied by using dopaminergic SH-SY5Y cells in culture and comparing its mechanisms with that of MPP$^+$. We [60] selected the SH-SY5Y cell line, a subclone derived from the human neuroblastoma cell line SK-N-SH, which expresses tyrosine hydroxylase, dopamine-β-hydroxylase, and DAT, which has unique properties similar to DA neurons in brain [61–63].

We [60] evaluated the mechanisms of 1BnTIQ in human dopaminergic SH-SY5Y cells and tested the neuroprotective action of SKF-38393, a dopamine receptor (D$_1$) agonist. 1BnTIQ dose-dependently decreased cell viability in dopaminergic SH-SY5Y cells, and the extent of cell death was more pronounced when compared to that caused by MPP$^+$. Similar to MPP$^+$, 1BnTIQ significantly decreased [^{3}H] dopamine uptake. 1BnTIQ significantly increased lipid peroxidation, Bax expression, and active caspase-3 formation. Furthermore, it decreased the expression of Bcl-xL, an antiapoptotic protein, in these cells. SKF-38393, a D$_1$ agonist, (in concentrations of 1 and 10 μM) completely prevented cell death and significantly increased cell viability. These results strongly suggest that 1BnTIQ induces dopaminergic cell death by apoptosis, and dopamine receptor agonists may be useful neuroprotective agents against 1BnTIQ toxicity [60].

PD is characterized by an extensive loss of DA neurons in the SN and the appearance of Lewy bodies in the remaining small number of surviving neurons, which results in the loss of dopamine-induced signals in the brain. Langston et al. [44] reported that MPTP produces parkinsonism in humans and in experimental animals by causing degeneration of DA neurons, specifically, in the nigrostriatal region. Furthermore, it was discovered that MPTP is converted into MPP$^+$ by MAO-B, which is present in glial cells. MPP$^+$ is actively taken up into DA neurons via DAT and gradually induces apoptosis by inhibition of complex I [44,63,65] and by decreasing ATP production. In addition, endogenously produced MPTP-like toxins have been suspected to be involved in the etiology of PD. Understanding the cause of dopaminergic cell death in the SN may clarify the etiopathogenesis of PD.

In the mouse brain and human CSF, 1BnTIQ was detected as a novel endogenous amine and its administration to rodents and primates induced symptoms similar to those seen in parkinsonism [53,59]. In our study, we [60] investigated the mechanisms of 1BnTIQ-induced neurotoxicity in human dopaminergic SH-SY5Y cells and compared its mechanisms with that of MPP$^+$, another neurotoxin. 1BnTIQ dose-dependently induced cell death in dopaminergic SH-SY5Y cells, similar to MPP$^+$. However, at the given concentrations, the extent of cell death was more pronounced with 1BnTIQ than with MPP$^+$. This was evident by MTT cell-proliferation assay. Fluorescent microscopic studies revealed an extensive loss of viable cells at higher concentrations of 1BnTIQ [66]. 1BnTIQ also dose-dependently decreased [^{3}H]dopamine uptake in these cells. The inhibition of dopamine uptake by 1BnTIQ suggests that it may affect the function of DATs in dopaminergic cells.

Isoquinoline derivatives, structurally related to MPTP, are known to inhibit [^{3}H]dopamine uptake into striatal synaptosomes [67,68]. It has been reported that 1BnTIQ derivatives, especially 3,4-dihydroxy-1-BnTIQ and 6,7-dihydroxy-1BnTIQ utilize DATs to actively accumulate in DA neurons. However, a recent report by Park et al. [67], indicated that neurotoxicants generating free radicals, such as ONOO$^-$, can also inactivate human dopamine transporter (hDAT) by modifying the cysteine residue at the 342 position in the transporter. Furthermore, 1BnTIQ significantly induces lipid peroxidation in these dopaminergic cells, suggesting that this compound induces oxidative stress, as observed in the SN of PD patients.

In our study [60], we specifically evaluated the nature of cell death by 1BnTIQ. It has been shown that apoptotic death of neurons occurs in the brain of PD patients [65]. Dopaminergic cells showed an apoptotic mode of cell death when treated with MPP$^+$, with specific apoptotic characteristics, such as increased reactive oxygen species (ROS) production, caspase-3 activation, and cleavage of PARP, with DNA condensation and fragmentation [71]. We have observed that 1BnTIQ dose-dependently increased proapoptotic protein (BAX) and decreased antiapoptotic protein Bcl-xL when analyzed by Western immunoblot technique. We also identified increased formation of active caspase-3 protein fragments (18 kDa) by 1BnTIQ. These results strongly support the possibility that apoptotic-related pathways are involved in 1BnTIQ-induced cell death in DA neurons [60].

Evidence is now emerging that some endogenous amines could act as antagonists for dopamine receptors and inhibit receptor function. Kawai et al. [69] demonstrated that 1BnTIQ and 6,7-dihydroxy-1BnTIQ administration reduced locomotor activity in mice. Several studies indicate that dopamine agonists have neuroprotective properties not only through acting on dopamine receptors but also by serving as antioxidants. Kitamura et al. [70] reported that pramipexole, a D_2/D_3 dopamine receptor agonist, has a neuroprotective action against MPP$^+$-induced toxicity in dopaminergic cells. Furthermore, a recent report from our laboratory [71] also suggested that SKF-38393, a dopamine D_1 agonist attenuated MPTP-induced neurotoxicity *in vivo* by preventing the depletion of glutathione and dopamine and by enhancing the activity of the antioxidant enzyme superoxide dismutase [4]. In our study [60], we evaluated the neuroprotective effects of SKF-38393 against 1BnTIQ-induced cell death. SKF-38393 dose-dependently (1 to 10 μM) prevented 1BnTIQ-induced cell death and increased cell viability in these dopaminergic cells. These results suggest that 1BnTIQ may act as an antagonist on dopamine receptors, and SKF-38393 could prevent cell death by blocking the interaction between 1BnTIQ and D_1 receptors. These results also explain that SKF-38393 receptor mediated signals are necessary for the protection of the dopaminergic cells against 1BnTIQ toxicity.

CoQ$_{10}$ protects against SAL-induced DNA fragmentation and enhanced expression of α-Syn (Figure 12.3). Further, selegiline attenuated SAL-induced apoptosis (Figure 12.4).

In conclusion, our study [60] strongly indicates that 1BnTIQ is toxic to dopaminergic cells and 1BnTIQ-induced cell death is apoptotic in nature. Furthermore, 1BnTIQ could be an endogenous neurotoxin involved in the pathogenesis

Coenzyme Q_{10} inhibits Salsolinol-induced α-Synuclein expression in dopaminergic cells

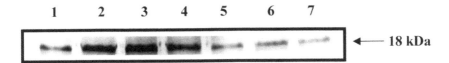

FIGURE 12.3 Effects of CoQ_{10} on SAL-induced α-Syn expression in dopaminergic SH-SY5Y cells as determined by Western immunoblot analysis. Cells were treated with SAL/CoQ_{10} for 24 h. The proteins were separated and transferred onto nitrocellulose membrane and probed with α-Syn antibody. Lane 1: Control; Lanes 2, 3, and 4: SAL 10, 20, and 50 μM, respectively; Label 5: SAL + CoQ_{10} (10 μM; Lane 6: SAL + CoQ_{10} (20 μM); Lane 7: CoQ_{10} (100 μM).

of PD, and dopamine receptor agonists may be useful neuroprotective compounds in protecting DA neurons against 1BnTIQ toxicity [71].

12.5 METALLOTHIONEIN (MT) ATTENUATES OXIDATIVE STRESS IN DA NEURONS

Although, the exact etiopathogenesis of PD remains unknown, it has been hypothesized that the neuronal demise of nigrostriatal DA neurons could occur due to the production of endogenous neurotoxins, such as tetrahydroisoquinolines, or by exposure to various environmental neurotoxins, such as rotenone [72]. These neurotoxins produce a significant downregulation of mitochondrial complex I, as observed in the majority of PD patients. Furthermore, significantly reduced glutathione in the SN enhances the risk of free radical [mainly ·OH and ·NO] overproduction, leading to neuronal damage in PD.

NO plays a critical role in mediating the neurotoxicity associated with various neurological disorders, such as stroke, PD, HIV dementia [73–77], and multiple sclerosis [78]. In the SN of PD patients, a significant increase in the density of glial cells expressing tumor necrosis factor-α, interleukin-1β, and interferon-γ has been observed. Although CD23 was not detectable in the SN of control subjects, it was found in both astroglial and microglial cells of parkinsonian patients, indicating the existence of cytokine/CD23-dependent activation pathways of iNOS and of proinflammatory mediators in glial cells and their involvement in the pathophysiology of PD [79]. In addition to NO, accumulation of iron in the SN has been implicated in the death of DA neurons in PD [79–81]. ONOO ions, generated in the mitochondria by Ca^{2+}-dependent NOS activation during oxidative and nitrative stresses, readily react with lipids, aromatic amino acids, or metalloproteins, inhibiting respiratory complexes and, hence, are thought to be involved in the etiopathogenesis of many diseases, including PD [82].

SELEGILINE ATTENUATES SALSOLINOL APOPTOSIS

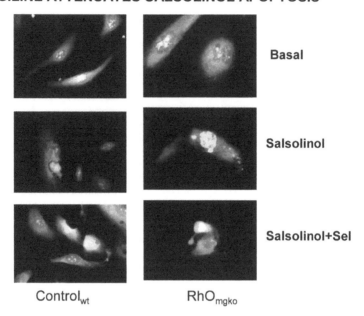

FIGURE 12.4 Multiple fluorochrome digital fluorescence imaging microscopic analysis of SAL-induced apoptosis in SK-N-SH neurons illustrating nuclear DNA fragmentation and condensation particularly in aging RhO$_{mgko}$ neurons. Selegiline pretreatment inhibits SAL-induced apoptosis by enhancing MTs expression. Upper left panel: Control$_{wt}$ SK-N-SH neurons exhibit bipolar spindle-shaped appearance with structurally intact plasma membranes, mitochondria, and nuclear DNA. Upper right panel: Aging RhO$_{mgko}$ neurons with rounded appearance, mitochondrial aggregation in the perinuclear region, and some evidence of nuclear DNA condensation. Middle left panel: Overnight exposure to SAL (100 μM) induces apoptosis characterized by phosphatidyl serine externalization, perinuclear aggregation of mitochondria and nuclear DNA fragmentation and condensation. Middle right panel: SAL induces severe apoptosis characterized by enhanced mitochondrial damage and nuclear DNA fragmentation and condensation. Lower left panel: pretreatment with selegiline inhibits SAL-induced plasma membrane phosphatidyl serine externalization, mitochondrial damage, and nuclear DNA fragmentation and condensation. Lower right panel: Selegiline-induced neuroprotection in response to SAL is compromised in RhO$_{mgko}$ neurons, suggesting that downregulation of mitochondrial genome renders the DA neurons highly susceptible to endogenously synthesized THIQ, SAL. The digital fluorescence images were captured individually in the beginning and subsequently merged to assess the overall structural and functional damage at the plasma membrane, mitochondrial membranes, and nuclear DNA. Flourescence images were captured by SpotLite digital camera and analyzed by ImagePro computer software. Blue: DAPI; Green: FITC annexin V; Red: Mitochondrial membrane potential marker, JC-1. Triple fluorochrome analysis was done to examine the extent of damage induced by SAL on the plasma membrane, mitochondria, and nuclear DNA simultaneously. [DAPI preferentially stains the structurally intact nuclear DNA, FITC annexin V stains phosphatidyl serine, and JC-1 is a molecular marker of mitochondrial membrane potential, $\Delta\psi$.

Although NO has been shown to possess both apoptogenic and apoptostatic properties, its overproduction during oxidative and nitrative stresses could induce deleterious consequences on mitochondrial complex I activity [83,84]. We have discovered recently that MT gene overexpression in MT transgenic (MT$_{trans}$) mouse brain inhibited the MPTP-induced nitration of α-Syn, and preserved mitochondrial CoQ$_{10}$ levels, affording neuroprotection against nitrative and oxidative stress in the aging brain [85,86]. In addition, MT isoforms were able to suppress 6-hydroxy-dopamine (6-OHDA)-induced ·OH radical generation. We have also, reported that selegiline, a MAO-B inhibitor, provides neuroprotection via MT gene overexpression [87,88].

As the involvement of oxidative and nitrative stresses is now implicated in the etiopathogenesis of PD, a detailed study was needed to explore the exact molecular mechanism of NO-mediated neurodegeneration and MT-induced neuroprotection in PD. 3-Morpholinosydnonimine (SIN-1), a vasorelaxant, a soluble guanylyl cyclase stimulator, and a potent ONOO⁻ generator, produced not only oxidative but also nitrative stresses in DA neurons and hence, promises to play an important role in clarifying the etiopathogenesis of PD [72].

Sharma and Ebadi [88] have examined SIN-1-induced apoptosis in control and MT-overexpressing DA neurons, with a primary objective of determining the neuroprotective potential of MT against ONOO⁻-induced neurodegeneration in PD. SIN-1 induced lipid peroxidation and triggered plasma membrane blebbing. In addition, it caused DNA fragmentation, α-Syn induction, intramitochondrial accumulation of metal ions (copper, iron, zinc, and calcium), and enhanced the synthesis of 8-OH-2dG. Furthermore, it downregulated the expression of Bcl-2 and PARP polymerase but upregulated the expression of caspase-3 and Bax in DA (SK-N-SH) neurons.

SIN-1 induced apoptosis in aging RhO$_{mgko}$ cells, α-Syn-transfected cells, MT double knock out cells, and in caspase-3-overexpressed DA neurons. SIN-1-induced changes were attenuated with selegiline or in MT-transgenic striatal fetal stem cells. SIN-1-induced oxidation of dopamine to dihydroxyphenylacetalde-hyde was attenuated in MT-transgenic fetal stem cells and in cells transfected with a mitochondrial genome, and enhanced in aging RhO$_{mgko}$ cells, in MT double knock out cells, and in caspase-3 gene-overexpressing DA neurons.

Selegiline, melatonin, ubiquinone, and MT suppressed SIN-1-induced down-regulation of mitochondrial genome and upregulation of caspase-3 as determined by reverse transcription-polymerase chain reaction. The synthesis of mitochondrial 8-OH-2dG and apoptosis-inducing factors were increased following exposure to MPP⁺ ion or rotenone [110–113]. Pretreatment with selegiline or MT suppressed 1-methyl, 4-phenyl tetrahydropyridinium ion-, 6-hydroxydopamine-, and rotenone-induced increases in mitochondrial 8-OH-2dG accumulations. Transfection of aging RhO$_{mgko}$ neurons with mitochondrial genome encoding complex I or melanin attenuated the SIN-1-induced increase in lipid peroxidation. SIN-1 induced the expression of α-Syn, caspase-3, and 8-OH-2dG and augmented protein nitration. MT attenuated SAL-induced apoptosis (Figure 12.5). These effects were attenuated by MT gene overexpression. These studies provide

METALLOTHIONEIN ATTENUATES SALSOLINOL APOPTOSIS

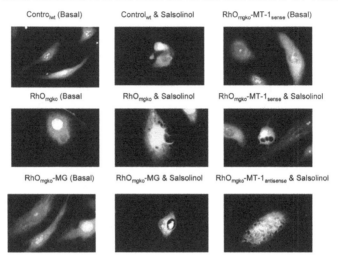

FIGURE 12.5 Multiple fluorochrome digital fluorescence imaging microscopic analysis of SAL-induced apoptosis in control$_{wt}$, aging RhO$_{mgko}$, RhO$_{mgko}$-MT-1$_{sense}$, and RhO$_{mgko}$-MT$_{antisense}$ oligonucleotide-transfected SK-N-SH neurons. Aging RhO$_{mgko}$ neurons were prepared by exposing SK-N-SH neurons to 5 µg/l of the DNA intercalating agent, ethidium bromide for 6 to 8 weeks in Dulbecco's modified Eagle's medium (DMEM) supplemented with 10% fetal bovine serum, high glucose, and glutamine. The RhO$_{mgko}$ neurons were authenticated by complete suppression of complex I gene by RT-PCR analysis and by estimating superoxide dismutase and catalase. RhO$_{mgko}$ neurons were transfected with either sense or antisense oligonucleotides to metallothionein-I using Effectene transfection reagent and the DNA enhancer as per the manufacturer's recommendations. Upper left panel: Control$_{wt}$ SK-N-SH neurons exhibiting spindle-shaped bipolar appearance with structurally intact plasma membranes, mitochondria, and nuclear DNA. Upper middle panel: Overnight exposure to SAL induces spherical or elliptical appearance with condensations of plasma membrane, mitochondrial aggregation, and nuclear DNA condensation. Upper right panel: Aging RhO$_{mgko}$ neurons exhibit rounded appearance with mitochondrial aggregation in the perinuclear region. Middle central panel: RhO$_{mgko}$ neurons following overnight exposure to SAL (100 µM). SAL induces apoptosis characterized by phosphatidyl serine externalization and blebbing, mitochondrial aggregation, and nuclear DNA condensation. Middle right panel: RhO$_{mgko}$ neurons, transfected with MT-1$_{sense}$ oligonucleotides exhibit relatively reduced SAL apoptosis. Lower left panel: RhO$_{mgko}$ neurons, transfected with mitochondrial complex I exhibit normal appearance without any evidence of plasma membrane or nuclear DNA damage. Lower central panel: SAL induces apoptosis even in RhO$_{mgko}$ neurons, transfected with mitochondrial complex I gene, indicating its deleterious effects on the mitochondrial as well as nuclear DNA. Lower right panel: RhO$_{mgko}$ neurons, transfected with MT-1$_{antisense}$ oligonucleotides and exposed overnight to SAL exhibit severe apoptosis, characterized by plasma membrane blebbing, perforations, mitochondrial aggregation, nuclear DNA condensation and fragmentation. SAL induced severe apoptosis in aging RhO$_{mgko}$ and MT-1$_{antisense}$ oligonucleotide-transfected neurons. SAL-induced apoptosis was attenuated in mitochondrial genome-transfected aging RhO$_{mgko}$

evidence that NOS activation and ONOO ion overproduction may be involved in the etiopathogenesis of PD, and MT gene induction may provide neuroprotection [88,110–113].

12.6 THE NEUROPROTECTIVE EFFICACY OF CoQ_{10} IN TREATING PD

A *neuroprotectant* is an agent (e.g., CoQ_{10}) or a condition (e.g., hypothermia or hyperbaric oxygen) that is used to prevent neuronal death by inhibiting one or more of the pathophysiologic steps in the processes that follow injury to the CNS or ischemia due to occlusion of an artery or hypoxia due to any cause. This definition has now been extended to include diseases (e.g., PD) or neurotoxins (e.g., MPTP). CoQ_{10} is a mitochondrial stabilizer and an antioxidant, and in higher concentrations than is available in the diet is a neuroprotectant [89] and a neurorestorative agent.

Crane [90], in describing the biochemical functions of CoQ_{10}, stated that it is well defined as a critical component of the oxidative phosphorylation process in mitochondria, which converts the energy from carbohydrates and fatty acids into ATP to drive cellular machinery and synthesis functions. New roles for CoQ_{10} in other cellular functions are only now coming to be recognized. The new aspects have developed from the realization that CoQ_{10} can undergo oxidation/reduction reactions in other cell membranes, such as lyosome, Golgi, or plasma membranes. In mitochondria and lysosomes, CoQ_{10} undergoes reduction/oxidation cycles during which it transfers protons across the membrane to form a proton gradient. The presence of high concentrations of quinol in all membranes provides a basis for antioxidant action either by direct reaction with radicals or by regeneration of tocopherol and ascorbate. Evidence for a function in redox control of cell signaling and gene expression is developing from studies on CoQ_{10} stimulation of cell growth, inhibition of apoptosis, control of thiol groups, formation of hydrogen peroxide, and control of membrane channels. Deficiency of CoQ_{10}, causing failure of biosynthesis has been demonstrated by gene mutation, inhibition of biosynthesis of HMG-CoA reductase, or in aging. Correction of the deficiency requires supplementation with higher levels of CoQ_{10} than is available in the diet [90].

FIGURE 12.5 (continued) and MT-1_{sense} oligonucleotide-transfected neurons, indicating that MT's induction provides neuroprotection. Triple fluorochrome analysis was done to examine the extent of damage induced by SAL on the plasma membrane, mitochondria, and nuclear DNA simultaneously. The digital fluorescence images were captured individually in the beginning and subsequently merged to assess the overall structural and functional damage at the plasma membrane, mitochondrial membranes, and nuclear DNA. Fluorescence images were captured by SpotLite digital camera and analyzed by ImagePro computer software. Blue: DAPI; green: FITC annexin V; red: mitochondrial membrane potential marker, JC-1.

The pathophysiology of mitochondrial DNA (mtDNA) diseases is caused by increased cell death and dysfunction due to the accumulation of mutations to mtDNA. Although the disruption of oxidative phosphorylation is central to mtDNA disease, many other factors, such as Ca^{2+} dyshomeostasis, increased oxidative stress, and defective turnover of mitochondrial proteins, may also contribute [91].

Studies by Shults et al. [92] have shown that the activities of complex I and complex II/III in platelet mitochondria are reduced in patients with early, untreated PD. CoQ_{10} is the electron acceptor for complex I and complex II (see Figure 12.1). They found that the level of CoQ_{10} was significantly lower in mitochondria from parkinsonian patients than in mitochondria from age- and sex-matched control subjects, and that the levels of CoQ_{10} and the activities of complex I and complex II/III were significantly correlated. Furthermore, studies by Chan et al. [93] have shown that the cytotoxicity caused due to complex I inhibition by rotenone, but not by antimycin, can be prevented by CoQ_{10} or menadione.

Studies by Blandini et al. [94] showed that the activity of mitochondrial enzyme complex I is reduced in the SN of patients with PD. A less pronounced decrease in the activity of this enzyme has also been reported in platelets of PD patients. Furthermore, these investigators studied platelet complex I in 16 patients and 16 age-matched controls, using a newly developed technique based on the binding of [^{3}H]dihydrorotenone ([^{3}H]DHR), an analog of the pesticide rotenone, to complex I. They also investigated the inhibitory effect of MPP$^+$ on [^{3}H]DHR-specific binding to platelet complex I. PD patients and their controls showed similar levels of [^{3}H]DHR-specific binding. Preincubation of platelets with MPP$^+$ caused the same degree of inhibition of [^{3}H]DHR-specific binding in the two groups. In PD patients, they observed a direct correlation between MPP$^+$-induced inhibition of [^{3}H]DHR-specific binding and the daily intake of levodopa, which may be related to drug-induced changes in the transport of MPP$^+$ into the platelet or in its binding to complex I. These findings demonstrate that the reported reduction in complex I activity in platelets of PD patients cannot be accounted for by an abnormality at the level of the rotenone binding sites, although they do not exclude differences in complex I activity between PD patients and the control subjects.

Vitamin E (VE) and **coenzyme Q** (CoQ) are essential for maintaining function and integrity of mitochondria, and high concentrations of these compounds are found in the inner membranes of mitochondria. Ibrahim et al. [95] conducted a study to examine the interaction between exogenously administered CoQ_{10} and VE in rats. Male Sprague-Dawley rats (12 months old) were fed a basal diet (containing 10 IU VE or 6.7 mg *RRR*-α-tocopherol equivalent) supplemented with either 0 to 500 mg CoQ_{10} or 0, 100, or 1310 IU VE/kg diet for 14 or 28 d. Liver, spleen, heart, kidney, skeletal muscle, brain, and serum were analyzed for the levels of CoQ_{10}, CoQ_9, and VE. CoQ_{10} supplementation significantly ($p < .05$) increased CoQ_{10} concentration in the liver and spleen (total and mitochondrial) and serum, but not in the other organs. Interestingly, rats supplemented with CoQ_{10} plus 100 IU VE/kg diet had significantly higher CoQ_{10} levels in the liver and spleen, whereas those supplemented with CoQ$_{10}$ plus 1310 IU VE/kg diet had lower levels, compared with those that received CoQ$_{10}$ alone. As expected, dietary VE supplementation increased VE content in all the organs analyzed in a dose-dependent manner. However, rats fed the basal diet supplemented with CoQ$_{10}$ had significantly higher VE levels in the liver (total and mitochondrial) than those not receiving CoQ$_{10}$ supplementation. CoQ$_9$ levels were higher in the liver and skeletal muscle and unaltered in brain, serum, heart, and kidney of rats supplemented with CoQ$_{10}$ compared with the controls. These data provide direct evidence for an interactive effect between exogenously administered VE and CoQ$_{10}$ in terms of tissue uptake and retention, and for a sparing effect of CoQ$_{10}$ on VE. The data also suggests that dietary VE plays a key role in determining the tissue retention of exogenous CoQ$_{10}$.

CoQ acts as an antioxidant in respiring mitochondria by regenerating α-tocopherol from its phenoxyl radical. Lass et al. [96] attempted to find out whether CoQ has a similar sparing effect on α-tocopherol *in vivo*. Mice were administered either CoQ$_{10}$ (123 mg/kg/d) or α-tocopherol (200 mg/kg/d) alone for 13 weeks, after the amounts of CoQ$_{10}$, CoQ$_9$, and α-tocopherol were determined by HPLC in the serum, as well as the levels of homogenates in the mitochondria of the liver, kidney, heart, upper hind limb skeletal muscle, and brain. Administration of CoQ$_{10}$ and α-tocopherol, alone or together, increased the corresponding levels of CoQ$_{10}$ and α-tocopherol in the serum. Supplementation with CoQ$_{10}$ also elevated the amounts of the predominant homologue, CoQ$_9$, in the serum and the mitochondria. A notable effect of CoQ$_{10}$ intake was the enhancement of α-tocopherol in mitochondria. Administration of α-tocopherol resulted in an elevation of the α-tocopherol content in the homogenates of nearly all tissues and their mitochondria. Results of this study, thus, indicate that relatively long-term administration of CoQ$_{10}$ or α-tocopherol can result in an elevation of their concentrations in the tissues of the mouse. More importantly, CoQ$_{10}$ intake has a sparing effect on α-tocopherol in mitochondria *in vivo*.

Matthews et al. [97] and Beal [99] have shown that CoQ$_{10}$ exerts neuroprotection by increasing brain mitochondria, they demonstrated that feeding with CoQ$_{10}$ increased its cerebral cortex concentrations in 12- and 24-month-old rats. In 12-month-old rats, administration of CoQ$_{10}$ resulted in significant increases in cerebral cortex concentrations. Oral administration of CoQ$_{10}$ markedly attenuated striatal lesions produced by systemic administration of 3-nitropropionic acids and significantly increased life span in a transgenic mouse of familial amyotrophic lateral sclerosis. These results show that oral administration of CoQ$_{10}$ increases both brain tissue and brain mitochondrial concentrations of CoQ$_{10}$ and provide further evidence that CoQ$_{10}$ can exert a neuroprotective effect that might be useful in the treatment of neurodegenerative diseases.

Beal [97–100] and Beal and Matthews [97] have advocated use of CoQ$_{10}$ for treatment of neurodegenerative disease. Fariello et al. [101] studied the regional distribution of ubiquinones in mouse brain and found that the lowest level was found in the SN. Beal et al. [102] showed that CoQ$_{10}$ is able to attenuate MPTP-induced loss of dopamine and dopaminergic axons in aged mice. The beneficial effects of CoQ_{10} in the treatment of PD have been described in several clinical reports [103–109]. Similar benefits have been shown in patients with Huntington's disease [108,109]. CoQ_{10} is well absorbed in parkinsonian patients and caused a tendency for increased complex I activity. These data suggest that CoQ_{10} and MT isoforms [110–113], may play roles in the cellular dysfunction found in PD and may be potential protective agents for treating parkinsonian patients.

ACKNOWLEDGMENTS

The authors express their heartfelt appreciation to Joann Lynn Johnson for her excellent secretarial skills in typing this review. The studies reported have been supported by grants from the USPHS provided by the National Institute of Aging AG 17059-06 (ME) and National Institute of Neurological Disorders and Stroke NS 34566-09 (ME).

REFERENCES

1. Ebadi, M. and Pfeiffer, R. (2005) *Parkinson's Disease,* CRC Press, Boca Raton.
2. Sangchot, P., Sharma, S., Fugere, N., Porter, J., Ebadi, M. (2002). Iron-induced oxidative stress, mitochondrial aggregation, and α-synuclein translocation in SK-N-SH cells in culture. *Soc Neurosci Abstr* 28: 594–619.
3. Sharma, S., Shavali, S., ElRefaey, H., Ebadi, M. (2002). Inhibition of α-synuclein nitration and perinuclear aggregation by antioxidants in metallothionein transgenic and aging RhO (mgko) dopaminergic neurons. *FASEB J* 16: 696–711.
4. Ebadi, M., Srinivasan, S. K., Baxi, M. (1996). Oxidative stress and antioxidants therapy in Parkinson's disease. *Prog Neurobiol* 48: 1–19.
5. Ebadi, M., Sharma, S., Shavali, S., ElRefaey, H. (2002). Neuroprotective actions of Selegiline. *J Neurosci Res* 67: 285–289.
6. Ebadi, M., Sharma, S., Shavali, S., Sangchot, P., Brekke, L. (2002). Neuroprotective actions in Parkinson's disease. *45*th *Annual Meeting of the Western Pharmacological Society and XXV Congreso Nacional de Farmacologia* 45: 77.
7. Eve, D., Nisbet, A., Kingsbury, A., Hewson, E., Daniel, S., Lees, A., Marsden, C., Foster, O. (1998). Basal ganglia neuronal nitric oxide synthase mRNA expression in Parkinson's disease. *Mol Brain Res* 63: 62–71.
8. Ebadi, M., Hiramatsu, M. (2000). Glutathione and metallothionein in oxidative stress of Parkinson's disease. In: *Free Radicals in Brain Pathophysiology,* Eds., Poli, G., Cadenas, E., Packer, L. New York: Marcel Dekker, pp. 427–465.
9. Heales, S., Bolanos, J., Stewart, V., Brookes, P., Land, J., Clark, J. (1999). Nitric oxide, mithochondria and neurological disease. *Biochim Biophys Acta* 1410: 215–228.
10. Schulz, J., Matthews, R., Klockgether, T., Dichgans, J., Beal, M. (1997). The role of mitochondrial dysfunction and neuronal nitric oxide in animal models of neurodegenerative disease. *Mol Cell Biochem* 174: 193–197.
11. Dawson, V., Dawson, T. (1998). Nitric oxide in neurodegeneration. *Prog Brain Res* 118: 215–229.

12. Rose, S., MacKenize, G.M., Jenner, P. (1999). Nitric oxide and basal ganglia degeneration. *Adv Neurol* 80: 247–257.

13. Gerlach, M., Blum-Degen, D., Lan, J., Riederer, P. (1999). Nitric oxide in the pathogenesis of Parkinson's disease. *Adv Neurol* 80: 239–245.

14. Barthwal, M., Srivastava, N., Dikshit, M. (2001). Role of nitric oxide in a progressive neurodegeneration model of Parkinson's disease in the rat. *Redox Rep* 6: 297–302.

15. Riobo, N. A., Schopfer, F. J., Boveris, A. D., Cadenas, E., Poderoso, J. J. (2002). The reaction of nitric oxide with 6-hydroxydopamine: implications for Parkinson's disease. *Free Radic Biol Med* 32: 115–121.

16. Hunot, S., Boissiere, F., Faucheux, F., Brugg, B., Mouatt-Prigent, A., Agid, Y., Hirsch, E. C. (1996). Nitric oxide synthase and neuronal vulnerability in Parkinson's disease. *Neuroscience* 72: 355–363.

17. Tipton, K. F., Singer, T. P. (1993). Advances in our understanding of the mechanisms of the neurotoxicity of MPTP and related compounds. *J Neurochem* 61: 1191–1206.

18. Gainetdinov, R. R., Fumagali, F., Jones, S. R., Caron, M. G. (1997). Dopamine transporter is required for in vivo MPTP neurotoxicity: evidence from mice lacking the transporter. *J Neurochem* 69: 1322–1325.

19. Javitch, J. A., D'Amsto, R. J., Strittmatter, S. M., Snyder, S. (1985). Parkinsonism-inducing neurotoxin, N-methyl-4-phenyl-1,2,3,6-tetrahydropyridine uptake of the metabolite N-methyl-4-phenyl pyridium by dopamine neurons explains selective toxicity. *Proc Natl Acad Sci* 82: 2173–2177.

20. Amara, S. G., Kuhar, M. J. (1993). Neurotransimtter transporters: recent progress. *Annu Rev Neurosci* 16: 73–93.

21. Reith, M. E., Xu, C., Chen, N. H. (1997). Pharmacology and regulation of the neuronal dopamine transporter. *Eur J Pharmacol* 324: 1–10.

22. Uhl, G. R., Walther, D., Mash, D., Faucheux, B., Javoy-Agid, F. (1994). Dopamine transporter messenger RNA in Parkinson's disease and control substania nigra neurons. *Ann Neurol* 35: 494–498.

23. Harrington, K. A., Augood, S. J., Kingsbury, A. E., Foster, O. J., Emson, P. C. (1996). Dopamine transporter (DAT) and synaptic vesicle amine transporter (VMAT2) gene expression in the substantia nigra of control and Parkinson's disease. *Brain Res Mol Res* 36: 157–162.

24. Speciale, S. G., Liang, C. L., Sonsalla, P. K., Edwards, R. H., German, D. C. (1998). The neurotoxin 1-methyl-4-phenylpyridinium is sequestered within neurons that contain the vesicular momoamine transporter. *Neuroscience* 84: 1177–1185.

25. Chang, G. D., Ramirez, V. D. (1986). The mechanism of action of MPTP and MPP$^+$ on endogenous dopamine release from the rat corpus striatum superfused in vitro. *Brain Res* 368: 134–140.

26. Obata, T., Chiueh, C. C. (1992). *In vitro* trapping of hydroxyl free radicals in the striatum utilizing intracranial microdialysis perfusion of salicylate: effect of MPTP, MPDP$^+$, and MPP$^+$. *J Neural Transm Gen Sect* 89: 139–145.

27. Lotharius, J., O'Malley, K. L. (2000). The parkinsonism-inducing drug 1-methyl-4-phenylpyridinium triggers intracellular dopamine oxidation. A novel mechanism of toxicity. *J Biol Chem* 275: 38581–38588.

28. Przedborski, S., Jackson-Lewis, V. (1998). Mechanisms of MPTP toxicity. *Mov Disord* 13: 35–38.

29. Ramsay, R. R., Salach, J. I., Singer, T. P. (1986). Uptake of the neurotoxin 1-methyl-4-phenylpyridine (MPP$^+$) by mitochondria and its relation to the inhibition of the mitochondrial oxidation of NAD$^+$-linked substrates by MPP$^+$. *Biochem Biophys Res Commun* 134: 743–748.

30. Barc, S., Page, G., Fauconneau, B., Barrier, L., Huguet, F. (2001). A new *in vitro* approach for investigating the MPTP effect on DA uptake. *Neurochem Int* 38: 243–248.

31. Fonck, C., Baudry, M. (2001). Toxic effects of MPP+ and MPTP in PC12 cells independent of reactive oxygen species formation. *Brain Res* 905: 199–206.

32. Fonck, C., Baudry, M. (2003). Rapid reduction of ATP synthesis and lack of free radical formation by MPP$^+$ in rat brain synaptosomes and mitochondria. *Brain Res* 975: 214–221.

33. Storch, A., Ludolph, A. C., Schwarz, J. (1999). HEK-293 cells expressing the human dopamine transporter are susceptible to low concentration of 1-methyl-4-phenylpyridine (MPP$^+$) via impairment of energy metabolism. *Neurochem Int* 35: 393–403.

34. Chagkutip, J., Vaughan, R., Govitrapong, P., Ebadi, M. (2003). 1-Methyl-4-phenylpyridinium-induced down-regulation of dopamine transporter function correlates with a reduction in dopamine transporter cell surface expression. *Biochem Biophys Res Commun* 344: 49–54.

35. Kitayama, S., Mitsuhata, C., Davis, S., Wang, J. B., Sato, T., Morita, K., Uhl, G. R., Dohi, T. (1998). MPP$^+$ toxicity and plasma membrane dopamine transporter: study using lines expressing the wild-type and mutant rat dopamine transporter. *Biochim Biophys Acta* 1404: 305–313.

36. Ebadi, M., Govitrapong, P., Sharma. S., Muralikrishnan, D., Shavali, S., Pellett, L., Schafer, R., Albano. C., Eken. J. (2001). Ubiquinone (coenzyme Q_{10}) and mitochondria in oxidative stress of Parkinson's disease. *Biol Signals Recept* 10: 224–253.

37. Ebadi, M., Sharma, S. (2003). Peroxynitrite and mitochondrial dysfunction in the pathogenesis of Parkinson's disease. *Antioxid Redox Signal* 5: 319–335.

38. Vaughan, R., Huff, R. A., Uhl, G. R., Kuhar, M. J. (1997). Protein kinase C-mediated phosphorylation and functional regulation of dopamine transporters in striatal synaptosomes. *J Biol Chem* 272: 15541–15546.

39. Zhang, L., Coffey, L. L., Reith, M. E. (1997). Regulation of the functional activity of the human dopamine transporter by protein kinase C. *Biochem Pharmacol* 53: 677–688.

40. Pristupa, Z. B., McConkey, F., Lui, F., Man, H.Y., Lee, F. J., Wang, Y. T., Niznik, H. B. (1998). Protein kinase-mediated bi-directional trafficking and functional regulation of the human dopamine transporter. *Synapse* 30: 79–87.

41. Melikian, H. E., Buckley, K. M. (1999). Membrane trafficking regulates the activity of the human dopamine transporter. *J Neurosci* 19: 7699–7710.

42. Bauman, A. L., Apparsundaram, S., Ramamoorthy, S., Wadzinski, B. E., Vaughan, R., Blakely, R. D. (2000). Cocaine and antidepressant sensitive biogenic amine transporters exist in regulated complexes with protein phosphatase 2A. *J Neurosci* 20: 7571–7578.

43. Schapira, A. H. V., Mann, V. M., Cooper, J. M., Dexter, D., Daniel, S. F., Jenner, P., Clark, J. B., Marsden, C. D. (1990). Anatomic and disease specifically of NADH CoQ1 reductase (complex-I) deficiency in Parkinson's disease. *J Neurochem* 5: 2142–2145.

44. Langston, J. W., Balard, P., Tetrud, J. W., Irvin, I. (1983). Chronic Parkinsonism in humans due to a product of meperdine-analog synthesis. *Science* 219: 979–980.
45. Snyder, S. H., D'amato, R. J. (1986). MPTP a neurotoxin relevant to the pathophysiology of Parkinson's disease. *Neurology* 36: 250–258.
46. Irvin, I., Langston, J. W. (1993). MPTP and Parkinson's disease. In: Harvey, A. L., Ed. *Natural and synthetic neurotoxins.* New York: Academic Press: 225–256.
47. Mizuno, Y., Susuki, K., Sone, N., Saitoh, T. (1988). Inhibition of mitochondrial respiration by 1-methyl-4-phenyl-1,2,3,6-tetrahydropyridine (MPTP) in mouse brain *in vivo. Neurosci Lett* 91: 349–353.
48. Cleeter, M. W., Cooper, J. M., Schapira, A. H. (1992). Irreversible inhibition of mitochondrial complex-I by methyl-4-phenylpyridinium: evidence for free radical involvement. *J Neurochem* 58: 786–789.
49. Thomas, B., Muralikrishnan, D., Mohanakumar, K.P. (2000). *In vivo* hydroxyl radical generation in the striatum following systemic administration of 1-methyl-4-phenyl-1236-tetrahydropyridine in mice. *Brain Res* 852: 221–224.
50. Nagatsu, T. (1997). Isoquinoline neurotoxins in the brain and Parkinson's disease. *Neurosci Res* 29: 99–111.
51. Musshoff, F., Schmidt, P., Dettmeyer, R., Priemer, F., Witting, H., Madea, B. (1999). A systematic regional study of dopamine and dopamine-derived salsolinol and nonsalsolinol levels in human brain areas. *Forensic Sci Int* 105: 1–11.
52. Kotake, Y., Tasaki, Y., Makino, Y., Ohta, S., Hirobe, M. (1995). 1-Benzyl-1,2,3,4-tetrahydroisoquinoline as a Parkinsonism-inducing agent: a novel endogenous amine in mouse brain and Parkinsonian CSF. *J Neurochem* 65: 2633–2638.
53. Makino, Y., Ohta, S., Tachikawa, O., Hirobe, M. (1988). Presence of tetrahydroisoquinoline and 1-methyl-tetrahydroisoquinoline in foods: compounds related to Parkinson's disease. *Life Sci* 43: 373–378.
54. Niwa, T., Yoshozumi, H., Tatematsu, A., Matsuura, S., Nagatsu, T. (1989). Presence of tetrahydroisoquinoline, a Parkinsonism-related compound in food. *J Chromatogr* 493: 347–352.
55. Niwa. T., Takeda, N., Tatematsu, A., Matsuura, S., Yoshida, M., Nagatsu, T. (1988). Migration of tetrahydroisoquinoline, a possible Parkinsonian neurotoxin, into monkey brain from blood as proved by gas chromatography-mass spectrometry. *J Chromatogr* 452: 85–91.
56. Ohta, S., Tachikawa, O., Makino, Y., Tasaki, Y., Hirobe, O. M. (1990). Metabolism and brain accumulation of tetrahydroisoquinoline (TIQ): a possible Parkinsonism inducing substance in an animal model of a poor debrisoquine metabolizer. *Life Sci* 46: 599–605.
57. Kotake, Y., Tasaki, Y., Makino, Y., Ohta, S., Hirobe, M. (1995). 1-Benzyl-1,2,3,4-tetrahydroisoquinoline as a Parkinsonism-inducing agent: a novel endogenous amine in mouse brain and parkinsoonian CSF. *J Neurochem* 65: 2633-2638.
58. Kotake, Y., Yoshida, M., Ogawa, M., Tasaki, Y., Hirobe, M., Ohta, S. (1996). Chronic administration of 1-benzyl-1,2,3,4-tetrahydroisoquinoline, an endogenous amine in the brain, induces parkinsonism in a primate. *Neurosci Lett* 217: 69–71.
59. Antkiewicz-Michaluk, L., Mischaluk, J., Mokrosz, M., Romanska, I., Lorenc-Koci, E., Ohta, S., Betulani, J. (2001). Different action on dopamine catabolic pathways of two endogenase 1,2,3,4-tetrahydroisoquinolines with similar anti-dopaminergic properties. *J Neurochem* 78: 100–108.

60. Shavali, S., Ebadi, M. (2003). 1-Benzyl-1,2,3,4-tetrahydroisoquinoline (1BnTIQ), an endogenous neurotoxin causing Parkinson's induced dopaminergic cell death through apoptosis. *Neurotoxicology* 24: 417–424.

61. Lee, H. S., Park, C. W., Kim, Y. S. (2000). MPP⁺ increases the vulnerability to oxidative stress rather than directly medicating oxidative damage in human neuroblastoma cells. *Exp Neurol* 165: 164–171.

62. Heikkila, R. E., Manzino, L., Cabbat, F. S., Duvoisin, R. C. (1984). Protection against the dopaminergic neurotoxicity of 1-methyl-4-phentl-1,2,5,6-tetrahydropyridine by monoamine oxidase inhibitors. *Nature* 311: 467–469.

63. Javitch, J. A., Amato, R. J. D., Strittmatter, S. M., Snyder, S. H. (1985). Parkinsonism-inducing neurotoxin MPTP: uptake of the metabolite MPP⁺ by dopamine neurons explains selective toxicity. *Proc Natl Acad Sci U.S.A.* 82: 2173–2177.

64. McNaught, K. S., Thull, U., Carupt, P. A., Altomare, C., Cellamare, S., Carotti, A., Testa, B., Jenner, P., Marsden, C. D. (1996). Inhibition of [³H]dopamine uptake into striatal synaptosomes by isoquinoline derivates structurally related to 1-methyl-4-phenyl-1,2,3,6-tetrahydropyridine. *Biochem Pharmacol* 52: 29–34.

65. Barc, S., Page, G., Barrier, L., Hugnet, F., Fauconneau, B. Progressive alteration of neuronal dopamine transporter activity in a rat injured by an intranigral injection of MPP⁺. *Brain Res* 2002; 941: 72–81.

66. Kawai, H., Kotake, Y., Ohta, S. (2000). Inhibition of dopamine receptors by endogenous amines: binding to striatal receptors and pharmacological effects on locomotor activity. *Bioorg Med Chem Lett* 15: 1669–1671.

67. Park, S. U., Ferrer, J. V., Javitvh, J. A., Kuhn, D. M. (2002). Peroxynitrite inactivates the human dopamine transporter by modification of cysteine 342: potential mechanism of neurotoxicity in dopamine neurons. *J Neurosci* 22: 4399–4405.

68. Mochizuki, H., Goto, K., Mori, H., Mizuno, Y. (1996). Histochemistry detection of apoptotis in Parkinson's disease. *J Neurol Sci* 137: 120–123.

69. Kawai, H., Makino, Y., Hirobe, M., Ohta, S. (1998). Novel endogenous 1,2,3,4-tetrahydroisoquinoline derivatives: uptake by dopamine transporter and activity to induce Parkinsonism. *J Neurochem* 70: 745–751.

70. Kitamura, Y., Kosaka, T., Kakimura, J.I., Matsuoka, Y., Kohno, Y., Nomura, Y., Taniguchi, T. (1998). Protective effects of the antiparkinsonian drugs talipexole and pramipexole against 1-methyl-4-phenylpyridinium-induced apoptotic death in human neuroblastoma SH-SY5Y cells. *Mol Pharmacol* 54: 1046–1054.

71. Muralikrishnan, D., Ebadi, M. (2001). SKF-38393, a dopamine receptor agonist, attenuates 1-methyl-4-phenyl-1-2-3-6-tetrahydropyridine-induced neurotoxicity *Brain Res* 892: 241–247.

72. Ebadi, M., Govitrapong, P., Sharma, S., Muralikrishnan, D., Shavali, S., Pellett, L., Schafer, R., Albano, C., Eken J. (2001) Ubiquinine (coenzyme Q_{10}) and mitochondria in oxidation stress of Parkinson's disease. *Biol Signals Recept* 10: 224–253.

73. Beal, M.F. (1998). Excitotoxicity and nitric oxide in Parkinson's disease pathogenesis. *Ann Neurol* 44: S110–S114.

74. Bredt, D.S. (1999). Endogenous nitric oxide synthesis: biological functions and pathophysiology. *Free Radic Res* 31: 544–596.

75. Dawson, V. L., Dawson, T. M. (1998). Nitric oxide in neurodegeneration. *Prog Brain Res* 118: 215–229.

76. Ghafourifar, P., Bringhold, U., Klein, S. D. Richter, C. (2001). Mitochondrial nitric oxide synthase, oxidative stress and apoptosis. *Biol Signals Recept* 10: 56–65.

77. Hirsch, E. C., Hunot, S. (2000). Nitric oxide, glial cells and neuronal degeneration in parkinsonism. *Trends Pharmacol Sci* 21: 163–165.

78. Smith, K. J., Kapoor, R., and Felts, P. A. (1999). Demyelination: the role of reactive oxygen and nitrogen species. *Brain Pathol* 9: 69–92.

79. Imam, S. Z., el-Yazal, J., Newport, G. D., Itzhak, Y., Cadet, J. L., Slikker, W., Ali, S. F. (2002). Methamphetamine-induced dopamine neurotoxicity: role of peroxynitrite and neuroprotective role of antioxidants and peroxynitrite decomposition catalysts. *Ann NY Acad Sci* 939: 366–380.

80. Youdim, M. B., Ben-Shachar, D., Eshel, G., Finberg, J. P., Reiderer, P. (1993). The neurotoxicity of iron and nitric oxide: relevance to the etiology of Parkinson's disease. *Adv Neurol* 60: 259–266.

81. Youdim, M. B., Lavie, L., Riederer, P. (1994). Oxygen free radicals and neurodegeneration in Parkinson's disease: a role for nitric oxide. *Ann NY Acad Sci* 738: 64–68.

82. Torreilles, F., Salman-Tabcheh, S., Guerin, M., Torreilles, J. (1999). Neurodegenerative disorders: the role of peroxynitrite. *Brain Res Rev* 30: 153–163.

83. Bringfold, U., Ghafourifar, P., Richter, C. (2000). Peroxynitrite formed by mitochondrial NO synthase mitochondrial Ca^{2+} release. *Free Radic Biol Med* 29: 343–348.

84. Grunewald, T., Beal, M.F. (1999). NOS knockouts and neuroprotection, *Nat Med* 5: 1354–1355.

85. Sharma, S., Sangchot, P., Ebadi, M. (2002). MT gene manipulation influences striatal mitochondrial ubiquinones and MPTP-induced neurotoxicity in dopaminergic neurons. *World Congress of Pharmacology* 106–107.

86. Sharma, S., Shavali, S., ElRafaey, H., Ebadi, M. (2002). Inhibition of α-Syn nitration and perinuclear aggregation by antioxidants in metallothionein transgenic and aging RhO (mgko) dopaminergic neurons. *FASEB J* 16: 686–711.

87. Ebadi, M., Hiramatsu, M., Burke, W. J., Folks, D. G., and el-Sayed, M. A. (1998). MT isoforms provide neuroprotection against 6-hydroxy-dopamine-generated hydroxyl radicals and superoxide anions. *Proc West Pharmacol Soc* 41: 155–158.

88. Sharma, S., Ebadi, M. (2003). Metallothioneon attenuates 3-morpholinosydnonimine (SIN-1)-induced oxidative stress in dopaminergic neurons. *Antioxidants Redox Signaling* 5: 251–263.

89. Dawson, T., and Dawson, V. (2002). Neuroprotective and neurorestorative strategies for Parkinson's disease. *Nat Neurosci* 5: 1058–1061.

90. Crane, F. (2001). Biochemical Functions of Coenzyme Q10. *J Am Coll Nutr* 20: 591–598.

91. James, A., Murphy, M. (2002). How mitochondrial damage affects cell function. *J Biomed Sci* 9: 475–487.

92. Shults, C. W., Haas, R. H., Passov, D., Beal, M.F. (1997). Coenzyme Q10 levels correlate with the activities of complexes I and II/III in mitochondria from parkinsonian and nonparkinsonian subjects. *Ann Neurol* 42: 261–264.

93. Chan, T., Teng, S., Wilson, J., Galatt, G., Khan, S., O'Brien, P. (2002). Coenzyme Q cytoprotective mechanisms for mitochondrial complex I cytopathies involves NAD(P)H: Quinone oxidoreductase 1(NQ01). *J Free Radic Res* 36 (4): 421–427.

94. Blandini, F., Nappi, G., Greenamyre, T. (1998). Quantitative study of mitochondrial complex I in platelets of parkinsonian patients. *Mov Disord* 13: 11–15 Movement Disorder Society.

95. Ibrahim, W., Bhagavan, H., Chopra, R., Chow, C. (2000). Dietary coenzyme Q10 and vitamin E alter the status of these compounds in rat tissue and mitochondria. *J Nutr* 130: 2343–2348.

96. Lass, A., Forster, M., Sohal, R. (1999). Effects of coenzyme Q_{10} and α-tocopherol administration on their tissue levels in the mouse: elevation of mitochondrial α-tocopherol by coenyzme Q_{10}. *Free Radic Biol Med* 26: 1375–1382.

97. Matthews, R., Yang, L., Browne, S., Baik, M., Flint-Beal, M. (1998). Coenzyme Q10 administration increases brain mitochondrial concentrations and exerts neuroprotecive effects. *Proc Natl Acad Sci* 95: 8892–8897.

98. Beal, F.M. (1999). Coenzyme Q_{10} administration and its potential for treatment of neurodegenerative disease. *Biofactors* 9: 261–266.

99. Beal, F.M. (2002) Coenzyme Q_{10} as a possible treatment for neurodegeneration diseases. *Free Radic Res* 36 (4): 455–460.

100. Beal, F.M., Matthews, R. (1997). Coenzyme Q_{10} in the central nervous system and its potential usefulness in the treatment of neurodegenerative disease. *Mol Aspects Med* 18: 169–179.

101. Fariello, R. G., Ghilardi, O., Peschechera, A., Ramacci, M. T., Angelucci, L. (1988). Regional distribution of ubiquinones and tocopherol in the mouse brain: lowest-content of ubiquinols in the substantia nigra. *Neuropharmacology* 27: 1077–1080.

102. Beal, F.M., Matthews, R., Tieleman, A., Shults, W. (1998). Coenzyme Q_{10} attenuates the 1-methyl-4-phenyl-1,2,3,6-tetrahydropyridine (MPTP) induced loss of striatal dopamine and dopaminergic axons in aged mice. *Brain Res* 783: 109–114.

103. Shults, C.W., Haas, R., Flint-Beal. M. (1999). A possible role of coenzyme Q_{10} in the etiology and treatment of Parkinson's disease. *Biofactors* 9: 267–272.

104. Shults, C. W., Oakes, D., Kieburtz, K., Beal, M. F., Haas, R., Plumb, S., Juncos, J. I., Nutt, J., Shoulson, I., Carter, J., Kompoliti, K., Perlmutter, J. S., Reich, S., Stern, M., Watts, R. L., Kurlan, R., Molho, E., Harrison, M., Lew, M., Parkinson Study Group. (2002). Effects of coenzyme Q_{10} in early Parkinson disease: evidence on slowing of the functional decline. *Arch Neurol* Oct. 59 (10): 1541–1550.

105. Strijks, E., Kremer, H. P. H., Horstink, M., W., I., M. (1997). Q_{10} therapy in patients with idiopathic Parkinson's disease. *Mol Aspects Med* 18: 237–240.

106. Götz, M. E., Gerstner, A., Harth, R., Dirr, A., Janetzky, B., Kuhn, W., Riederer, P., Gerlach, M. (2000). Altered redox state in platelet coenzyme Q_{10} in Parkinson's disease. *J Neural Transm* 107: 41–48.

107. Jiménéz-Jiménéz, F., J., Molina, A., de Bustos, F., García-Redondo, A., Gómez-Escalonilla, C., Martínez-Salio, A., Berbel, A., Camacho, A., Zurdo, M., Barcenilla, B., Enríquez de Salamanca, R., Arenas, J. (2000). Serum levels of coenzyme Q10 in patients with Parkinson's disease. *J Neural Transm* 107: 177–181.

108. Schilling, G., Coonfield, M., Ross, C., Borchelt, D. (2001). Coenzyme Q_{10} and remacemide hydrochloride ameliorate motor deficits in a Huntington's disease transgenic mouse model. *Neurosci Lett* 315: 149–153.

109. Ferrante, R., Andreassen, O., Dedeoglu, A., Ferrante, K., Jenkins, B., Hersch, S., Flint-Beal, M. (2002). Therapeutic effects of coenzyme Q_{10} and remacemide in transgenic mouse models of Huntington's disease. *J Neurosci* 22(5): 1592–1599.

110. Ebadi, M., Kumari, M.V.R., Hiramatsu, M., Hao, R., Pfeiffer, R.F., Rojas, P. (1998). Metallothionein, neurotrophins and selegiline in providing neuroprotection in Parkinson's disease. *Restor Neurol Neurosci* 12: 103–111.

111. Kumari, M.N.R., Hiramatsu, M., Ebadi, M. (1998). Free radicals scavenging actions of metallothionein isoforms I and II. *Free Radical Res* 29: 93-101.

112. Ebadi, M., Hiramatsu, M., Burke, W.J., Folks, D.G., El Sayed, M.A. (1998). Metallothionein isoforms provide neuroprotection against 6-hydroxydopamine-generated hydroxyl radicals and superoxide anions. *Proc West Pharmacol Soc* 41: 155–158.

113. Ebadi, M., Hiramatsu, M., Kumari, M.V.R., Hao, R., Pfeiffer, R.F. (1999). Metallothionein in oxidative stress of Parkinson's disease. In: Metallothionein IV (C.D. Klaassen, Ed.) *Advances in Life Sciences* 341–349.

13 The *In Vitro* and *In Vivo* Molecular Mechanisms of Neuroprotection by the Major Green Tea Polyphenol, (–)-Epigallocatechin-3-gallate, with Special Reference to Parkinson's Disease

Orly Weinreb, Silvia Mandel, Tamar Amit, and Moussa B.H. Youdim
Technion–Israel Institute of Technology, Haifa, Israel

CONTENTS

13.1 OVERVIEW

Tea consumption is changing its status from a mere ancient beverage and a lifestyle habit to a nutrient endowed with prospective neurobiological–pharmacological actions beneficial to human health. Accumulating evidence suggest that oxidative stress resulting in reactive oxygen species (ROS) generation and inflammation plays a pivotal role in neurodegenerative diseases, supporting the implementation of radical scavengers, transition metals (e.g., iron and copper) chelators, and nonvitamin natural antioxidant polyphenols in the clinic. These observations are in line with the current view that polyphenolic dietary supplementation may have an impact on cognitive deficits in old age. As a consequence, green tea polyphenols are now being considered as therapeutic agents in well-controlled epidemiological studies, aimed at altering brain aging processes and as possible neuroprotective agents in progressive neurodegenerative disorders such as Parkinson's disease (PD) and Alzheimer's disease (AD). In particular, literature on the putative, novel neuroprotective mechanism of the major green tea polyphenol, (−)-epigallocatechin-3-gallate (EGCG), will be examined and discussed in this review.

13.2 INTRODUCTION

Polyphenols are natural substances present in beverages obtained from plants, fruits, and vegetables such as olive oil, red wine, and tea. Flavonoids form the largest group of polyphenols, which is mainly divided into anthocyanins, glycosylated derivative of anthocyanidin, present in colorful flowers and fruits; and anthoxantins, colorless compounds further divided into several categories; including flavones, isoflavones, flavans, and flavonols [1] (Figure 13.1). Flavonoids consist of an aromatic ring, which is condensed to a heterocyclic ring and attached to a second aromatic ring. The abundant phenolic hydroxyl groups on the aromatic ring confer the antioxidant capacity, and the 3-OH is essential for the iron-chelating activity of these compounds [2].

The importance of polyphenolic flavonoids in enhancing cell resistance to oxidative stress (OS) goes beyond the simple scavenging activity and is mostly interesting in those pathologies in which OS plays an important role. Numerous studies in the past 10 yr have shown that polyphenols have *in vitro* and *in vivo* activity in preventing and reducing the deleterious effects of oxygen-derived free radicals associated with several chronic and stress-related human and animal diseases. Several lines of evidence suggest that oxidative stress resulting in ROS generation and inflammation plays a pivotal role in clinical disorders such as arteriosclerosis, ischemia-reperfusion injury, cancer, stroke, and neurodegenerative disorders [3,4]. Special interest has been assigned to the therapeutic role of antioxidants in neurodegenerative diseases, PD and AD [5,6]. Oxidative damage to neuronal biomolecules and increased accumulation of iron in specific brain areas are considered major pathological aspects of PD and AD [7]. Although the etiology of both disorders and their respective dopaminergic or cholinergic neuron

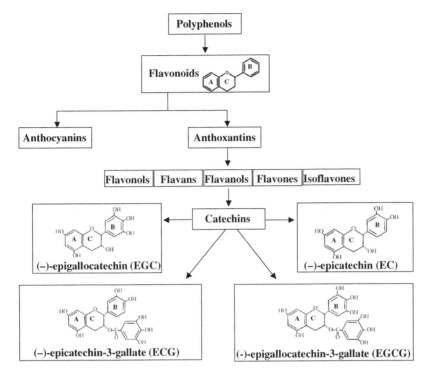

FIGURE 13.1 A diagram of the natural polyphenol classification and the chemical structure of green tea catechins.

degeneration remains elusive, the chemical pathology of PD shows many similarities to AD, involving increase in iron concentration, release of cytochrome c, alpha-synuclein aggregation, OS, loss of tissue, reduced glutathione (GSH), reduction in mitochondrial complex I activity, and increased lipid peroxidation [8–10].

Large investigation of PD found a moderate risk reduction in tea consumers compared to non-tea-drinkers [11]. The favorable properties ascribed to tea consumption are believed to rely on its bioactive components, catechins and their derivatives, demonstrated to act directly as radical scavengers and exert indirect antioxidant effects through activation of transcription factors and antioxidant enzymes (for reviews see References 12 and 13). In line with this evidence, particular attention has been paid on studying the neuroprotective action of antioxidants, iron-chelating and antiinflammatory agents, tea flavonoids, and especially the major component of green tea, (−)-epigallocatechin-3-gallate (EGCG) [14–16]. The revelation of novel molecular targets, possibly implicated in their neuroprotective action, include calcium homeostasis [17], the extracellular mitogen-activated protein kinases (MAPK) [18], protein kinase C (PKC) [19], antioxidant enzymes [20], antioxidant response element (ARE) [21], survival genes [22], and the processing of the amyloid precursor protein (APP) pathway [19].

The specific mechanisms by which green tea polyphenols exert their neuroprotective action are not clearly defined. But recent evidence indicates that besides their antioxidant and iron-chelating properties, polyphenols have a profound effect on cell survival and death genes and signal transduction. This review will focus on the most recent studies relating to the mechanism underlying the biochemical and molecular effects of EGCG, the major green tea polyphenol component, with particular relevance to PD.

13.3 GREEN TEA POLYPHENOLS

Green tea is a drink made from the steamed and dried leaves of the *Camellia sinesis* plant, a shrub native to Asia. It is a widely consumed beverage in Japan, China, and other Asian nations and is becoming more popular in the West. Recently, green tea has attracted attention for its health benefits, particularly with respect to its potential for preventing and treating cancer, cardiovascular diseases, inflammatory diseases, and neurodegenerative diseases in humans [23–26].

Green tea contains a number of bioactive chemicals; it is particularly rich in flavonoids (30% of the dry weight of a leaf), including catechins and their derivatives (Figure 13.1). The most abundant polyphenolic compound is EGCG, thought to contribute to the beneficial effects attributed to green tea, such as its anticancer and antioxidant properties. Catechins have been found to be more efficient radical scavengers than vitamin E and vitamin C [27,16]. Relative antioxidant activities among tea catechins (Figure 13.1) have been found to be EGCG=(–)-epicatechin-3-gallate (ECG) > (–)-epigallocatechin (EGC) > (–)-epicatechin (EC) [28]. EGCG accounts for more than 10% of the extract dry weight (30 to 130 mg per cup of tea), followed by EGC > EC ≥ ECG [29].

The metabolism of green tea catechins has been studied in various animals including humans [30,31]. Catechin orally administered to humans is absorbed, metabolized, and excreted within 24 h [32]. Study of healthy green tea consumers revealed levels of EGCG, EGC, and EC in the plasma in dose-dependent concentrations varying between 0.2% and 2% of the ingested amount, with maximal concentration 1.4 to 2.4 h after ingestion [33,34]. In addition, catechin intake does not alter endogenous antioxidant levels in human blood plasma [34]. The half-life for EGCG is about 5 h, and for EGC and EC it varies between 2.4 and 3.4 h [35]. Several reports indicated that tea polyphenols can reach the brain and exert neuroprotective effect simply by drinking. Recently, it was reported that EC metabolites (epicatechin glucuronide and 3'-O-methylated epicatechin glucuronide) formed after oral ingestion of EC by rats had gained entry to the brain [36]. Furthermore, a study with labeled EGCG demonstrated a wide distribution of radioactivity in mouse organs including brain after oral administration and a small amount of [^{3}H] EGCG excretion in the urine after direct administration into the stomach [37].

13.4 EGCG NEUROPROTECTIVE ACTIVITIES IN RELATION TO PD AND AD

Owing to the lack of evidence on the activity of tea polyphenols in the human brain and well-controlled clinical trials, the effect of these compounds in the progression of neurodegenerative disorders has not been performed on a large scale. Recent epidemiological studies have shown reduced risk of PD associated with consumption of 2 cups per day or more of tea [38], and that the prevalence of PD is much lower in Chinese populations than in white people [39,40]. Clinical trials with AD patients did not report any significant result relative to tea consumption [41], although treatment with an extract of another natural flavonoid, *Gingko biloba* leaves, improved cognitive performance of AD patients [42]. Nonetheless, a study on a French community of 65-year-old people showed an opposite connection between moderate wine drinking and AD incidence [43].

13.4.1 NEUROPROTECTION *IN VITRO* AND IN ANIMAL MODELS

Neuroprotective *in vivo* studies employing *N*-methyl-4-phenyl-1,2,3,6-tetrahydro-pyridine (MPTP) have shown that both green tea extract and EGCG possess highly potent capabilities in preventing mice striatal dopamine depletion and substantia nigra dopaminergic neuron loss induced by the parkinsonism-inducing neurotoxin [20]. One possible mechanism underlying the effectiveness of green tea and EGCG against MPTP neurotoxicity may involve its catechol-like structure because it is known that catechol-containing compounds are potent radical anti-oxidants and chelators of ferric ion [44]. In agreement, the iron chelators, radical scavengers, and catechol derivative compounds — R-apomorphine (R-APO), dopamine (DA) receptor agonist, and its S-isomer — induced neuroprotection in animal models of PD [45,46]. The catechol structural resemblance may account for a recently reported inhibitory effect of green tea polyphenols on [^{3}H] DA uptake by presynaptic transporters. This inhibition was suggested to block the metabolic product of MPTP, the neurotoxin 1-methyl-4-phenylpyridinium (MPP$^+$) uptake (because of competition for the vesicular transporter), thereby protecting DA-containing neurons against MPP$^+$-induced injury [47]. *In vitro* studies also demonstrated inhibition of MPP$^+$ and 6-hydroxydopamine (6-OHDA)-induced neurotoxicity by EGCG [22]. Furthermore, EGCG inhibited the activity of the enzyme catechol-*O*-methyltransferase (COMT) in rat liver cytosol homogenates at a low IC$_{50}$ concentration (0.2 μM) [48]. DA and related catecholamines, are physiological substrates of COMT. The COMT inhibitors entacapone and tolca-pone, clinically prescribed to PD-affected individuals, dose-dependently inhibit the formation of the major metabolite of levodopa, 3-*O*-methyldopa, thereby improving its bioavailability in the brain [49]. The implication of the pivotal role of EGCG in neuroprotection as an iron chelator has been strengthened by the observations that both MPTP and 6-OHDA significantly increased iron in substantia nigra pars compacta (SNPC) of mice, rats and monkeys treated with these neurotoxins [50–53]. Iron accumulation has been implicated in a range of neurodegenerative disorders [54], and iron has been reported to accumulate in the neurons in SN of patients with PD.

A considerable evidence points to an amyloid cascade of events in the pathogenesis of AD where amyloid precursor protein (APP) is processed to amyloid β-peptide (Aβ) which spontaneously self-aggregates, in the presence of divalent metals (Fe^{2+}, Cu^{2+}), into neurotoxic amyloid fibrils in the neocortex [55]. APP can be processed via two pathways: (1) a nonamyloidogenic secretory pathway that involves cleavage of APP to soluble APP (sAPP) by a putative α-secretase within the sequence of Aβ peptide, thus precluding the formation of Aβ, and (2) a formation of amyloidogenic Aβ peptides that is regulated by the sequential action of β and γ-secretases [55–57]. As the proportion of APP processed by β-secretase vs. α-secretase may affect the amount of Aβ produced, the regulation of these two pathways might be critically important to the pathogenesis of AD.

Although AD case-control studies did not report any significant outcome relative to tea consumption, *in vitro* observations show that EGCG inhibits induced oxidative stress, neurotoxicity, [19,58] and Aβ-fibrils formation [59]. In addition, EGCG is able to regulate the proteolytic processing of APP under *in vivo* and *in vitro* conditions [19], suggesting that green tea polyphenols might be potentially promising therapeutic agents not only for PD but also for AD. EGCG promoted the nonamyloidogenic α-secretase pathway of APP neuronal cell cultures. The increase was dose-dependent and the stimulatory effect of EGCG on sAPPα secretion was inhibited by the hydroxamic acid-based metalloprotease inhibitor Ro31-9790, indicating that this effect was mediated via α-secretase processing. Also, long-term treatment of mice with EGCG resulted in decreases in cell-associated, full-length APP levels, as well as increases in sAPP levels in the hippocampus. In addition, inhibition of PKC activity, whose involvement in sAPP release is well established [56,60], prevented EGCG-induced sAPPα release, indicating a key role for PKC in mediating EGCG effect. Although it is not known which isoenzyme of PKC plays a major role in modulating APP processing, several lines of evidence suggest the involvement of PKCα and PKCε in APP processing [61,62]. Previous studies in brains of AD patients demonstrated reduction of PKCε activity in the membrane fraction [63]. In agreement with these findings, repeated administration of EGCG for one or two weeks caused significant increases in the protein expression of PKC isoenzymes α and ε in mice hippocampus [19].

13.5 EGCG ANTIOXIDATIVE AND IRON-CHELATING MECHANISMS OF ACTION

The protective effect of EGCG against neuronal diseases may involve its radical scavenging and iron-chelating activities and regulation of antioxidant protective enzymes.

Tea polyphenols have been found to be potent scavengers of free radicals, such as singlet oxygen, superoxide anions, hydroxyl radicals, and peroxyl radicals, in a number of *in vitro* systems [15,16,64]. In the majority of these studies,

EGCG was shown to be more efficient as a radical scavenger than its counterparts ECG, EC, and EGC, which may be attributed to the presence of the trihydroxyl group on the B ring (Figure 13.1) and the gallate moiety at the 3 position in the C ring [16]. 3-Hydroxykynurenine (3-HK) is an endogenous metabolite of tryptophan in the kynurenine pathway and is a potential neurotoxin in several neurodegenerative disorders. EGCG attenuated cell death, presumably via its antioxidant activity, in neuronal culture after treatment with 3-HK-induced increase in the concentration of ROS as well as caspase-3 activity [65]. In brain tissue, green tea and black tea extracts were shown to inhibit lipid peroxidation promoted by iron-ascorbate in homogenates of brain mitochondrial membranes (IC_{50}: 2.44 and 1.40 μmol/l, respectively) [66]. A similar effect was also reported using brain synaptosomes, in which the four major polyphenol catechins of green tea were shown to inhibit iron-induced lipid peroxidation [44]. In this regard, it has been shown that EGCG attenuated paraquat-induced microsomal lipid peroxidation and increased the survival rate of paraquat-poisoned mice [67]. The herbicide paraquat is a strong redox agent, contributes to the formation of ROS, and induces toxicity of the nigrostriatal dopaminergic system, and therefore is used as a model for parkinsonism *in vivo* [68]. EGCG inhibited paraquat-induced malondialdehyde production in rat liver microsome system containing $FeSO_4$ by two possible mechanisms. One may be scavenging of superoxide radicals, which are responsible for the reduction of ferric to ferrous catalyzed by the Fenton reaction. The other is iron-chelating activity, given that the inhibition disappeared when an excessive amount of $FeSO_4$ was added to the reaction, which indicates that EGCG inhibits iron-driven lipid peroxidation by pulling out available iron in the mixture.

The ability of green tea polyphenols, and catechins in particular, to chelate metal ions such as iron and copper may contribute to their antioxidant and neuroprotective activity by inhibiting transition metal-catalyzed free radical formation. The two points of attachment of transition metal ions to the flavonoid molecule are the *o*-diphenolic groups in the 3,4-dihydroxy positions in the B ring and the keto structure *4-keto,* 3-hyroxy or *4-keto* and 5-hydroxy in the C ring of the flavonols [2,69]. The ability of green tea polyphenols to act as relatively potent metal chelators [44,70] may be of major significance for the treatment of neurodegenerative diseases, where accumulation of iron has been shown to happen at brain areas in which neurodegeneration occurs. The localization of iron and ferritin in PD patients is restricted to specific brain areas [7,71,72] in the SNPC and not the reticulata, even though the latter region has higher iron content than the SNPC [71]. Similarly, AD pathogenesis is also associated with iron accumulation and is linked to the characteristic neocortical Aβ deposition, phosphorylation of tau and tangle formation, which may be mediated by abnormal interaction with excess of free-chelatable iron. Ionic iron can, in turn, participate in Fenton reaction with subsequent generation of ROS, which is thought to initiate the processes of OS and inflammatory cascade resulting in production of cytotoxic cytokines (tumor necrosis-alpha (TNF-α), interluekine-1 and -6) in the microglia and the surrounding neurons [73–76], and activation of transcription factors and nuclear factor-kappa B (NF-κB) [77,78]. Indeed, a 70-fold increase in NF-κB immunoreactivity was found in the nucleus of melanized dopaminergic neurons of the parkinsonian SNPC compared to normal brains [79]. EGCG was found to inhibit the nuclear translocation of NF-κB in *in vitro* systems: immunofluorescence and electromobility shift assays showed that introduction of green tea extract before 6-OHDA-induced OS, inhibited both NF-κB nuclear translocation and binding activity in neuroblastoma SH-SY5Y cells [66]. Furthermore, the reduced activity of NF-κB by EGCG and the theaflavin-3,3-digallate polyphenol from black tea was associated with inhibition of lipopolysaccharid (LPS)-induced TNF-α production [24] and the enzyme-inducible nitric oxide synthase (iNOS) [77,80] in activated macrophages. This enzyme is an inducible form of the nitric oxide synthase (NOS), which is responsible for the production of the short-lived free radical, nitric oxide, functioning as a signaling molecule.

The inhibition of enzymes, whose activity may promote oxidative stress and an increase of antioxidant enzyme activities, might have a beneficial significance to neuroprotection. EGCG was found to elevate the activity of two major antioxidant enzymes, superoxide dismutase (SOD) and catalase, in mice striatum [20]. Furthermore, EGCG was found to activate ARE, which has been found in promoters of genes of the cellular defensive enzymes, glutathione-*s*-transferase and heme-oxygenase 1 [21], and this was correlated with increased activity and nuclear binding of the transcription factors Nrf1 and Nrf2 and stimulation of the MAPK pathway [81].

13.6 THE MECHANISMS UNDERLYING EGCG-INDUCED CELL SURVIVAL AND CELL DEATH

13.6.1 MODULATION OF CELL-SIGNALING PATHWAYS

Emerging evidence suggests that the antioxidant activity of green tea polyphenols cannot be the sole mechanism responsible for their neuroprotective action, but rather, their ability to alter signaling pathways may significantly contribute to the cell survival effect. The modulation of cell survival and signal transduction pathways has significant biological consequences that are important in understanding the various pharmacological and toxicological responses of antioxidant drugs. A number of intracellular signaling pathways have been described to play central functions in EGCG-promoted neuronal protection against a variety of extracellular insults such as the MAPK [21,82,83], PKC [22,84], and phosphatidylinositol-3-kinase (PI-3 kinase)-AKT [85–87] pathways, as described in Figure 13.2. Given the critical role of MAPK pathways in regulating cellular processes that are affected in neurodegenerative diseases, the importance of MAPKs as transducers of extracellular stimuli into a series of intracellular phosphorylation cascades in disease pathogenesis is being increasingly recognized. OS seems to be a major stimulus for the MAPK cascade, which might lead to cell survival or cell death. Among the MAPKs, the extracellular signal-regulated kinases (ERK1/2) are mainly activated by mitogen and growth factors [88], whereas p38 and c-jun-N-terminal kinase (JNK) respond to stress stimuli [89,90].

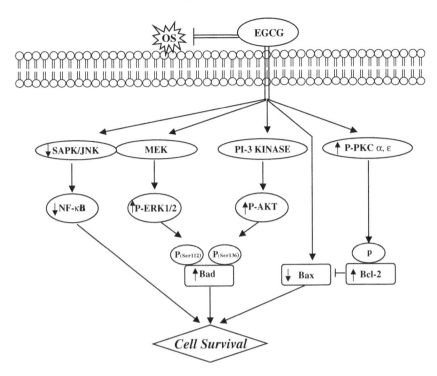

FIGURE 13.2 A model for EGCG protective mechanism of action indicating potential pathways of action with respect to its proposed modulative effect in neuronal and extra-neuronal tissues.

Previous *in vitro* studies [21], as discussed above, demonstrated the potency of EGCG to induce ARE-mediated defensive genes and MAPK pathways including ERK, JNK, and p38 MAPK [81] that enhanced cell survival and beneficial homeostatic response. The role of ERK1/2 signaling seems to be connected to attenuation of neuronal death and cellular injury by OS [91]. Moreover, different antioxidant molecules modulate, in a diverse way, kinases downstream to ERK1/2 and JNK [92]. In addition, EGCG counteracted the decline in ERK1/2 induced by 6-OHDA in neuroblastoma cells, whereas neither EGCG nor catechin at their neuroprotective concentrations (1–10 µM) affected the levels of ERK1/2 phosphorylation by themselves, in the absence of any exogenous damage, in neuronal cell line and primary neuron cultures [22,93]. Conversely, resveratrol, a natural flavonoid occurring in grapes and wine, was shown to directly induce phosphorylation of ERK1/2 in SH-SY5Y neuroblastoma cells [94]. This discrepancy may have resulted from structural differences between the catechins and flavonoids affecting the specific molecular interaction, which is only partially related to antioxidant capability.

EGCG neuroprotective activity also involves the intracellular signaling mediator PKC [19], thought to have an essential role in the regulation of cell survival and programmed cell death [95,96]. PKCs belong to the family of serine and threonine kinases consisting of 11 isoforms: conventional ($\alpha, \beta_{II}, \beta_{II}$, and γ), novel ($\delta, \epsilon, \theta, \eta$, and μ) and atypical ($\iota/\lambda, \zeta$). A rapid loss of neuronal PKC activity is a common consequence of several brain damages [97,98]. A previous study [99] demonstrated a significant reduction of phorbol ester binding, a direct activator of PKC, in the substantia nigra of PD patients compared to controls. The induction of PKC activity in neurons by EGCG is thought to be a prerequisite for neuro-protection against several neurotoxins, such as Aβ [19], serum withdrawal [87], and 6-OHDA [22]. Inhibition of PKC phosphorylation completely abolished the protection induced by EGCG and by the PKC activator, phorbol 12-myristate 13-acetate (PMA). These *in vitro* results were supported by a recent study [19], which showed that EGCG oral administration to mice caused significant increase of the PKC isoenzymes α and ϵ protein levels in the membrane and cytosolic fractions of hippocampus. These isoforms play a crucial role in cell survival and differentiation of pathways [100,101] and may be involved in APP processing elevated pathogenesis of AD [61,62].

The mechanism by which PKC activation leads to neuroprotection and cell survival has not been clearly defined. Studies with extraneuronal tissues support a role for PKCα as a kinase of the antiapoptotic Bcl-2, probably through direct or indirect phosphorylation of this cell survival protein [102]. Moreover, overex-pression of PKCϵ in a hematopoietic cell line resulted in increased expression of the mitochondrial protein Bcl-2 [100]. Recent study in human epidermal keratinocytes, taken from EGCG-treated skin of healthy subjects, indicated that EGCG promoted cell survival by increasing the ratio of Bcl-2 to proapoptotic Bax and phosphorylation of another proapoptotic Bcl-2 family member, Bad, through ERK and AKT signaling pathways [103]. Using mitogen-activated protein kinase 1 (MEK1) inhibitor (PD98059), EGCG induced only the phosphorylation of Ser136 of Bad, and when the authors used PI-3 kinase inhibitor (LY294002), EGCG induced only the phosphorylation of Ser112. These results indicated that EGCG is affecting not only the ERK pathway involved in phosphorylation of Ser112 but also the PI-3 Kinase/AKT pathway involved in phosphorylation of Ser136 of Bad.

Nonetheless, a study with high concentrations of EGCG reported cell prolif-eration arrest of tumor cells and inhibition of ERK1/2 and AKT phosphorylation, which was associated with reduced phosphorylation of Bad [104]. In line with these results, a study in human prostate carcinoma LNCaP cells [105] reported that high concentrations of EGCG induced apoptosis via negative regulation of the activity of the transcription factor NF-κB, thereby decreasing the expression of Bcl-2 and upregulation of the transcriptional activity of p53, resulting in activation of its downstream targets, cyclin-dependent kinase inhibitor p21/WAF1 and Bax. The latter caused a change in the ratio of Bax/Bcl-2 in favor of cell death, followed by activation of caspase-9, -8 and -3. This biphasic mode of biological activity of EGCG relies on its concentration-dependent window of pharmacological action: EGCG exhibits prooxidant and proapoptotic activity at high concentrations, that are responsible for the anti-cancer-cell death effect, whereas at low doses it is neuroprotective against a wide spectrum of neurotoxic compounds. A biphasic mode of action has been described for most of the typical radical scavengers and antioxidants such as ascorbic acid (vitamin C) [106], iron chelators such as DA receptor agonist R-APO [107], and also for green tea polyphenols [66].

13.6.2 EFFECT ON CELL SURVIVAL AND APOPTOTIC GENE EXPRESSION

Studies based on customized cDNA and quantitative real-time RT-PCR were recently conducted to verify the molecular mechanisms involved in the cell survival and cell death action of EGCG [22,108,109]. In these studies, the gene expression profile of EGCG was compared to that of three other neuroprotective and antioxidant drugs, DA, R-APO (both catechol derivatives), and the pineal indoleamine hormone melatonin, at low and high concentrations [110]. EGCG (1 μM), DA (10 μM), and R-APO (1 μM) behaved as potent neuroprotective agents, decreasing the expression of the proapoptotic genes bax, bad, gadd45, and fas ligand. However, EGCG did not affect the expression of the antiapoptotic bcl-w, bcl-2, and bcl-xL, whereas the other three compounds increased them. In addition, when cell viability of neuroblastoma SH-SY5Y cells was challenged with 6-OHDA, low micromolar concentration of EGCG abolished the induction of proapoptotic-related mRNAs and the decrease in bcl-2, bcl-w, and bcl-xL. EGCG neuroprotective effect is thought to be mediated through downregulation of proapoptotic genes, as shown for mdm2, caspase-1, cyclin-dependent kinase inhibitor p21 and TNF-related apoptosis-inducing ligand TRAIL, rather than upregulation of antiapoptotic genes. These findings support the assumption that complementary mechanisms, in addition to radical scavenging, are involved in the neuronal survival effect. In contrast to the antiapoptotic effect observed with low concentrations (<10 μM), a proapoptotic pattern of gene expression is observed with high concentrations of EGCG, DA, R-APO, and melatonin (>20 μM). It includes the expression of bax, gadd45, caspase family members (3, 6, and 10), and TNF receptor family member fas and fas-ligand mRNAs. The results revealed a significant functional-group homology between these drugs in genes coding for signal transducers, transcriptional repressors, and growth factors, which may account for their mechanism of action as illustrated in Figure 13.3. This might be predictable, given that EGCG, DA, and R-APO share similar attributes, being catechol-like derivatives, iron chelators and offering protection against neurotoxicity induced by 6-OHDA or MPTP [111,112].

13.7 CONCLUSIONS AND PERSPECTIVES FOR THE FUTURE

To date, an accumulated evidence on the health properties of green tea polyphenols on pathologies, such as neurodegenerative diseases, in particular PD and AD, is associated with oxidative damage and iron accretion. However, the effect

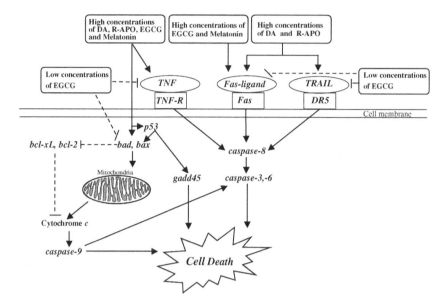

FIGURE 13.3 Schematic overview demonstrating possible gene targets involved in the antiapoptotic and proapoptotic actions of low and high concentrations of EGCG, respectively, compared with other antioxidants. Solid arrows and dotted lines indicate induction and inhibition of gene expression, respectively. Full explanation is given in the text.

of green tea and its constituents have not been evaluated enough in these pathologies. Most studies are carried out in animal models and cell culture to evaluate the effect of acute and chronic administration of the compounds. EGCG appears to affect the mortality of neuronal cells. Given the central role of the mitochondria in OS-induced apoptosis, it may be speculated that EGCG-mediated inhibition of apoptosis might implicate mitochondrial targets. This may be a consequence of the blockade of mitochondrial permeability transition pore (mPTP) opening, because EGCG has an effect on the mitochondrial protein expression, the Bcl-2 family members such as Bax and Bad. Similarly, the anti-Parkinson neuroprotective antiapoptotic drug, rasagiline, was recently shown to prevent the fall in mitochondrial membrane potential and the opening of mitochondrial voltage-dependent anion channel via the increase in Bcl-2 and Bcl-xl mRNAs and their proteins [113,114]. In addition, the toxicity of the metabolic product of MPTP, MPP+, which is also a mitochondrial complex I inhibitor [115] and inducer of oxidative stress, iron signaling, and α-synuclein expression [116], is attenuated by EGCG probably via its metal-chelating ability. Consistent with this idea, a recent study reported a novel iron-responsive element (IRE-type II) within the 5-untranslated region of the Alzheimer APP transcript that can be regulated by metal chelators [117] like EGCG. In the light of this accumulated information, our group strongly advocates the use of brain-permeable iron chelators as neuroprotective drugs to "iron out iron" from those brain areas where it preferentially accumulates in neurodegenerative diseases.

Future efforts in the understanding of the protective effect mechanism of action of green tea polyphenols must concentrate on deciphering the cell targets affected by these compounds and other neuroprotectants. This will allow a more specific therapy design, especially when a multipharmacological action drug cocktail is considered for formulation, which can be more effective when used in clinical treatment of PD and AD.

ACKNOWLEDGMENTS

We thank the National Parkinson Foundation (Miami), Stein Foundation (Philadelphia), and Rappoport Family Research, Technion–Israel Institute of Technology for support of this study.

REFERENCES

1. Butterfield D, Castegna A, Pocernich C, Drake J, Scapagnini G, Calabrese V. Nutritional approaches to combat oxidative stress in Alzheimer's disease. *J Nutr Biochem* 2002; 13: 444–461.
2. van Acker SA, van den Berg DJ, Tromp MN, Griffioen DH, van Bennekom WP, van der Vijgh WJ, Bast A. Structural aspects of antioxidant activity of flavonoids. *Free Radic Biol Med* 1996; 20: 331–342.
3. Gotz ME, Kunig G, Riederer P, Youdim MB. Oxidative stress: free radical production in neural degeneration. *Pharmacol Ther* 1994; 63: 37–122.
4. Halliwell B. Role of free radicals in the neurodegenerative diseases: therapeutic implications for antioxidant treatment. *Drugs Aging* 2001; 18: 685–716.
5. Gotz ME, Freyberger A, Riederer P. Oxidative stress: a role in the pathogenesis of Parkinson's disease. *J Neural Transm Suppl* 1990; 29: 241–249.
6. Halliwell B. Reactive oxygen species and the central nervous system. *J Neurochem* 1992; 59: 1609–1623.
7. Riederer P, Sofic E, Rausch WD, Schmidt B, Reynolds GP, Jellinger K, Youdim MBH. Transition metals, ferritin, glutathione, and ascorbic acid in parkinsonian brains. *J Neurochem* 1989; 52: 515–520.
8. Linazasoro G. Neuroprotection in Parkinson's disease: love story or mission impossible? *Expert Rev Neurother* 2002; 2: 403–416.
9. Perry G, Cash AD, Smith MA. Alzheimer Disease and Oxidative Stress. *J Biomed Biotechnol* 2002; 2: 120–123.
10. Perry G, Taddeo MA, Nunomura A, Zhu X, Zenteno-Savin T, Drew KL, Shimohama S, Avila J, Castellani RJ, Smith MA. Comparative biology and pathology of oxidative stress in Alzheimer and other neurodegenerative diseases: beyond damage and response. *Comp Biochem Physiol C Toxicol Pharmacol* 2002; 133: 507–513.
11. Hellenbrand W, Seidler A, Boeing H, Robra BP, Vieregge P, Nischan P, Joerg J, Oertel WH, Schneider E, Ulm G. Diet and Parkinson's disease. I: A possible role for the past intake of specific foods and food groups. Results from a self-administered food-frequency questionnaire in a case-control study. *Neurology* 1996; 47: 636–643.
12. Higdon JV, Frei B. Tea catechins and polyphenols: health effects, metabolism, and antioxidant functions. *Crit Rev Food Sci Nutr* 2003; 43: 89–143.

13. Wiseman SA, Balentine DA, Frei B. Antioxidants in tea. *Crit Rev Food Sci Nutr* 1997; 37: 705–718.

14. Slikker W, Youdim MB, Palmer GC, Hall E, Williams C, Trembly B. The future of neuroprotection. *Ann N Y Acad Sci* 1999; 890: 529–533.

15. Salah N, Miller NJ, Paganga G, Tijburg L, Bolwell GP, Rice-Evans C. Polyphenolic flavanols as scavengers of aqueous phase radicals and as chain-breaking antioxidants. *Arch Biochem Biophy* 1995; 322: 339–346.

16. Nanjo F, Goto K, Seto R, Suzuki M, Sakai M, Hara Y. Scavenging effects of tea catechins and their derivatives on 1,1-diphenyl-2-picrylhydrazyl radical. *Free Radic Biol Med* 1996; 21: 895–902.

17. Ishige K, Schubert D, Sagara Y. Flavonoids protect neuronal cells from oxidative stress by three distinct mechanisms. *Free Radic Biol Med* 2001; 30: 433–446.

18. Schroeter H, Boyd C, Spencer JP, Williams RJ, Cadenas E, Rice-Evans C. MAPK signaling in neurodegeneration: influences of flavonoids and of nitric oxide. *Neurobiol Aging* 2002; 23: 861–880.

19. Levites Y, Amit T, Mandel S, Youdim MBH. Neuroprotection and Neurorescue Against Amyloid beta Toxicity and PKC-Dependent Release of Non-Amyloidogenic Soluble Precusor Protein by Green Tea Polyphenol (–)-Epigallocatechin-3-gallate. *Faseb J* 2003; 17: 952–954.

20. Levites Y, Weinreb O, Maor G, Youdim MBH, Mandel S. Green tea polyphenol (–)-Epigallocatechin-3-gallate prevents *N*-methyl-4-phenyl-1,2,3,6-tetrahydropyridine-induced dopaminergic neurodegeneration. *J Neurochem* 2001; 78: 1073–1082.

21. Chen C, Yu R, Owuor ED, Kong AN. Activation of antioxidant-response element (ARE), mitogen-activated protein kinases (MAPKs) and caspases by major green tea polyphenol components during cell survival and death. *Arch Pharm Res* 2000; 23: 605–612.

22. Levites Y, Amit T, Youdim MBH, Mandel S. Involvement of protein kinase C activation and cell survival/cell cycle genes in green tea polyphenol (–)-epigallocatechin-3-gallate neuroprotective action. *J Biol Chem* 2002; 277: 30574–30580.

23. Hollman PC, Feskens EJ, Katan MB. Tea flavonols in cardiovascular disease and cancer epidemiology. *Proc Soc Exp Biol Med* 1999; 220: 198–202.

24. Yang F, de Villiers WJ, McClain CJ, Varilek GW. Green tea polyphenols block endotoxin-induced tumor necrosis factor-production and lethality in a murine model. *J Nutr* 1998; 128: 2334–2340.

25. Tedeschi E, Menegazzi M, Yao Y, Suzuki H, Forstermann U, Kleinert H. Green tea inhibits human inducible nitric-oxide synthase expression by down-regulating signal transducer and activator of transcription-1alpha activation. *Mol Pharmacol* 2004; 65: 111–120.

26. Weisburger JH, Chung FL. Mechanisms of chronic disease causation by nutritional factors and tobacco products and their prevention by tea polyphenols. *Food Chem Toxicol* 2002; 40: 1145–1154.

27. Pannala AS, Rice-Evans CA, Halliwell B, Singh S. Inhibition of peroxynitrite-mediated tyrosine nitration by catechin polyphenols. *Biochem Biophys Res Commun* 1997; 232: 164–168.

28. Nanjo F, Mori M, Goto K, Hara Y. Radical scavenging activity of tea catechins and their related compounds. *Biosci Biotechnol Biochem* 1999; 63: 1621–1623; Guo Q, Zhao B, Shen S, Hou J, Hu J, Xin W. ESR study on the structure-antioxidant activity relationship of tea catechins and their epimers. *Biochim Biophys Acta* 1999; 1427: 13–23.

29. Wang ZY, Huang MT, Lou YR, Xie JG, Reuhl KR, Newmark HL, Ho CT, Yang CS, Conney AH. Inhibitory effects of black tea, green tea, decaffeinated black tea, and decaffeinated green tea on ultraviolet B light-induced skin carcinogenesis in 7,12-dimethylbenz[a]anthracene-initiated SKH-1 mice. *Cancer Res* 1994; 54: 3428–3455.

30. Pietta PG, Simonetti P, Gardana C, Brusamolino A, Morazzoni P, Bombardelli E. Catechin metabolites after intake of green tea infusions. *Biofactors* 1998; 8: 111–118.

31. Li C, Lee MJ, Sheng S, Meng X, Prabhu S, Winnik B, Huang B, Chung JY, Yan S, Ho CT, et al. Structural identification of two metabolites of catechins and their kinetics in human urine and blood after tea ingestion. *Chem Res Toxicol* 2000; 13: 177–184.

32. Harada M, Kan Y, Naoki H, Fukui Y, Kageyama N, Nakai M, Miki W, Kiso Y. Identification of the major antioxidative metabolites in biological fluids of the rat with ingested (+)-catechin and (–)-epicatechin. *Biosci Biotechnol Biochem* 1999; 63: 973–977.

33. Nakagawa K, Miyazawa T. Chemiluminescence-high-performance liquid chromatographic determination of tea catechin, (–)-epigallocatechin 3-gallate, at picomole levels in rat and human plasma. *Anal Biochem* 1997; 248: 41–49.

34. Nakagawa K, Okuda S, Miyazawa T. Dose-dependent incorporation of tea catechins, (–)-epigallocatechin-3-gallate and (–)-epigallocatechin, into human plasma. *Biosci Biotechnol Biochem* 1997; 61: 1981–1985.

35. Yang CS, Kim S, Yang GY, Lee MJ, Liao J, Chung JY, Ho CT. Inhibition of carcinogenesis by tea: bioavailability of tea polyphenols and mechanisms of actions. *Proc Soc Exp Biol Med* 1999; 220: 213–217.

36. Abd El Mohsen MM, Kuhnle G, Rechner AR, Schroeter H, Rose S, Jenner P, Rice-Evans CA. Uptake and metabolism of epicatechin and its access to the brain after oral ingestion. *Free Radic Biol Med* 2002; 33: 1693–1702.

37. Suganuma M, Okabe S, Oniyama M, Tada Y, Ito H, Fujiki H. Wide distribution of [3H](–)-epigallocatechin gallate, a cancer preventive tea polyphenol, in mouse tissue. *Carcinogenesis* 1998; 19: 1771–1776.

38. Checkoway H, Powers K, Smith-Weller T, Franklin GM, Longstreth WT, Swanson Jr. PD. Parkinson's disease risks associated with cigarette smoking, alcohol consumption, and caffeine intake. *Am J Epidemiol* 2002; 155: 732–738.

39. Li SC, Schoenberg BS, Wang CC, Cheng XM, Rui DY, Bolis CL, Schoenberg DG. A prevalence survey of Parkinson's disease and other movement disorders in the People's Republic of China. *Arch Neurol* 1985; 42: 655–657.

40. Zhang ZX, Roman GC. Worldwide occurrence of Parkinson's disease: an updated review. *Neuroepidemiology* 1993; 12: 195–208.

41. Forster DP, Newens AJ, Kay DW, Edwardson JA. Risk factors in clinically diagnosed presenile dementia of the Alzheimer type: a case-control study in northern England. *J Epidemiol Community Health* 1995; 49: 253–258.

42. Le Bars PL, Katz MM, Berman N, Itil TM, Freedman AM, Schatzberg AF. A placebo-controlled, double-blind, randomized trial of an extract of Ginkgo biloba for dementia. North American EGb study group. *JAMA* 1997; 278: 1327–1332.

43. Orgogozo JM, Dartigues JF, Lafont S, Letenneur L, Commenges D, Salamon R, Renaud S, Breteler MB. Wine consumption and dementia in the elderly: a prospective community study in the Bordeaux area. *Rev Neurol* (Paris) 1997; 153: 185–192.

44. Guo Q, Zhao B, Li M, Shen S, Xin W. Studies on protective mechanisms of four components of green tea polyphenols against lipid peroxidation in synaptosomes. *Biochim Biophys Acta* 1996; 1304: 210–222.

45. Grunblatt E, Mandel S, Maor G, Youdim MBH. Effects of R-apomorphine and S-apomorphine on MPTP-induced nigro-striatal doamine neuronal loss. *J Neurochem* 2001; 77: 146–156.

46. Grunblatt E, Mandel S, Berkuzki T, Youdim MBH. Apomorphine protects against MPTP-induced neurotoxicity in mice. *Mov Disord* 1999; 14: 612–618.

47. Pan T, Fei J, Zhou X, Jankovic J, Le W. Effects of green tea polyphenols on dopamine uptake and on MPP+ -induced dopamine neuron injury. *Life Sci* 2003; 72: 1073–1083.

48. Lu H, Meng X, Yang CS. Enzymology of methylation of tea catechins and inhibition of catechol-O-methyltransferase by (–)-epigallocatechin gallate. *Drug Metab Dispos* 2003; 31: 572–579.

49. Deleu D, Northway MG, Hanssens Y. Clinical pharmacokinetic and pharmacodynamic properties of drugs used in the treatment of Parkinson's disease. *Clin Pharmacokinet* 2002; 41: 261–309.

50. Monteiro HP, Winterbourn CC. 6-Hydroxydopamine releases iron from ferritin and promotes ferritin-dependent lipid peroxidation. *Biochem Pharmacol* 1989; 38: 4177–4182.

51. Mochizuki H, Imai H, Endo K, Yokomizo K, Murata Y, Hattori N, Mizuno Y. Iron accumulation in the substantia nigra of 1-methyl-4-phenyl-1,2,3,6-tetrahydropyridine (MPTP)-induced hemiparkinsonian monkeys. *Neurosci Lett* 1994; 168: 251–253.

52. Oestreicher E, Sengstock GJ, Riederer P, Olanow CW, Dunn AJ, Arendash GW. Degeneration of nigrostriatal dopaminergic neurons increases iron within the substantia nigra: a histochemical and neurochemical study. *Brain Res* 1994; 660: 8–18.

53. Temlett JA, Landsberg JP, Watt F, Grime GW. Increased iron in the substantia nigra compacta of the MPTP-lesioned hemiparkinsonian African green monkey: evidence from proton microprobe elemental microanalysis. *J Neurochem* 1994; 62: 134–146.

54. Senior K. New genes reveal major role for iron in neurodegeneration. *Lancet* 2001; 358: 302.

55. Bush AI. The metallobiology of Alzheimer's disease. *Trends Neurosci* 2003; 26: 207–214.

56. Checler F. Processing of the beta-amyloid precursor protein and its regulation in Alzheimer's disease. *J Neurochem* 1995; 65: 1431–1444.

57. Nunan J, Small DH. Regulation of APP cleavage by alpha-, beta- and gamma-secretases. *FEBS Lett* 2000; 483: 6–10.

58. Choi YT, Jung CH, Lee SR, Bae JH, Baek WK, Suh MH, Park J, Park CW, Suh SI. The green tea polyphenol (–)-epigallocatechin gallate attenuates beta-amyloid-induced neurotoxicity in cultured hippocampal neurons. *Life Sci* 2001; 70: 603–614.

59. Ono K, Yoshiike Y, Takashima A, Hasegawa K, Naiki H, Yamada M. Potent anti-amyloidogenic and fibril-destabilizing effects of polyphenols *in vitro*: implications for the prevention and therapeutics of Alzheimer's disease. *J Neurochem* 2003; 87: 172–181.

60. Mills J, Reiner PB. Regulation of amyloid precursor protein cleavage. *J Neurochem* 1999; 72: 443–460.

61. Slack BE, Nitsch RM, Livneh E, Kunz GM, Jr., Breu J, Eldar H, Wurtman RJ. Regulation by phorbol esters of amyloid precursor protein release from Swiss 3T3 fibroblasts overexpressing protein kinase C alpha. *J Biol Chem* 1993; 268: 21097–21101.

62. Benussi L, Govoni S, Gasparini L, Binetti G, Trabucchi M, Bianchetti A, Racchi M. Specific role for protein kinase C alpha in the constitutive and regulated secretion of amyloid precursor protein in human skin fibroblasts. *Neurosci Lett* 1998; 240: 97–101.

63. Matsushima H, Shimohama S, Chachin M, Taniguchi T, Kimura J. Ca2+-dependent and Ca2+-independent protein kinase C changes in the brain of patients with Alzheimer's disease. *J Neurochem* 1996; 67: 317–323.

64. Morel I, Lescoat G, Cogrel P, Sergent O, Pasdeloup N, Brissot P, Cillard P, Cillard J. Antioxidant and iron-chelating activities of the flavonoids catechin, quercetin and diosmetin on iron-loaded rat hepatocyte cultures. *Biochem Pharmacol* 1999; 45: 13–19.

65. Jeong JH, Kim HJ, Lee TJ, Kim MK, Park ES, Choi BS. Epigallocatechin 3-gallate attenuates neuronal damage induced by 3-hydroxykynurenine. *Toxicology* 2004; 195: 53–60.

66. Levites Y, Youdim MBH, Maor G, Mandel S. Attenuation of 6-hydroxydopamine (6-OHDA)-induced nuclear factor-κB (NF-κB) activation and cell death by tea extracts in neuronal cultures. *Biochem Pharmacol* 2002; 63: 21–29.

67. Higuchi A, Yonemitsu K, Koreeda A, Tsunenari S. Inhibitory activity of epigallocatechin gallate (EGCG) in paraquat-induced microsomal lipid peroxidation — a mechanism of protective effects of EGCG against paraquat toxicity. *Toxicology* 2003; 183: 143–149.

68. Liou HH, Chen RC, Chen TH, Tsai YF, Tsai MC. Attenuation of paraquat-induced dopaminergic toxicity on the substantia nigra by (–)-deprenyl *in vivo*. *Toxicol Appl Pharmacol* 2001; 172: 37–43.

69. Thompson M, Williams CR, Elliot GE. Stability of flavonoid complexes of copper(II) and flavonoid antioxidant activity. *Anal Chim Acta* 1976; 85: 375–381.

70. Grinberg LN, Newmark H, Kitrossky N, Rahamim E, Chevion M, Rachmilewitz EA. Protective effects of tea polyphenols against oxidative damage to red blood cells. *Biochem Pharmacol* 1997; 54: 973–978.

71. Jellinger K, Paulus W, Grundke-Iqbal I, Riederer P, Youdim MBH. Brain iron and ferritin in Parkinson's and Alzheimer's diseases. *J Neural Transm Park Dis Dement Sect* 1990; 2: 327–340.

72. Sofic E, Paulus W, Jellinger K, Riederer P, Youdim MBH. Selective increase of iron in substantia nigra zona compacta of parkinsonian brains. *J Neurochem* 1991; 56: 978–982.

73. Mogi M, Harada M, Riederer P, Narabayashi H, Fujita K, Nagatsu T. Tumor necrosis factor-alpha (TNF-alpha) increases both in the brain and in the cerebrospinal fluid from parkinsonian patients. *Neurosci Lett* 1994; 165: 208–210.

74. Blum-Degen D, Muller T, Kuhn W, Gerlach M, Przuntek H, Riederer P. Interleukin-1 beta and interleukin-6 are elevated in the cerebrospinal fluid of Alzheimer's and de novo Parkinson's disease patients. *Neurosci Lett* 1995; 202: 17–20.

75. McGeer PL, McGeer EG. The inflammatory response system of brain: implications for therapy of Alzheimer and other neurodegenerative diseases. *Brain Res Brain Res Rev* 1995; 21: 195–218.

76. Sakaguchi S, Furusawa S, Yokota K, Sasaki K, Takayanagi M, Takayanagi Y. The enhancing effect of tumour necrosis factor-alpha on oxidative stress in endotoxemia. *Pharmacol Toxicol* 1996; 79: 259–265.

77. Lin YL, Lin JK. (–)-Epigallocatechin-3-gallate blocks the induction of nitric oxide synthase by down-regulating lipopolysaccharide-induced activity of transcription factor nuclear factor-kappaB. *Mol Pharmacol* 1997; 52: 465–472.

78. Schreck R, Rieber P, Baeuerle PA. Reactive oxygen intermediates as apparently widely used messengers in the activation of the NF-kappaB transcription factor and HIV-1. *EMBO J* 1991; 10: 2247–2258.

79. Hunot S, Brugg B, Ricard D, Michel PP, Muriel MP, Ruberg M, Faucheux BA, Agid Y, Hirsch EC. Nuclear translocation of NF-kappaB is increased in dopaminergic neurons of patients with parkinson disease. *Proc Natl Acad Sci USA* 1997; 94: 7531–7536.

80. Lin YL, Tsai SH, Lin-Shiau SY, Ho CT, Lin JK. Theaflavin-3,3'-digallate from black tea blocks the nitric oxide synthase by down-regulating the activation of NF-kappaB in macrophages. *Eur J Pharmacol* 1999; 367: 379–388.

81. Owuor ED, Kong AN. Antioxidants and oxidants regulated signal transduction pathways. *Biochem Pharmacol* 2002; 64: 765–770.

82. Xia Z, Dickens M, Raingeaud J, Davis RJ, Greenberg ME. Opposing effects of ERK and JNK-p38 MAP kinases on apoptosis. *Science* 1995; 270: 1326–1331.

83. Singer CA, Figueroa-Masot XA, Batchelor RH, Dorsa DM. The mitogen-activated protein kinase pathway mediates estrogen neuroprotection after glutamate toxicity in primary cortical neurons. *J Neurosci* 1999; 19: 2455–2463.

84. Cordey M, Gundimeda U, Gopalakrishna R, Pike CJ. Estrogen activates protein kinase C in neurons: role in neuroprotection. *J Neurochem* 2003; 84: 1340–1348.

85. Kaplan DR, Miller FD. Neurotrophin signal transduction in the nervous system. *Curr Opin Neurobiol* 2000; 10: 381–391.

86. Koh SH, Kim SH, Kwon H, Park Y, Kim KS, Song CW, Kim J, Kim MH, Yu HJ, Henkel JS, et al. Epigallocatechin gallate protects nerve growth factor differentiated PC12 cells from oxidative-radical-stress-induced apoptosis through its effect on phosphoinositide 3-kinase/Akt and glycogen synthase kinase-3. *Brain Res Mol Brain Res* 2003; 118: 72–81.

87. Mandel S, Reznichenko L, Amit T, Youdim M. Green tea polyphenol (–)-epigallocatechin-3-gallate protects rat PC12 cells from apoptosis induced by serum withdrawal independent of P13-Akt pathway. *Neurotoxocity Research* 2003; 5: 419–424.

88. Vaudry D, Stork PJ, Lazarovici P, Eiden LE. Signaling pathways for PC12 cell differentiation: making the right connections. *Science* 2002; 296: 1648–1649.

89. Harris CA, Deshmukh M, Tsui-Pierchala B, Maroney AC, Johnson Jr. EM. Inhibition of the c-Jun N-terminal kinase signaling pathway by the mixed lineage kinase inhibitor CEP-1347 (KT7515) preserves metabolism and growth of trophic factor-deprived neurons. *J Neurosci* 2002; 22: 103–113.

90. Johnson GL, Lapadat R. Mitogen-activated protein kinase pathways mediated by ERK, JNK, and p38 protein kinases. *Science* 2002; 298: 1911–1912.

91. Satoh T, Nakatsuka D, Watanabe Y, Nagata I, Kikuchi H, Namura S. Neuroprotection by MAPK/ERK kinase inhibition with U0126 against oxidative stress in a mouse neuronal cell line and rat primary cultured cortical neurons. *Neurosci Lett* 2000; 288: 163–166.

92. Della Ragione F, Cucciolla V, Criniti V, Indaco S, Borriello A, Zappia V. Antioxidants induce different phenotypes by a distinct modulation of signal transduction. *FEBS Lett* 2002; 532: 289–294.

93. Schroeter H, Spencer JP, Rice-Evans C, Williams RJ. Flavonoids protect neurons from oxidized low-density-lipoprotein- induced apoptosis involving c-Jun N-terminal kinase (JNK), c-Jun and caspase-3. *Biochem J* 2001; 358: 547–557.

94. Miloso M, Bertelli AA, Nicolini G, Tredici G. Resveratrol-induced activation of the mitogen-activated protein kinases, ERK1 and ERK2, in human neuroblastoma SH-SY5Y cells. *Neurosci Lett* 1999; 264: 141–144.

95. Dempsey EC, Newton AC, Mochly-Rosen D, Fields AP, Reyland ME, Insel PA, Messing RO. Protein kinase C isozymes and the regulation of diverse cell responses. *Am J Physiol Lung Cell Mol Physiol* 2000; 279: L429–438.

96. Maher P. How protein kinase C activation protects nerve cells from oxidative stress-induced cell death. *J Neurosci* 2001; 21: 2929–2938.

97. Cardell M, Wieloch T. Time course of the translocation and inhibition of protein kinase C during complete cerebral ischemia in the rat. *J Neurochem* 1993; 61: 1308–1314.

98. Busto R, Globus MY, Neary JT, Ginsberg MD. Regional alterations of protein kinase C activity following transient cerebral ischemia: effects of intraischemic brain temperature modulation. *J Neurochem* 1994; 63: 1095–1103.

99. Nishino N, Kitamura N, Nakai T, Hashimoto T, Tanaka C. Phorbol ester binding sites in human brain: characterization, regional distribution, age-correlation, and alterations in Parkinson's disease. *J Mol Neurosci* 1989; 1: 19–26.

100. Gubina E, Rinaudo MS, Szallasi Z, Blumberg PM, Mufson RA. Overexpression of protein kinase C isoform epsilon but not delta in human interleukin-3-dependent cells suppresses apoptosis and induces bcl-2 expression. *Blood* 1998; 91: 823–829.

101. Whelan RD, Parker PJ. Loss of protein kinase C function induces an apoptotic response. *Oncogene* 1998; 16: 1939–1944.

102. Ruvolo PP, Deng X, Carr BK, May WS. A functional role for mitochondrial protein kinase Calpha in Bcl2 phosphorylation and suppression of apoptosis. *J Biol Chem* 1998; 273: 25436–25442.

103. Chung JH, Han JH, Hwang EJ, Seo JY, Cho KH, Kim KH, Youn JI, Eun HC. Dual mechanisms of green tea extract-induced cell survival in human epidermal keratinocytes. *FASEB J* 2003; 17: 1913–1915.

104. Sah JF, Balasubramanian S, Eckert RL, Rorke EA. Epigallocatechin-3-gallate inhibits epidermal growth factor receptor signaling pathway: evidence for direct inhibition of ERK1/2 and AKT kinases. *J Biol Chem* 2004; 279: 12755–12762.

105. Hastak K, Gupta S, Ahmad N, Agarwal MK, Agarwal ML, Mukhtar H. Role of p53 and NF-kappaB in epigallocatechin-3-gallate-induced apoptosis of LNCaP cells. *Oncogene* 2003; 22: 4851–4859.

106. Halliwell B. Vitamin C: antioxidant or pro-oxidant *in vivo*? *Free Radic Res* 1996; 25: 439–454.

107. Gassen M, Gross A, Youdim MB. Apomorphine enantiomers protect cultured pheochromocytoma (PC12) cells from oxidative stress induced by H2O2 and 6-hydroxydopamine. *Mov Disord* 1998; 13: 242–248.

108. Mandel S, Grunblatt E, Maor G, Youdim MBH. Early and late gene changes in MPTP mice model of Parkinson's disease employing cDNA microarray. *Neurochemical Research* 2002; 27: 1231–1243.

109. Weinreb O, Mandel S, Youdim MB. Gene and protein expression profiles of anti- and pro-apoptotic actions of dopamine, R-apomorphine, green tea polyphenol (-)-epigallocatechine-3-gallate, and melatonin. *Ann NY Acad Sci* 2003; 993: 351–361; discussion: 87–93.

110. Weinreb O, Mandel S, Youdim MBH. CDNA gene expression profile homology of antioxidants and their anti-apoptotic and pro-apoptotic activities in human neuroblastoma cells. *Faseb J* 2003; 17: 935–937.

111. Gassen M, Youdim MBH. Free radical scavengers: chemical concepts and clinical relevance. *J Neural Transm Suppl* 1999; 56: 193–210.

112. Mandel S, Grunblatt E, Riederer P, Gerlach M, Levites Y, Youdim MBH. Neuroprotective strategies in Parkinson's disease: an update on progress. *CNS Drugs* 2003; 17: 729–762.

113. Maruyama W, Takahashi T, Youdim MBH, Naoi M. The anti-parkinson drug, rasagiline, prevents apoptotic DNA damage induced by peroxynitrite in human dopaminergic neuroblastoma SH-SY5Y cells. *J Neural Transm* 2002; 109: 467–481.

114. Maruyama W, Akao Y, Youdim MBH, Boulton AA, Davis BA, Naoi M. Transfection-enforced Bcl-2 overexpression and an anti-Parkinson drug, rasagiline, prevent nuclear accumulation of glyceraldehyde-3 phosphate dehydrogenase induced by an endogenous dopaminergic neurotoxin, N-methyl(R)salsolinol. *J Neurochem* 2001; 78: 727–735.

115. Bates TE, Heales SJ, Davies SE, Boakye P, Clark JB. Effects of 1-methyl-4-phenylpyridinium on isolated rat brain mitochondria: evidence for a primary involvement of energy depletion. J Neurochem 1994; 63: 640-8.

116. Kalivendi SV, Cunningham S, Kotamraju S, Joseph J, Hillard CJ, Kalyanaraman B. α-Synuclein up-regulation and aggregation during MPP+-induced apoptosis in neuroblastoma cells: intermediacy of transferrin receptor iron and hydrogen peroxide. *J Biol Chem* 2004; 279: 15240–15247.

117. Rogers JT, Randall JD, Cahill CM, Eder PS, Huang X, Gunshin H, Leiter L, McPhee J, Sarang SS, Utsuki T. An iron-responsive element type II in the 5′-untranslated region of the Alzheimer's amyloid precursor protein transcript. *J Biol Chem* 2002; 277: 45518–45528.

14 Kainic-Acid-Induced Neurotoxicity: Involvement of Free Radicals

Akitane Mori
Okayama University, Okayama, Japan

Toshiki Masumizu
JEOL Ltd., Tokyo, Japan

Isao Yokoi
Oita University, Oita, Japan

Lester Packer
University of Southern California, Los Angeles, California

CONTENTS

14.1 INTRODUCTION

Systemic or intracerebral administration of kainic acid (kainate, KA), a glutamate receptor agonist, generates severe and stereotyped behavioral convulsions and brain damage syndrome, and is used widely in rodent models of human temporal lobe epilepsy and status epilepticus [1–3] and also for Huntington's chorea [4]. Excellent reviews are available on the neuropathology of KA seizures [2,3,5,6]. Generally, excessive stimulation of excitatory amino acid (EAA) neurotransmitter receptors is known to give rise to an excitotoxic effect leading to neuronal cell death [7], although the precise mechanism for the neurodegenerative change by KA is not yet clearly elucidated. Evidence suggesting involvement of reactive oxygen species (ROS) in KA-induced neurotoxicity have been accumulating. Here, we review KA-induced neurotoxicity from the point of view of free radical injury.

At present, many neuroscientists are investigating KA-induced neuropathology but few of them know who discovered KA. Therefore, as an introduction, we briefly describe the career of Tsunematsu Takemoto, the discoverer of KA. This was made possible by a valuable document kindly supplied recently by Noboru Shoji, his son-in-law.

14.2 TSUNEMATSU TAKEMOTO: DISCOVERER OF KAINIC ACID

Tsunematsu Takemoto (Figure 14.1) was born on January 28, 1913, in Tadaoka, Osaka, Japan. He graduated from the Osaka Pharmaceutical College in 1933 and then studied in the Tokyo Imperial University (now Tokyo University). He received his Ph.D. from this university in 1945. In 1947, he was appointed as a professor of organic chemistry in Osaka Pharmaceutical College, and here he started his work on ascariasis and ascarides, especially on the active principles of *Digenea simplex* (Wulf.) C. Aq., a sea weed (Figure 14.2). He became a professor in the Osaka University, after its merger with Osaka Pharmaceutical College. He and his colleagues continued their investigations, and in 1953 discovered digenic acid in *Digenea simplex* Aq. [8]. The name of the acid was, of course, derived from the name of the plant. However, it has been changed to "kainic acid" to avoid a confusion with "diginin," which was a registered for another substance. "Kainic" refers to *kainin-so*, the Japanese name of *Digenea simplex* [9]. In that era, ascariasis was a very common disease in Japan, but it had been eradicated by KA treatment. He was given many awards for his discoveries. In 1960 he moved to Tohoku University, and 16 yr later retired from the university in 1976. He then served as a professor and dean of Tokushima Bunri University from 1976 to 1988. He died on January 23, 1989, at the age of 76. The list of the compounds researched by Dr. Takemoto and his coworkers, which was published on the occasion of his retirement from Tohoku University in 1976, contains 334 compounds, in which we can find very-well-known substances discovered by him and his colleagues, e.g., kainic acid, α-allo-kainic acid,

FIGURE 14.1 Tsunematsu Takemoto botanizing on a mountain. From a handwritten note of his: "Take care every day, and take great care of the plants botanized on a mountain. Treat them in all ways with much care. Any may have medical usefulness. Tsunematsu."

domoic acid, tricholomic acid, ibotenic acid, quisqualic acid [10]. All of these substances are heteropentacyclic compounds, and all have been shown to be EAA after extensive studies by neuroscientists [11]. He published more than 400 scientific papers. His specialty covered amino acids, phytoecdysones, cardiac glycosides, terpens, alkaloids, phenolic compounds, etc. Although most of his original papers were published in Japanese, one can find many important articles in *Yakugaku Zasshi*, the official journal of the Japan Pharmaceutical Society (volume 70–93, 1953–1973). A review article in English, "Isolation and structural identification of naturally occurring excitatory amino acids" by Takemoto (1978) is necessary reading for KA researchers.

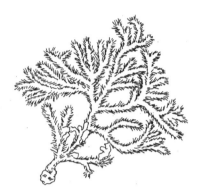

FIGURE 14.2 *Digenea simplex* (Wulf.) C. Aq. (Adapted from Chuuyaku-Daijiten, 5th ed., 1985, Vol 1, p. 223. With permission from Shougakkan, Tokyo.)

14.3 KA-INDUCED NEUROTOXICITY AS A GROUNDWORK OF FREE RADICAL GENERATION

The potent excitatory property of KA was first demonstrated by Shinozaki and Konishi in 1970 in mammalian neurons microelectrophoretically administered with KA [12]. Glutamate-like neurotoxic action of KA was demonstrated in the mouse brain with KA-induced convulsion by Olney et al. [13]. KA has been shown to cause selective degeneration of neurons, especially in the strial and hippocampal areas of the brain after intraventricular or intracerebral injection [4,14].

Generally, the mechanism for KA-induced neurotoxicity is thought to involve EAA receptors; i.e., KA activates not only postsynaptic KA receptors but also presynaptic KA receptors to induce the Ca^{2+}-dependent release of endogenous glutamate [7,15–18]. Glutamate release acts postsynaptically on non-NMDA receptors to depolarize the membrane and relieve the voltage-dependent block by magnesium of NMDA receptors, which can then contribute to neuronal damage [19]. The excessive stimulation of NMDA receptors accelerates Ca^{2+} influx [18] and may lead to calcium events (Figure 14.3). KA has been shown to stimulate Ca^{2+} influx into rat brain cells [20], cultured neuronal cells [21], synapstosomes [22], and perfused hippocampus [23].

KA-induced calcium events may involve NMDA-receptor-mediated stimulation of Ca^{2+}-dependent phospholipase A_2 (PLA_2), which releases arachidonic acid (AA) [24,25]. The liberated AA may become the source of $O_2^{\cdot-}$. The KA-induced stimulation of AA release, the Ca^{2+} influx, the accumulation of lipid peroxides and their decomposition products (e.g., 4-hydroxynonenal), along with alterations in cellular redox state and ATP depletion may each play an important role in KA-induced cell death [26]. The depletion of ATP may lead to the degradation of adenosine nucleotide and to the generation of $O_2^{\cdot-}$ [27–30].

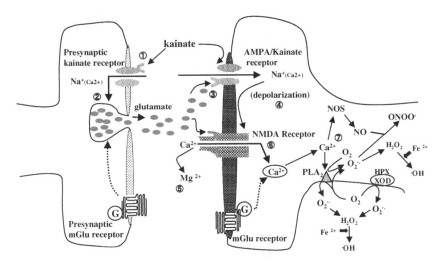

FIGURE 14.3 Possible mechanism for KA-induced neurotoxicity. (1) KA activates the postsynaptic KA receptor and (2) induces Ca^{2+}-dependent release of glutamate. (3) The released glutamate acts on postsynaptic non-NMDA receptors (4) to depolarize the membrane and (5) relieve the voltage-dependent block by Mg^{2+} of NMDA receptor. (6) The excess stimulation of NMDA receptor by glutamate accelerates Ca^{2+} influx and (7) leads to calcium events. These reactions are modulated by stimulation of metabotropic glutamate receptors.

Another important effect of increased Ca^{2+} may be the activation of a protease, which converts xanthine dehydrogenase to xanthine oxidase, and generates $O_2^{\cdot-}$ [30], as described in the following text.

The neurotoxic action of KA is known to also involve metabotropic glutamate receptors (mGluR), which are G-protein-coupled receptors. Usually, mGluRs are divided into three subgroups based on sequence homology of the cloned receptors, group-selective agonists, and second messenger pathways, i.e., Group I: mGluR1 and 5, group II: mGluR2 and 3, and group III: mGluR4, 6, 7, and 8. Activation of group I mGluRs (phosphatidylinositol-linked) increases neuronal excitability and facilitates NMDA-dependent long-term potentiation [31,32], presumably by releases of Ca^{2+} from intracellular stores and potentiation of ionotropic receptors. Moreover, a selective agonist for group I mGluRs induces seizures in mice [6,33], whereas an antagonist for Group I mGluRs limits KA-induced hippocampal dysfunction [34]. An agonist for both group I and group II mGluRs induces seizures and neuronal damage [35–37]. However, mGluRs activation on neurotoxicity may not be uniform but may vary with their subtypes and also with their doses; e.g., L-CCG1, a preferential agonist for group II mGluRs and a lesser agonist for group I and group III mGluRs, induces seizures with high doses, though low doses antagonize 3,5-DHPG-induced seizures in mice [38]. On the other hand, agonists for group II or group III mGluRs have been shown to be anticonvulsive and neuroprotective [38,39]. Low doses of DVG IV, a specific and potent agonist for group II mGluRs, show anticonvulsive and neuroprotective actions in the KA-administered rat, although they are unaffected by higher doses [40]. Pharmacological actions of mGluR agonists and antagonists are of interest from the point of view of anticonvulsants and/or neuroprotective agents for treatment of epilepsy and other neurodegenerative diseases, although a more exact mechanism for the regulation of KA-induced neurotoxicity by mGluRs still remains to be clarified.

14.4 INVOLVEMENT OF FREE RADICALS IN KA-INDUCED NEUROTOXICITY

ROS are known to play a role in KA-induced neuronal damage. Free radical formation was confirmed by Sun et al. [30] using an electron spin resonance (ESR) technique with a spin-trapping agent, phenyl-t-butylnitrone (PBN), for detecting the spin-trapped signal of a carbon-centered radical in the KA-treated gerbil brain. By analyzing thiobarburic acid reactive substances and conjugated diene formation [30], they also observed increased levels of lipid peroxidation in the brain after KA treatment. The presence of hydroxyl radical was demonstrated in cultured rat retinal neurons exposed to KA also by an ESR spin-trapping method [41].

Increase in superoxide dismutase (SOD) [42–44], malondialdehyde (MDA) [42,45], protein oxidation [42], lipid radicals [46], and iron ion [47] in the brain after KA treatment, all suggest that KA could generate ROS in $vivo$.

Generation of $O_2^{\cdot-}$ by KA may occur principally via calcium events induced by excess stimulation of the NMDA receptor [18]. KA is known to activate PLA_2 and to liberate AA. Then, AA generates $O_2^{\cdot-}$ through its subsequent metabolism by lipo-oxygenases and cyclo-oxygenases to form eiconasoids [48]. Otherwise, xanthine dehydrogenase is known to be converted to xanthineoxidase and to generate $O_2^{\cdot-}$ under conditions of energy failure and elevated intracellular Ca^{2+} such as ischemia [49]. KA toxicity in cerebellar neurons suggests the formation of cytotoxic $O_2^{\cdot-}$ [29].

Meanwhile, KA-induced neurotoxicity is prevented by inhibiting xanthine oxidase, a cellular source of cytotoxic $O_2^{\cdot-}$ [29]. KA-induced neurotoxicity can be attenuated also by treatment with synthetic radical scavengers both in $vitro$ and in $vivo$, e.g., 21-aminosteroids [50] and desferrioxamine–manganese complex [51]. Systemically administered antioxidants, such as ascorbate [19], melatonin, vitamin E, and estrogen, show protective effects against KA-induced neuronal damage [52]. SOD transgenic mice, being SOD rich, were more resistant than wild mice to KA-induced damage [53]. These evidences clearly show the involvement of ROS in the mechanism of KA-induced neurotoxicity.

Nitric oxide (NO) has a possible role in convulsive phenomenon, i.e., the excess synthesis and release of NO occurs in the stimulation of the NMDA

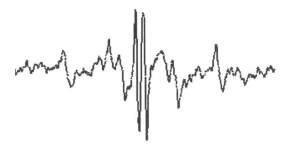

FIGURE 14.4 ESR spectrum of ascorbyl radical detected in the mouse brain after convulsive seizures induced by KA. (Adapted from Reference 74.)

receptor [54,55]. Stimulation of KA receptors is known to activate NO synthesis [56–58]. NO production activates guanylcyclase, and the generated cyclic GMP is related to initiation and propagation of seizures [59]. Excitatory amino acid receptor activation by glutamate or NMDA is known to produce NO and also other ROS, i.e., $O_2^{\cdot-}$, H_2O_2, $\cdot OH$ [60–62]. Nitric oxide synthase (NOS) is able to generate $O_2^{\cdot-}$, H_2O_2, and $\cdot OH$, instead of NO, in the brain subjected to highly excitotoxic conditions [62].

ONOO$^-$ formed from NO and $O_2^{\cdot-}$, is a potent oxidant, and may play a important role in KA-induced neurotoxicity. There are a number of reports showing the proconvulsant effect of NO in KA-induced seizure mechanisms [63–65], and NOS inhibitors protect against KA-induced excitotoxicity [66]. On the other hand, the anticonvulsant effect of NO has been documented in other studies [67–70]. Excess NO generated by accelerated NOS activity inhibits glutamate binding to the NMDA receptor in a negative feedback manner [71,72], i.e., it contributes to the termination of seizures. These conflicting findings may be based on the different experimental conditions used, i.e., experimental animal models, methods for inducing convulsions, methods for administration of NOS inhibitors, stage of seizure observed — preconvulsive stage, during convulsion, or postconvulsive stage.

14.5 ASCORBYL RADICAL GENERATION IN RODENT HIPPOCAMPUS BY KA: A NEW FINDING

Ascorbate (Asc) is generally a physiological antioxidant of major importance in a hydrophilic phase for protection against oxidative stress. Asc is oxidized to ascorbyl radical (Asc$^{\cdot-}$) and then to dehydroascorbate. We observed Asc$^{\cdot-}$ in the circulating blood of living rats, using ESR spectrometry [73]. Recently we successfully demonstrated specific ESR spectra in the hippocampal slices of the rodent after KA-induced seizures by ESR [74]. In this experiment, KA (25 mg/kg body weight) was intraperitoneally injected into male Wistar rats (6 to 8 weeks old) and male ddY mice (6 to 8 weeks old). Animals were killed 15 to 30 min after KA injection. Convulsive seizures were observed before this stage in all experimental animals. The hippocampus was cut out from the brain, and the slices were placed on a specially prepared quartz cell for tissue samples (LTC-10, Labotec, Tokyo) to which 10 μl of DMPO, a spin trapping reagent, was applied without dilution. The ESR spectrum were recorded with an JES FA-200ESR spectrometer (JEOL Ltd., Japan).

The results of the experiment are: no ESR signal of Asc$^{\cdot-}$ was observed in the hippocampal slices of the normal control or before seizure initiation. The typical Asc$^{\cdot-}$ signal was detected clearly in the hippocampal slices after KA-induced seizures. This observation shows obvious intervention of free radicals at the earlier stage of KA-induced neurotoxicity. The ESR signal of Asc$^{\cdot-}$ was detectable in the hippocampal slices only when DMPO was present. This finding suggests that Asc$^{\cdot-}$ may be produced from Asc mainly by reaction with DPMO hydroxyl radical as shown Figure 14.5. In this experiment, the anticonvulsant zonisamide (25 mg/kg body weight), administrated 30 min before KA i.p. injection, completely inhibited the appearance of Asc$^{\cdot-}$. Zonisamide is known to scavenge free radicals and to inhibit NOS activity induced by NMDA [75,76]. The evidence suggests that formation of Asc$^{\cdot-}$ by KA may be inhibited by synergistic effects of zonisamide, i.e., an anticonvulsive effect and an antioxidant effect.

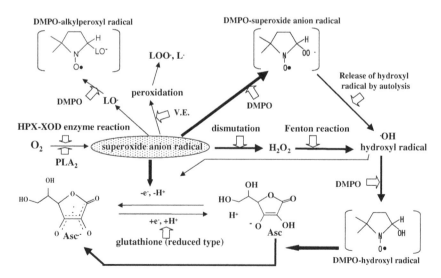

FIGURE 14.5 Possible mechanism for ascorbyl radical formation from ascorbate. Super-oxide anion radical may form DMPO hydroxyl radical in coexisting with DMPO, as shown in the right upper part of the figure, and then reacts with ascorba.

14.6 CONCLUSIONS

KA is an important tool for studying the mechanism of seizures and neurodegeneration related to human degenerative diseases such as epilepsy and Huntington's chorea. Accumulated evidence indicates that not only ROS but also reactive nitrogen species, e.g., NO and ONOO⁻, may participate in the mechanism of KA-induced neurotoxicity, although their detailed functions are not completely elucidated. We observed the typical signal of Asc$^{\bullet-}$ in the hippocampus after KA-induced seizures, using an ESR technique with a spin-trapping reagent, DMPO. Asc is a major endogenous antioxidant against oxidative stress in the hydrophilic phase. Therefore, Asc$^{\bullet-}$ could be an important endogenous indicator of oxidative stress. An easier estimation of Asc$^{\bullet-}$ may be very useful as an oxidative stress marker for studies in the field of free radical biology.

ACKNOWLEDGMENT

The authors thank Dr. Noboru Shoji, associate professor at Tokushima Bunri University, for the valuable information he provided on Tsunematsu Takemoto.

REFERENCES

1. Ben-Ari Y, Lagowska J, Tremblay E, LeGal Lalle G. A new model of focal status epilepticus: intra-amygdaloid application of kainic acid elicits repetitive secondarily generalized convulsive seizures. *Brain Res* 1979; 163: 176–179.
2. Ben-Ari Y. Limbic seizure and brain damage produced by kainic acid: mechanisms and relevance to human temporal lobe epilepsy. *Neuroscience* 1985; 14: 375–403.
3. Nedler JV. Minireview: kainic acid as a tool for the study of temporal lobe epilepsy. *Life Sci* 1981; 29: 2031–2042.
4. Coyle JT, Schwarz R. Lesion of striatal neurons with kainic acid provides a model for Huntington's chorea. *Nature* 1976; 263: 186–246.
5. Tanaka T, Tanaka S, Fujita T, Takano K, Fukuda H, Sako K, Yonemasu Y. Experimental complex partial seizures induced by a microinjection of kainic acid into limbic structures. *Prog Neurobiol* 1992; 38: 317–334.
6. Sperk G. Kainic acid seizures in the rats. *Prog Neurobiol* 1994; 42: 1–32.
7. Choi DW, Rothman SM. The role of glutamate neurotoxicity in hypoxic ischemic neuronal death. *Annu Rev Neurosci* 1990; 13: 171–182.
8. Murakami S, Takemoto T, Shimizu Z. Studies on the effective principles of Digenia simplex Aq. I.: separation of the effective fraction by liquid chromatography. *Yakugaku Zasshi* 1953; 73: 1026–1029.
9. Murakami S, Takemoto T, Shimizu Z, Daigo K. Proposed change of the name "digenic acid" to "kainic acid". *Yakugaku Zasshi* 1954; 74: 560.
10. The List of Achievements by Professor T. Takemoto: celebrating retirement from Tohoku University. Tohoku University, 1976.
11. Takemoto T. Isolation and structural identification of naturally occurring excitatory amino acids. In: McGeer G, Olney JW, McGeer PL, Eds., *Kainic Acid as a Tool in Neurobiology*. New York: Raven Press, 1978: 1–15.

12. Shinozaki H, Konishi S. Actions of several anthelmintics and insecticides on rat cortical neurons. *Brain Res* 1970; 24: 368–371.

13. Olney JW, Rhee V, Ho OL. Kainic acid: powerful neurotoxic analogue of glutamate. *Brain Res* 1974; 77: 507–512.

14. Coyle JT. Neurotoxic actions of kainic acid. *J Neurochem* 1988; 41: 1–11.

15. Ferkany JW, Zaczek R, Coyle JT. Kainic acid stimulates excitatory amino acid neurotransmitter release at presynaptic receptors. *Nature* 1982; 298: 757–759.

16. Connick JH, Stone TW. The effects of quinolinic, kainic and beta-kainic acids on the release of endogenous glutamate from rat brain slices. *Biochem Pharmacol* 1986; 35: 3631–3635.

17. Palmer AM, Reiter CT, Bostscheller M. Comparison of the release of exogenous and endogenous excitatory amino acids from rat cerebral cortex. *Ann N Y Acad Sci* 1992; 648: 361–364.

18. Coyle JT, Puttfarcken P. Oxidative stress, glutamate and neurodegenerative disorders. *Science* 1993; 262: 689–695.

19. MacGregor DG, Higgins MJ, Jones PA, Maxwell WL, Watson MW, Graham DI, Stone TW. Ascorbate attenuates the systemic kainate-induced neurotoxicity in the rat hippocampus. *Brain Res* 1996; 727: 133–144.

20. Berdichevsky E, Riveros N, Sanchez-Ormass S, Orre F. Kainate, N-methyl aspartate and other excitatory amino acid increases in calcium influx into rat brain cortex cells *in vitro*. *Neurosci Lett* 1983; 36: 75–80.

21. Wroblewski JT, Nicoletti F, Costa E. Different coupling of excitatory amino acid receptors with Ca^{2+} channels in primary culture of cerebellar granule cells. *Neuropharmacology* 1985; 24: 919–921.

22. Pastuszko A, Wilson DF, Erecuiska F. Effects of kainic acid in rat brain synaptosomes: the involvement of calcium. *J Neurochem* 1984; 43: 747–754.

23. Lazarewicz JW, Lehman A, Hagberg H, Hamberger A. Effects of kainic acid on brain calcium fluxes studied *in vivo* and *in vitro*. *J Neurochem* 1986; 46: 494–498.

24. Dumuis A, Sebben M, Haynes L, Pin JP, Bockaert J. NMDA receptors activate the arachidonic acid cascade system in striatal neuron. *Nature* 1988; 336: 68–70.

25. Lazarewicz JW, Wroblewski JT, Palmer ME, Costa E. Activation of N-methyl-D-aspartate-sensitive glutamate receptors stimulates arachidonic acid release in primary cultures of cerebellar granule cell. *Neuropharmacology* 1988; 27: 765–769.

26. Farooqui AA, Yi OW, Lu XR, Halliwell B, Horrocks, LA. Neurochemical consequences of kainate-induced toxicity in brain: involvement of arachidonic acid release and prevention of toxicity by phospholipase A(2) inhibitors. *Brain Res Brain Res Rev* 2001; 38: 61–78.

27. Biziere K, Coyle JT. Effects of kainic acid on ion distribution and ATP levels of striate slices incubated *in vitro*. *J Neurochem* 1978; 31: 513–520.

28. Retz KC, Coyle JT. The effect of kainic acid on high energy metabolites in the mouse striatum. *J Neurochem* 1982; 38: 196–203.

29. Dykens JA, Stern A, Trenkner E. Mechanism of kainate toxicity to cerebellar neurons *in vitro* is analogous to reperfusion tissue injury. *J Neurochem* 1987; 49: 1222–1228.

30. Sun AY, Cheng Y, Bu Q, Oldfield F. The biochemical mechanism of the excitotoxicity of kainic acid: free radical formation. *Mol Chem Neuropathol* 1992; 17: 51–63.

31. McGuiness N, Anwyl R, Rowan M. Trans ACPD enhances long-term potensiation in the hippocampus. *Eur J Pharmacol* 1991; 197: 231–232.

32. Behnish T, Reymann KG. Co-activation of metabotropic glutamate and N-methly-D-aspartate receptors is involved in mechanisms of long-term potensiation maintenance in rat hippocampal CA1 neurons. *Neuroscience* 1993; 54: 37–47.

33. Ito I, Kohda A, Tanabe S, Hirose E, Hayashi M, Mitunaga S, Sugiyama H. 3,5-Dihydroxyphenylglycine: a potent agonist of metabotropic glutamate receptors. *Neuro Rep* 1992; 3: 1013–1016.

34. Renaud J, Emond M, Meeilleur S, Psarropoulou C, Carmant L. AIDA, a class metabotropic glutamate-receptor antagonist limits kainate-induced hippocampal dysfunction. *Epilepsy Res* 2002; 43: 1306–1317.

35. McDonald JW, Fix AS, Tizzano JP, Schopp DD. Seizure and brain injury in neonatal rats induced by 1S,3R-ACPD, a metabotropic glutamate receptor agonist. *J Neurosci* 1978; 139: 381–383.

36. Sacaan A, SchoeppDD. Activation of hippocampal metabotropic excitatory amino acid receptors leads to seizures and neuronal damage. *Neurosci Lett* 1992; 139: 77–82.

37. TizzanoJP, Griffe I, Johnson JA, Fix AS, Hilton DO, Schoepp DD. Intracerebral 1S,3R-1-aminocyclopentane-1,3-dicarboxilic acid (1S,3R-ACPD) produces limbic seizures that are not blocked by ionotropic glutamate receptor antagonists. *Neurosci Lett* 1993; 162: 12–16.

38. Tizzano JP, Griffery KI, Schoepp DD. Induction or protection of limbic seizures in mice by mGluR subtype selective agonist. *Neuropharmacology* 1995; 34: 1063–1067.

39. Gereau RW, Conn PJ. Multiple presynaptic metabotropic glutamate receptors modulate excitatory and inhibitory transmission in hippocampal area CA1. *J Neurosci* 1995; 15: 6879–6889.

40. Miyamoto M, Ishida M, Shinozaki H. Anticonvulsive and neuroprotective actions of a potent agonist (DCG) for group metabotropic glutamate receptors against intraventricular kainate in the rat. *Neuroscience* 1997; 77: 131–140.

41. Dutrait N, Culcasi M, Cazevieille C, Pietri S, Tordo P, Bonne C, Muller A. Calcium-dependent free radical generation in cultured retinal neurons injured by kainate. *Neurosci Lett.* 1995; 198: 13–16.

42. Bruce AJ, Braudry M. Oxygen free radicals in rat limbic structures after kainate-induced seizures. *Free Radic Biol Med* 1995; 18: 993–1002.

43. Kim H-C, Bing G, Jhoo W-K, Ko KH, Suh J-H, Kim S-J, Kato K, Hong J-S. Changes of hippocampal Cu/Zn-superoxide dismutase after kainate treatment in the rat. *Brain Res* 2000; 853: 215–226.

44. Liang LP, Ho YS, Patel M. Mitochondrial superoxide production in kainate-induced hippocampal damage. *Neuroscience* 2000; 101: 563–570.

45. Waterfall AH, Singh G, Fry JR, Marsden CA. Detection of the lipid peroxidation product malonaldehyde in the rat brain *in vivo. Neurosci Lett* 1995; 200: 69–72.

46. Ueda Y, Yokoyama H, Niwa R, Konaka R, Ohya-Nishiguchi H, Kamada H. Generation of lipid radicals in the hippocampal extracellular space during kainic acid-induced seizures in rats. *Epilepsy Res* 1997; 26: 329–333.

47. Wang XS, Ong WY, Connor JR. Increase in ferric and ferrous iron in the rat hippocampus with time after kainate-induced excitotoxic injury. *Exp Brain Res* 2002; 143: 137–148.

48. Chan PH, Fishman RA. Transient formation of superoxide radicals in polyunsaturated fatty acid-induced brain swelling. *J Neurochem* 1980; 35: 1004–1007.

49. McCord JM. Oxygen-derived free radicals in postischemic tissue injury. *N Engl J Med* 1985; 312: 159–163.

50. Monyer H, Hartley DM, Choi DW. 21-Aminosteroids attenuate excitotoxic neuronal injury in cortical cell cultures. *Neuron* 1990; 5: 121–126.

51. Bruce A, Najm I, Malfroy B, Baudry M. Effects of desferrioxamine/manganase complex, a superoxide dismutase-mimic, on kainate-induced pathology in rat brain. *Neurodegeneration* 1992; 1: 265–271.

52. Oriz GG, Sanchez-Ruiz MY, Tan D-X, Reiter RJ, Benítez-King G, Beas-Zárate C. Melatonin, vitamine E, and estrogen reduce damage induced by kainic acid in the hippocampus: potassium-stimulated GABA release. *J Pineal Res* 2001; 31: 62–67.

53. Hirata H, Cadet JL. Kainate-induced hippocampal DNA damage is attenuated in superoxide dismutase transgenic mice. *Brain Res Mol Brain Res* 1997; 48: 145–148.

54. Garthwaite J, Garthwaite G, Palmer RM, Moncada S. NMDA receptor activation induces nitric oxide synthesis from arginine in rat brain slices. *Eur J Pharmacol* 1989; 172: 413–416.

55. Dowson VL, Dowson TM, London ED, Bred DS, Snyder SH. Nitric oxide mediates glutamate neurotoxicity in primary cortical cell cultures. *Proc Natl Acad Sci USA* 1991; 88: 6368–6371.

56. Garthwaite J, Southan E, Anderton M. A kinate receptor linked to nitric oxide synthesis from arginine. *J Neurochem* 1989; 53: 1952–1954.

57. Balcioglu A, Maher TJ. Determination of kainic acid-induced release of nitric oxide using a novel hemoglobin trapping technique with microdialysis. *J Neurochem* 1993; 61: 2311–2313.

58. Kashihara K, Sakai K, Marui, Shohmori T. Kainic acid may enhance hippocampal NO generation of awake rats in a seizure stage related fashion. *Neurosci Res* 1998; 32: 189–94.

59. Ferrendelli JA, Blant AC, Gross RA. Relationship between seizure activity and cyclic nucleotide levels in brain. *Brain Res* 1980; 200: 93–103.

60. Gunasekar PG, Kanthasam AG, Borowitz JL, Isom GE. NMDA receptor activation produces concurrent generation of nitric oxide and reactive oxygen species: implication for cell death. *J Neurochem* 1995; 65: 2016–2021.

61. Lafon-Cazal M, Pietri S, Culcasi M, Bockaert J. NMDA-dependent superoxide production and neurotoxicity. *Nature* 1995; 364: 534–537.

62. Lancelot R, Lecanue L, Revaud M-L, Boulu RG, Plotkine M, Callebert J. Glutamate induces hydroxyl radical formation *in vivo* via activation opg nitric oxide synthase in Sprague-Dawley rats. *Neurosci Lett* 1998; 242: 131–134.

63. De Sarro GB, Donato Di Paola E, De Sarro A, Vidal MJ. Role of nitric oxide in the genesis of excitatory amino acid-induced seizures from the deep prepiriform cortex. *Fundam Clin Pharmacol* 1991; 5: 503–511.

64. Mülsch A, Busse R, Mordvintcev PI, Vanin AF, Nielsen EO, Scheel-Krüger J, Olesen S-P. Nitric oxide promotes seizure activity in kainate-treated rats. *Neuro Rep* 1994; 5: 2325–2328.

65. Yasuda H, Fujii M, Fujisawa H, Ito H, Suzuki M. Changes in nitric oxide synthesis and epileptic activity in the contralateral hippocampus of rats following intra hippocampal kainate injection. *Epilepsia* 2001; 42: 13–20.

66. Jones PA, Smith RA, Stone TW. Nitric oxide synthase inhibitor L-NAME and 7-nitroindazole protect rat hippocampus against kainate-induced excitotoxicity. *Neurosci Lett* 1998; 249: 7578.

67. Rondouin G, Bockaert J, Lener-Natoli M. L-nitroarginine, an inhibitor of NO synthase, dramatically worsens limbic epilepsy in rats. *Neuro Rep* 1993; 4: 1187–1190.

68. Pregalinski E, Baran L, Siwanowicz J. The role of nitric oxide in the kainate-induced seizures in mice. *Neurosci Lett* 1994; 170: 74–76.

69. Penix LRP, Davis W, Subramaniam S. Inhibition of NO synthase increases the severity of kainic acid-induced seizures in rodents. *Epilepsy Res* 1994; 18: 177–184.

70. Maggio R, Fumagalli F, Donati E, Barbier P, Racagni G, Corsini GU, Riva M. Inhibition of nitric oxide synthase dramatically potentiates seizures induced by kainic acid and pilocarpine in rats. *Brain Res* 1995; 679: 184–187.

71. Manzoni O, Prezeau L, Martin P, Deshager S, Bockaert L, Fagni L. Nitric oxide-induced blockade of NMDA receptors. *Neuron* 1992; 8: 653–662.

72. Lipton SA. Distinctive chemistries of NO-related species. *Neurochem Int* 1996; 29: 111–114.

73. Wang X, Liu J, Yokoi I, Kohno M, Mori, A. Direct detection of circulating free radicals in the rat using electron spin resonance spectrometry. *Free Radic Biol Med* 1992; 12: 121–126.

74. Masumizu T, Noda Y, Yokoi I, Mori A. Ascorbyl radical generation in the rodent hippocampus during kainate-induced seizures: an electron spin resonance (ESR) study. *Free Radic Biol Med* 2003; 35(Suppl. 1): 159.

75. Mori A, Noda Y, Packer L. The anticonvulsant zonisamide scavenges free radicals. *Epilepsy Res* 1998; 30: 153–158.

76. Noda Y, Mori A, Packer L. Zonisamide inhibits nitric oxide synthase activity induced by N-methyl-D-aspartate and buthionine sulfoximine in the rat. *Res Commun Mol Pathol* 1999; 105: 23–33.

15 Prevention of Cerebral Oxidative Stress Using Traditional Chinese Medicines: A Model of Antioxidant-Based Composite Formula

Tetsuya Konishi
NUPALS, Niigata, Japan

CONTENTS

15.1 INTRODUCTION

Oxidative stress is now accepted as a ubiquitous condition involved in the initiation and progression of many diseases. Free radicals also play a critical role in brain injury caused by stroke or ischemia condition and, thus, radical-scavenging medicines such as Edaravone are clinically used to prevent progression of brain damage after the injury. However, the process leading to neuronal cell death is associated with multiple disturbances of cellular reaction such as cell signaling and damage repair. Therefore, composite therapy will be more appropriate to

189

treat or prevent the brain damage associated with oxidative stress because multiple agents will cooperatively participate in oxidative damage protection by not only scavenging radicals but also manipulating cellular defense or repair processes.

In this sense, traditional Chinese medicines (TCM) will be an attractive model of composite formula. The ancient theories of Chinese medicine recognized that physical health is maintained by the balance of five elements, and the strategy for controlling disease is to restore balance by adding what is deficient and subtracting excess. Although the significance of these elements and their use in combinations have not been clarified in Western terms as yet, TCM prescriptions are usually formulas comprising several herbal components [1].

Moreover, in the ancient theory of TCM, there is one specific pathological condition defined as a predisease condition, that is, in the early stages of what may become a more serious condition. It has been recognized that the treatment of this condition to prevent diseases was more important to treat the specific endpoint diseases. Therefore, the prevention of this predisease condition was the basic strategy of TCM. The predisease condition has not been fully understood by Western medicine but seems related to complex pathological conditions such as autonomic dysfunction, immune suppressive condition, and other disease conditions to which Western medicine fails to give an appropriate diagnostic disease name. Therefore, the TCM strategy against diseases is finding more favor in current preventive medicine (Figure 15.1).

Oxidative stress is a typical reaction caused by various stressors. Because these aforementioned complex conditions are associated with oxidative stress, the predisease condition could be partly defined as an oxidatively stressed condition. We, thus, focused our attention on antioxidant TCM that can be used to treat cerebral oxidative injury [2–4].

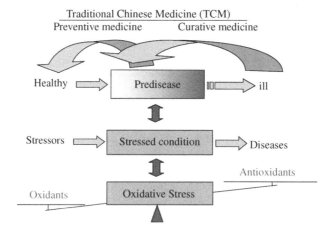

FIGURE 15.1 Predisease condition and oxidative stress as a target of TCM.

TABLE 15.1
Relative Radical-Scavenging Activities
of TCM *In Vitro*

O_2^{-a}	%	•OH[b]	%
0.7 unit SOD	100	20 mM Mannitol	100
SMS	60		208
DTS	132		145
LGSGT	132		90
XMT	131		88
GZFLW + YY	152		127

[a] 10 mg herbs/ml solution.
[b] 100 mg herbs/ml solution.

15.2 ANTIOXIDANT POTENTIAL OF SEVERAL TCM FORMULATIONS PRESCRIBED FOR CONDITIONING BRAIN FUNCTION

The brain is considered one of the organs most sensitive to oxidative abuse because of its high content of oxidizable unsaturated fatty acids, low physiological antioxidant defense level, and high oxygen consumption. We, therefore, studied first the *in vitro* antioxidant potential of several Kampo and TCM formulations prescribed for treating conditions related to brain function. They are sheng mai san (SMS), dia teng san (DTS), ling gui zhu gan tang (LGSGT), xu ming tang (XMT), and bui zhi fu ling wan + yi yi ren (GZFLW + YY). These prescriptions comprise three SMS to nine XMT herbal components but no herb is common to all prescriptions. When hydroxyl and superoxide radical-scavenging activities were measured for these TCMs by spin trapping ESR using DMPO, a characteristic feature of SMS became apparent. SMS showed a rather higher scavenging potential against the hydroxyl radical than the superoxide radical, whereas all other TCMs examined had stronger superoxide scavenging activity than hydroxyl radical (Table 15.1). So we further studied the preventive effect of SMS on cerebral oxidative injury caused by ischemia–reperfusion.

15.3 PREVENTION OF CEREBRAL OXIDATIVE INJURY BY SMS IN RAT [2,4,5]

In order to examine the protective efficiency of SMS against cerebral oxidative injury, rats (6-month-old male Wistar rats) were injected with SMS directly into the jejunum. Two hours after the SMS administration, ischemic condition was given by occluding both carotid arteries for 85 min and, then, blood circulation

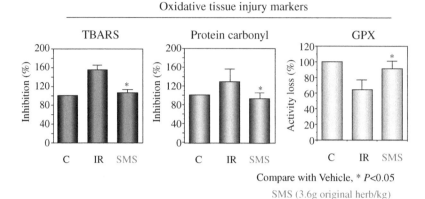

Compare with Vehicle, * P<0.05

SMS (3.6g original herb/kg)

FIGURE 15.2 Prevention of cerebral oxidative damage induced by ischemia–reperfusion in rat by shengmai san. Shengmai San was administered directly into rat duodenum 2 h before ischemia–reperfusion treatment. Cerebral oxidative injury was evaluated by lipid peroxidation, protein carbonyl formation, and glutathione peroxidase loss.

was recovered for a 45 min reperfusion treatment. The brain was removed and homogenized for biochemical measurement of oxidative injury. As shown in Figure 15.2, both lipid peroxidation and protein carbonyl formations were significantly increased after ischemia–reperfusion treatment, but in the rats administered SMS before the ischemia–reperfusion, the increase of these oxidative markers was almost completely suppressed. Glutathione peroxidase (GPX) activity, which was measured as a marker of tissue injury, decreased after ischemia–reperfusion but the enzyme level was found to be maintained as high as in the untreated control for the SMS-treated rats. The protective effect of SMS against cerebral oxidative injury was clearly demonstrated when the brain slices were stained by TTC as shown in Figure 15.3. A large infarction observed in the oxidatively damaged brain almost disappeared in the rats treated with SMS. The

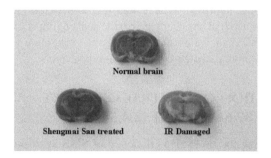

FIGURE 15.3 Protection of cerebral ischemia–reperfusion damage by shengmai san in rat. Brain slices were stained by TTC. Large infarction is observed in the brain damaged after ischemia–reperfusion.

preventive effect of SMS was dose dependent, and the damage was almost completely prevented with the SMS dose being more than 2.4 g of original herbs per rat. The dose-dependent effect was slightly different in TBARS formation and GPX loss. These results clearly showed that SMS demonstrated a rather strong hydroxyl-radical-scavenging activity and can prevent brain damage caused by oxidative stress.

The protective effect of SMS on cerebral oxidative injury was also demonstrated in the mouse model of Parkinson's disease induced by MPTP [5]. By treating mice with MPTP, both lipid peroxidation and protein carbonyl levels were significantly elevated up to 190% of control in midbrain, including substantia nigra and striatum. Such an increase in oxidative injury was not observed in other brain areas such as the cortex. The oxidative injury was effectively prevented by orally administered SMS and LGSGT for 5 d before MPTP injection. At the same time, an immunohistochemical study indicated a significant loss of adrenergic neurons in the substantia nigra. MPTP-induced loss of the adrenergic neurons were also prevented by SMS and LGSGT. These observations strongly suggest that the antioxidant potential of TCMs could be a reliable marker to evaluate their preventive potential against neuronal oxidative injury.

In order to learn if SMS acts as a radical scavenger in the brain, L-band ESR study was carried out using MC-Proxyl radical as a radical spin probe, which becomes ESR-silent when reduced by hydroxyl radicals generated in the brain. The MC-Proxyl radical signal decay in the brain significantly slowed down in the SMS-administered rats indicating that SMS (or its ingredients) was incorporated into the brain and functioned as a radical scavenger.

15.4 RELATIONSHIP BETWEEN *IN VITRO* AND *IN VIVO* ANTIOXIDANT ACTIVITY OF TCM [6]

Because TCM formulations are usually prescribed with several herbal components, complex interactions such as synergism and antagonism are predicted to occur among the component herbs or herbal ingredients. Due to this mixing effect, the therapeutic potential of TCM prescriptions is usually difficult to predict based on the contents of certain marker molecules present in the preparation. It is thus necessary to evaluate their therapeutic potential as a mixture *in vitro*. Because oxidative stress is the basic pathology involved in many disease conditions mentioned previously, antioxidant activity could be a reliable target by which we can evaluate the therapeutic potential of TCM as a mixture. In order to know how the *in vitro* antioxidant activities correlate to the preventive potential of cerebral oxidative injury, we prepared several types of SMS-related decoctions, including those from component herb itself and their combination and determined their antioxidant activity *in vitro* and *in vivo*. Radical-scavenging activity was determined toward hydroxyl radical, superoxide radical, and DPPH radical. Antioxidant activity was also evaluated by crocin bleaching and TBARS assay. In all assays, the antioxidant activity was essentially related to the *in vivo* inhibitory

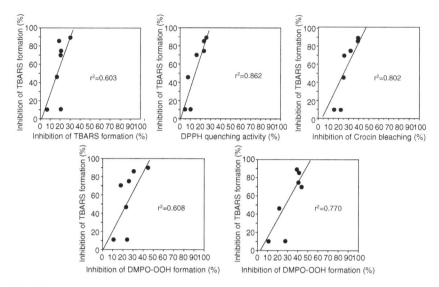

FIGURE 15.4 Correlation of *in vitro* antioxidant activity of shengmai san to *in vivo* protective potential against cerebral oxidative injury in rat. Shengmai san component herbs and their combinations were studied for their antioxidant activity by several *in vitro* assay systems and the activities were correlated to their *in vivo* protective potential of brain oxidative stress.

potential of cerebral oxidative damage caused by ischemia–reperfusion even among different decoctions. (Figure 15.4) Among the antioxidant assays examined, DPPH-radical-scavenging activity showed the best *in vitro–in vivo* correlation, followed by superoxide-radical-scavenging and crocin-bleaching activities. It was also noted that antioxidant activity of SMS was mainly due to that of schisandra. On the other hand, the prevention of GPX loss was attained only by ophiopogon and complete SMS.

15.5 SMS MODULATES CELLULAR ANTIOXIDANT DEFENSE POTENTIAL [7]

Because SMS prevented GPX loss in an oxidatively damaged brain, it was suggested that SMS functions not only as a simple radical scavenger but also as a modulator of cellular antioxidant defense mechanism. When PC12 culture cells as a neuronal cell model were treated with SMS, the cells demonstrated improved cellular resistance against oxidative stress due to hydrogen peroxide when evaluated by Comet assay (DNA fragmentation) and also protein carbonyl formation. This improving effect of SMS on cellular antioxidant resistance was dependent on the time of incubation before hydrogen peroxide abuse (Figure 15.5). The longer the preincubation time, the more the oxidative damage of cellular components and cell viability was reduced. These observations strongly support the

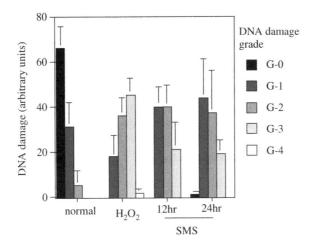

FIGURE 15.5 Protective effect of SMS on H_2O_2-induced DNA damage in PC12 cells. Cells ($5 \times 10^4/cm^2$) were preincubated with 0.66 mg/ml of SMS for either 12 hr or 24 hr and then treated with 0.3 mM H_2O_2 for 30 min. DNA damage was determined by Comet assay.

idea that SMS improved cellular antioxidant defense not only through its strong hydroxyl-radical scavenging activity but also manipulating inherent cellular antioxidant device such as induction of antioxidant enzymes. Multiple gene regulation by SMS was recently demonstrated in culture cells using DNA chip technology. Results showed that several stress-related genes were upregulated by SMS, but the SMS-dependent expression became more marked when the cells were oxidatively stressed, indicating an SMS-manipulated cellular response to the oxidative stress or damage.

15.6 CONCLUSION

SMS is a traditional Chinese herbal medicine prescribed for the treatment of coronary heart diseases. Although the brain was not included in the five element organs, defined in ancient TCM theory, which regulate the physiological activity of humans, brain function was considered to have a close relationship with the activities of heart, kidney, and liver. It was thus interesting to show that SMS was also effective in preventing cerebral oxidative injury. SMS is a formulation comprising three herbal constituents, *Panax ginseng*, *Ophiopogon japonicus*, and *Schisandra synensis*. Usually, in such mixed systems, the component–component interaction such as synergism or antagonism play an important role in their function. In SMS, some synergistic interactions among three components were observed in their antioxidant activity, for instance, the antioxidant activity of schisandra was enhanced when mixed with ophiopogon but was rather suppressed with ginseng. However, it is observed that SMS always showed the strongest antioxidant activity compared to the decoctions prepared by the component herbs and their combinations [6]. Therefore, rationale of the mixed formulation was recognized in terms of the antioxidant activity of SMS. Moreover, the *in vitro* antioxidant activity was essentially related to the *in vivo* efficiency in preventing cerebral oxidative stress. It was further shown that SMS manipulates cellular antioxidant defense mechanism. Although the mechanism of SMS function and the active ingredients involved in it were yet to be clarified, the present studies strongly suggest that the antioxidant-based composite therapy using a multifunctional formula will be a reliable approach for the prevention and treatment of the diseases associated with oxidative stress.

ACKNOWLEDGMENTS

This study was performed by collaboration with H. Nishida, H. Ichikawa, Wang X-J., Wang L., and other lab members. L-band ESR studies were done in collaboration with Lee M. of Kanagawa Dental Medical School. We thank Kotaro Kampo Pharmaceutical Co., Ltd., and Iskra Pharmaceutics Co., Ltd., for providing shengmai san and its component herbal decoctions.

REFERENCES

1. Cheng JT. Drug therapy in Chinese traditional medicine. *J Clin Pharmacol* 2000; 40: 445–450.
2. Wang XJ, Magara T, Konishi T. Prevention and repair of cerebral ischemia-reperfusion injury by Chinese herbal medicine, Shengmai San, in rats. *Free Radic Res* 1999; 31: 449–455.
3. Wang XJ, Ichikawa H, Konishi T. Antioxidant potential of Qizhu Tang, a Chinese herbal medicine, and the effect of cerebral oxidative damage after ischemia-reperfusion in rat. *Biol Pharm Bull* 2001; 24: 558–563.
4. Ichikawa H, Konishi T. *In vitro* antioxidant potentials of traditional Chinese medicine, Shengmai San and the relation to *in vivo* protective effect on cerebral oxidative damage in rats. *Biol Pharm Bull* 2002; 25: 898–903.
5. Xu B. Traditional Chinese medicines protect against MPTP-induced brain oxidative damage in C57BL/6 mice. *NUPALS Master Theses* (Manuscript in preparation) 2002.
6. Ichikawa H, Wang XJ, Konishi T. Role of component herbs in antioxidant activity of Shengmai San, a traditional Chinese medicine preventing cerebral oxidative damage in rats. *Am J Chin Med* 2002; 31: 509–521.
7. Wang L, Nishida H, Ogawa Y, Konishi T. Prevention of oxidative injury in PC12 cells by a traditional Chinese medicine, Shengmai San, as a model of an antioxidant-based composite formula. *Biol Pharm Bull* 2003; 26: 1000–1004.

16 Analysis of the Oxidative Stress State in Disease Development of Senescence-Accelerated Mouse

Seiichi Matsugo and Fumihiko Yasui
University of Yamanashi, Kofu, Japan

Kazuo Sasaki
Toyama University, Toyama, Japan

CONTENTS

16.1 OVERVIEW

We have reported previously that the level of lipid hydroperoxide in the brain tissue was significantly higher in SAMP8 than in SAMR1. To investigate the implication of oxidative stress in the development of learning and memory deficits observed in senescence-accelerated mouse (SAMP8), we analyzed the lipid hydroperoxide components and the amount of each in various brain parts (the olfactory center, septum, hippocampus, cortex, cerebellum, pons, and other parts including thalamus) using the HPLC-luminol chemiluminescence method. The levels of phosphatidylcholine hydroperoxide (PCOOH), phosphatidylethanolamine hydroperoxide (PEOOH), and phosphatidylserine hydroperoxide (PSOOH) in the hippocampus of SAMP8 were significantly higher than in those of SAMR1. In addition, the PCOOH level in the cortex was significantly higher in SAMP8 than in SAMR1. These results suggested that the severe oxidative stress observed in SAMP8 might be related to their learning and memory deficits.

16.2 INTRODUCTION

The SAM model, originally developed by Takeda and colleagues [1], has been used as an animal model for investigation of senescence acceleration and geriatric disorders. Of the many SAM substrains, senescence-accelerated prone 8 mice (SAMP8) exhibit deficits in learning and memory as an age-related disorder at an earlier stage in life [2].

Growing evidence indicates that oxidative stress is a major factor in the development of aging and its related disorders [3–5]. The brain tissue, especially, is vulnerable to oxidative damage, because this tissue contains a high level of polyunsaturated fatty acids (PUFAs). Elevated levels of lipid hydroperoxide or its decomposition products were observed in the brain tissue of SAMP8 [6] as well as patients with neurodegenerative disorders such as Alzheimer's and Parkinson's diseases [7–9]. We also reported previously that the level of lipid hydroperoxide in the brain of SAMP8 was significantly higher than that of SAMR1 after 2 months of age [10,11]. These results suggest the implication of oxidative stress in the development of learning and memory deficits in SAMP8. To understand the precise mechanism underlying the cognitive decline observed in SAMP8, we analyzed the lipid hydroperoxide components, including the amount of each, in various brain parts (the olfactory center, septum, hippocampus, cortex, cerebellum, pons, and other parts including thalamus), using HPLC–luminol chemiluminescence (CL) method.

16.3 MATERIALS AND METHODS

16.3.1 ANIMALS

Male SAMP8 and SAMR1 were caged at room temperature of $24 \pm 2°C$ with artificial light from 07.00 h to 19.00 h, in a 12-hourly light/dark cycle. Food and tap water were allowed *ad libitum*. At three months of age, the mice were used for the present study. The brain tissues of SAMP8 and SAMR1 were immediately dissected out after decapitation and separated into seven parts, comprising the olfactory center, septum, hippocampus, cortex, cerebellum, pons, and other parts including thalamus, at low temperature (0°C). The brain parts were frozen under liquid nitrogen and stored at −196°C until use. All experiments were carried out according to the guidelines for animal experimentation issued by Toyama University.

16.3.2 MEASUREMENT OF LIPID CONTENTS IN THE BRAIN

Each brain part was homogenized with 1.2 ml of PBS(−) using a polytetrafluoroethylene (PTFE)-glass homogenizer. The total lipid in the tissue homogenate

(1 ml) was extracted with 3 ml of $CHCl_3/CH_3OH$ (2:1, v/v) mixture containing 0.01% (w/v) of butylated hydroxytoluene. After centrifugation at 2000 r/min for 2 min, the $CHCl_3$ layer was transferred into a round-bottomed flask. The lipid was further extracted with 2 ml of $CHCl_3$. The combined $CHCl_3$ extract was evaporated at room temperature and the residue was redissolved in $CHCl_3$ (200 µl). An aliquot of the lipid solution (25 µl) was subjected to HPLC–CL analysis. The HPLC–CL system was set up according to the previous report [12]. By measuring the amount of protein in the homogenate with a protein assay kit (Bio-Rad), the level of lipid hydroperoxide was represented as pmol/mg protein. The data are expressed as the mean ± SEM (standard error of the mean). For statistical analysis, Student's or Welch's t-test was used for the detection of the significant difference between SAMP8 and SAMR1. Differences were considered significant at $p < .05$.

16.4 RESULTS AND DISCUSSION

We compared the weight of each brain part (the olfactory center, septum, hippocampus, cortex, cerebellum, pons, and other parts including thalamus) of SAMP8 with that of SAMR1 using Student's t-test, but no significant difference was observed in any of the brain parts.

Subsequently, we measured the levels of phospholipid hydroperoxides (PLOOHs) using the HPLC–CL method under normal-phase conditions. Comparing the retention times of authentic PLOOH samples with those of brain tissue samples, we detected PCOOH, PEOOH, and PSOOH in all brain parts of both SAMP8 and SAMR1. The levels of all these PLOOHs were markedly high in the olfactory center, septum, and hippocampus, whereas they were very low in the pons and other parts including thalamus. Also, in the cortex and cerebellum, the PLOOH levels were appropriate. In the olfactory center, the amounts of PCOOH, PEOOH, and PSOOH of SAMP8 were 128.97 ± 56.71, 1518.21 ± 589.04, and 192.25 ± 81.73 pmol/mg protein, respectively, whereas those of SAMR1 were 79.03 ± 32.61, 766.93 ± 337.87, and 103.23 ± 39.31 pmol/mg protein, respectively. The Student's t-test revealed no significant difference between SAMP8 and SAMR1 in any of the PLOOH components. In the septum, the amounts of PCOOH, PEOOH, and PSOOH of SAMP8 were 356.02 ± 151.78, 3351.45 ± 1047.139, 681.55 ± 281.47 pmol/mg protein, and those of SAMR1 were 150.00 ±68.04, 1977.96 ± 762.80, and 287.90 ± 110.72 pmol/mg protein. The Student's t-test also revealed no significant difference between SAMP8 and SAMR1 in any of the PLOOH components. In the hippocampus, the amounts of PCOOH, PEOOH, and PSOOH of SAMP8 were 84.36 ±16.13, 859.41 ± 236.48, and 74.02 ± 19.92 pmol/mg protein, whereas those of SAMR1 were 15.14 ± 5.72, 126.07 ± 44.89, and 11.60 ± 4.67 pmol/mg protein. We compared the levels of all PLOOHs of SAMP8 with those of SAMR1, using Student's t-test for PCOOH and Welch's t-test for PEOOH and PSOOH. The levels of PCOOH, PEOOH, and PSOOH in the hippocampus of SAM8 were significantly higher than those of SAMR1 ($**p < .01$ for PCOOH and PEOOH, $*p < .05$ for PSOOH). In the cortex, the amounts of PCOOH, PEOOH, and PSOOH of SAMP8 were

12.99 ± 3.04, 118.56 ± 57.64, and 6.24 ± 3.79 pmol/mg protein, whereas those of SAMR1 were 2.37 ±1.52, 23.05 ± 17.31, 0.91 ± 0.91 pmol/mg protein. The Student's t-test also revealed a significant difference in PCOOH, but not in PEOOH and PSOOH, between the two substrains. In the cerebellum, the amounts of PCOOH, PEOOH, and PSOOH of SAMP8 were 18.36 ± 7.13, 106.67 ± 48.33, and 7.35 ± 4.35 pmol/mg protein, and the PCOOH and PEOOH levels of SAMR1 were 5.37 ± 1.46 and 32.87 ± 13.22 pmol/mg protein, but the PSOOH level of SAMR1 was lower than the detection threshold of the HPLC–CL method. Welch's t-test did not reveal a significant difference in the three PLOOHs between SAMP8 and SAMR1. In the pons, the amounts of PCOOH, PEOOH, and PSOOH of SAMP8 were 2.34 ± 0.99, 16.82 ± 9.11, and 0.96 ± 0.96 pmol/mg protein, and those of SAMR1 were 13.7 ± 5.57, 78.24 ± 27.61, and 2.75 ± 1.32 pmol/mg protein. Neither Welch's nor Student's t-test showed a significant difference in the three PLOOHs between SAMP8 and SAMR1. In the residual parts including thalamus, the amounts of PCOOH, PEOOH, and PSOOH of SAMP8 were 3.09 ± 1.76, 6.08 ± 3.19, and 0.57 ± 0.34 pmol/mg protein, and those of SAMR1 were 1.19 ± 0.42, 1.74 ± 0.57, and 0.21 ± 0.21 pmol/mg protein. Neither Welch's nor Student's t-test showed a significant difference in the three PLOOHs between SAMP8 and SAMR1. Thus, we successfully detected the significant elevated level of PLOOHs in the hippocampus of SAMP8 using the HPLC–CL method; in other brain parts, these PLOOH levels — except the PCOOH level in the cortex — did not differ between SAMP8 and SAMR1. In addition, massive investigations reported the impairment of hippocampal function in SAMP8 [13]. Considering that the hippocampus is a major brain part involved in the learning and memory functions and that the cortex is an important brain part to store long-term memory, the aggravation of oxidative damage in the hippocampus and cortex of SAMP8 might play a crucial role in the development of cognitive decline in SAMP8.

REFERENCES

1. Takeda T, Hosokawa M, and Higuchi K. Senescence-accelerated mice (SAM): a novel murine model of accelerated senescence. *J Am Geriatr Soc* 1991; 39: 911–919.
2. Miyamoto M, Kiyota Y, Yamazaki N, Nagaoka A, Matsuo T, Nagawa Y, and Takeda T. Age-related changes in learning and memory in the senescence-accelerated mouse (SAM). *Physiol Behav* 1986; 38: 48–66.
3. Levine RL, Garland D, Oliver CN, Amici A, Climent I, Lenz AG, Ahn BW, Shaltiel S, and Stadtman ER. Determination of carbonyl content in oxidatively modified protein. *Methods Enzymol* 1994; 186: 464-487.
4. Mandavilli BS, Santos JH, and Van Houten B. Mitochondrial DNA repair and aging. *Mutat Res* 2002; 509: 127-151.
5. Townsend DM, Tew KD, and Tapiero H. The importance of glutathione in human disease. *Biomed Pharmacother* 2003; 57: 145–155.

6. Okatani Y, Wakatsuki A, Reiter RJ, and Miyahara Y. Melatonin reduces oxidative damage of neural lipids and proteins in senescence-accelerated mouse. *Neurobiol Aging* 2002; 23: 639–644.
7. DiCiero Miranda M, de Bruin VM, Vale MR, and Viana GS. Lipid peroxidation and nitrite plus nitrate levels in brain tissue from patients with Alzheimer's disease. *Gerontology* 2000; 46: 179–184.
8. Castellani RJ, Perry G, Siedlak SL, Nunomura A, Shimohama S, Zhang J, Montine T, Sayre LM, and Smith MA. Hydroxynonenal adducts indicate a role for lipid peroxidation in neocortical and brainstem Lewy bodies in humans. *Neurosci Lett* 2002; 319: 25–28.
9. Sayre LM, Zelasko DA, Harris PL, Perry G, Salomon RG, and Smith MA. 4-Hydroxynonenal-derived advanced lipid peroxidation end products are increased in Alzheimer's disease. *J Neurochem* 1997; 68: 2092–2097.
10. Matsugo S, Kitagawa T, Minami S, Esashi Y, Tokumaru S, Kojo S, Oomura Y, and Sasaki K. Age-dependent changes in lipid peroxide levels in peripheral organs, but not in brain, in senescence-accelerated mice. *Neurosci Lett* 2000; 278: 105–108.
11. Yasui F, Ishibashi M, Matsugo S, Kojo S, Oomura Y, and Sasaki K. Brain lipid hydroperoxide level increases in senescence-accelerated mice at early age. *Neurosci Lett* 2003; 350: 66–68.
12. Yamamoto Y, Brodsky MH, Baker JC, and Ames BN. Detection and characterization of lipid hydroperoxides at picomole levels by high-performance liquid chromatography. *Anal Biochem* 1987; 160: 7–13.
13. Armbrecht HJ, Boltz MA, Kumar VB, and Morley JE. Effect of age on calcium-dependent proteins in hippocampus of senescence-accelerated mice. *Brain Res* 1999; 842: 287–293.

Section III

Nutraceuticals, Functional Foods, Micronutrients, and Pharmacological Interventions

17 The Antioxidant Evolution: From Free Radical Scavenging to the Antioxidant Network and Gene Regulation by Flavonoids and Bioflavonoid-Rich Extracts from Pine Bark and Gingko Biloba Leaf

Lester Packer
University of Southern California, Los Angeles

CONTENTS

17.1 INTRODUCTION

Healthy aging and disease prevention by antioxidant nutrition is increasingly a subject of academic and public health interest and research. This chapter will provide an overview of oxidative stress and the antioxidant defense system. Antioxidants play an important role in maintaining the physiological redox status of cellular constituents. Antioxidants may quench free radicals, change their redox state, be targeted for destruction, regulate oxidative processes involved in signal transduction, and affect gene expression and pathways of cell proliferation, differentiation, and death. The action of antioxidants that interact with the redox antioxidant network will be highlighted, especially the bioflavonoid-rich botanical extracts from pine bark (Pycnogenol®) and *Gingko biloba* leaves (EGb 761).

17.2 FREE RADICALS AND NATURAL ANTIOXIDANTS

The chemical definition for free radicals (oxidants) is "atoms or groups of atoms with one or more unpaired electrons." This property makes them very unstable and highly reactive, trying to capture the needed electron from other compounds to gain stability. When the "attacked" molecule loses its electron, it becomes a free radical itself, beginning a chain reaction. Some free radicals, such as the toxic oxyradical species hydroxyl radical ($^{\bullet}OH$) or the less reactive superoxide radical ($O_2^{\bullet-}$), arise normally during mitochondrial oxidative metabolism. Also, the body's immune system cells purposefully generate free radicals to neutralize viruses and bacteria. Environmental factors such as pollution, radiation, cigarette smoke, and herbicides can also generate free radicals. Therefore, biological systems are continuously interacting with free radicals arising either from metabolism or from environmental sources, leading to a process called oxidation.

Free radicals have a precise chemical definition. However, the word antioxidant refers to a broad range of substances that have the ability to neutralize free radicals by donating one of their own electrons i.e., *antioxidation*. They act as scavengers, helping to prevent cell and tissue damage. Antioxidants display many different properties in biological systems. Consequently, it is difficult to find a precise description for the term *antioxidant* and many definitions abound.

In general, to be called antioxidant, a compound must be able to donate an electron and/or hydrogen atom and prevent or delay oxidation of an oxidizable substrate. They can act in different ways: by metal chelation (preventing free radical formation), scavenging free radicals, acting as chain-breakers (stopping propagation of the free radicals), being part of the redox antioxidant network, and regulating gene expression.

The major natural antioxidants from dietary sources include vitamins C (ascorbate) and E, polyphenols and bioflavonoids (redox-active substances that can be oxidized and reduced), and carotenoids, which act as free radical sinks. Other biofactors important in antioxidant defense are lipoic acid, co-enzyme Q10, and the various metals (micronutrients) essential for the activity of antioxidant enzymes, such as selenium, copper, zinc, manganese, and iron. The process of evolution has also equipped us with enzymes such as superoxide dismutases, catalases, glutathione, and thioredoxin peroxidases that neutralize or prevent the formation of free radicals.

When polyphenolic antioxidants, such as vitamin E or epigallocatechin gallate, donate an electron or hydrogen atom to quench a free radical, they become free radicals. However, because the unpaired electron is delocalized by resonance around the aromatic ring, this confers greater stability and, thus, less reactivity to natural antioxidants. Thus, they are less dangerous than the free radicals they have destroyed. The property of resonance stability by the free radical form of antioxidants is also exhibited by vitamin E (four different forms of tocopherols and tocotrienols) and vitamin C. For general references on antioxidants, see Reference 1 to Reference 5.

17.3 FLAVONOID ANTIOXIDANT DEFENSE MECHANISMS

Most studies have been with *in vitro* or *ex vivo* systems. Direct free radical scavenging action and metal chelating action preventing free radical formation have been extensively studied with *in vitro* systems. Often this is not relevant to their actions *in vivo* where little is known about bioavailability, tissue concentration, distribution, and metabolism [6].

The mechanisms most likely to account for effects of flavonoids *in vivo* direct reaction of flavonoids on molecular targets with high specificity, e.g., receptors, cell signaling pathways, transcription factors, and genes. Animal feeding studies with high-density oligonucleotide arrays demonstrate the presence of target genes, proteins, and enzymes [7].

The induction of phase 2 enzymes is of great importance in antioxidant defense, detoxification, and cancer prevention. Evidence of this *in vivo* response is from cell, animal, and human studies with Phase-2 conjugation enzymes including glutathione (as in GSH S-tranferases), glucuronic acid (as in UDP-glucuronosyltransferases), methylation (as with methyl transferases), and sulfation (as sulfotranferases).

In these studies flavonoid aglycones were found more effective antioxidants than their conjugated forms (e.g., quercetin, as compared with its conjugate quercetin-3-glucoside) [8].

17.4 OXIDATIVE STRESS AND THE ANTIOXIDANT NETWORK

Oxidants and antioxidants must be kept in balance to minimize molecular, cellular, and tissue damage because if the balance is upset in favor of the former, oxidative stress occurs. The term *oxidative stress* was first coined by Helmut Sies in 1986, referring to the imbalance that arises when exposure to oxidants changes the normal redox status of major tissue antioxidants, especially glutathione, the cell's primary preventative antioxidant [9,10]. Glutathione is usually present in the tissues in millimolar amounts in the aqueous compartments of cells and their organelles. Under normal conditions, glutathione exists primarily in its reduced form, as GSH. However, upon exposure to oxidants generated during flux through the respiratory chain, (cytochrome P450) electron transport reactions metabolism of foreign compounds, ligand–receptor interaction, immune system activation, or exposure to environmental oxidants, GSH is oxidized to glutathione disulfide (GSSG), thus changing the redox status of the glutathione system. This change in the reduced/oxidized ratio of glutathione is an example of oxidative stress. Recovery of the oxidative imbalance requires NADPH- and NADH-dependent reactions of reducing metabolism.

Oxidative stress often results in oxidative damage. This can arise from body metabolism, strenuous and/or traumatic exercise, exposure to environmental stressors such as ultraviolet irradiation and cigarette smoke, as well as from infection (microorganisms, viruses, parasites, etc.) and, of course, during aging. Molecular markers of oxidative damage to lipids (e.g., isoprostanes, age pigment, or lipofuscin), proteins (e.g., carbonyl and nitrotyrosine derivatives of protein), and DNA (products of DNA fragmentation and oxidized bases such as 8-hydroxy-2-deoxyguanosine) accumulate during aging and, therefore, tissues of aged individuals are more susceptible to disease. In addition, mild oxidative stress affects cell signaling pathways and thus gene expression.

All the redox-based antioxidants appear to interact with one another through the "antioxidant network" [11] consisting of nonenzymatic and enzymatic reactions (Figure 17.1). In this way, antioxidants are recycled or regenerated by biological reductants. For example, bioflavonoids and polyphenols interact with the vitamin C radical, thus lengthening its lifetime.

Oxidative stress can weaken the entire antioxidant system. Redox-active antioxidants from food or food supplements bolster antioxidant defenses and help in preventing oxidative damage. When redox antioxidants act and become free radicals, they are more reactive and can react with other radicals or one another and are thus depleted from the system. Thus, continuous replenishment of antioxidants from food or supplements is necessary. Carotenoids, though not redox-based antioxidants, also play an important role in bolstering the antioxidant defense, acting as sinks for quenching free radicals such as lipid peroxyl radicals. Presence of a double-bond system in their hydrocarbon chain enables them to take many hits from free radicals and to destroy them.

17.5 PHASE-2 ENZYME INDUCTION BY BIOFLAVONOID EXTRACTS

All cells possess elaborate antioxidant defense systems that consist of interacting micronutrients, enzymes, and other molecules. Oxidants as hydroperoxides (H_2O_2, CH_3OOH, $ROOH$), many natural antioxidant substances, and a broad range of chemicals reacting with sulfydryl groups induce phase-2 enzymes. Phase-2 enzymes protect against oxidants and electrophilic toxicity, i.e., they aid

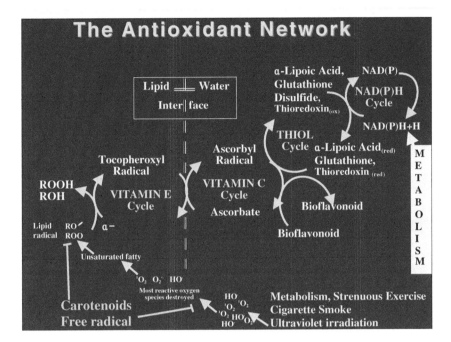

FIGURE 17.1 The redox antioxidant network.

antioxidant defense, detoxification, and cancer prevention. Phase-2 enzyme induction leads to a series of coordinated increases in the gene expression and the enzymatic activity of antioxidant defense system components [12].

Examples of phase-2 enzymes, proteins, and products important in antioxidant defense include glutamate-cysteine ligase, heme oxygenase-1, Mn superoxide dismutase, NAD(P)H: quinone reductase, dihydrodiol dehydrogenase, expoxide hydrolase, leukotriene B4 dehydrogenase, aflatoxin B1 dehydrogenase, and the major intracellular iron-binding protein ferritin. Upregulation of these systems result in the protection against reactive oxygen species and electrophile toxicity and therefore are very important in antioxidant defense, detoxification, and cancer prevention (Figure 17.2).

Evidence for phase-2 enzyme induction is readily observed from changes in the activity and levels of products of phase-2 enzyme action in cultured cell systems. Glutathione, the cell's primary preventative antioxidant, is often present in high concentration in cells and tissues, and may increase after being exposed to oxidative stress conditions and inducers of phase-2 enzymes. This results in synthesis of γ glutamate cysteinyl ligase (γGCL) a rate-limiting enzyme for glutathione biosynthesis. Cellular glutathione is present in the aqueous compartments, and thiolation reactions can lead to substantial reservoirs of protein-bound glutathione as conjugates. The latter, upon dethiolation can replenish aqueous glutathione levels. Often, enzyme proteins that participated in thiolation or dethiolation do not have their enzymatic activity altered by glutathione conjugation.

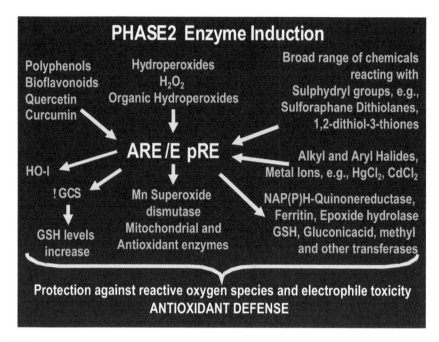

FIGURE 17.2 Phase-2 enzyme induction.

17.6 RESPONSES OF BRAIN GENES TO EGb 761

Herbal medicines and dietary supplements are widely used for health benefits, but the molecular basis of their therapeutic potential is often poorly defined. EGb 761 is a highly standardized bioflavonoid-rich preparation whose components interact in a redox antioxidant network (Figure 17.3). Because free radicals and other oxidants are known to modulate important cell signaling systems, it can be predicted that flavonoid-rich botanical extracts will exhibit gene modulatory activity. We hypothesized that *in vitro* and *in vivo* assays that allow quantitative analysis of gene expression profiles combined with targeted biochemical analysis could help identify their functional effects. This hypothesis was tested by application of high-density oligonucleotide microarrays to define mRNA expression in cell cultures [15–18] and brain tissue after EGb 761 treatment [7].

In vitro assays of gene expression analysis of botanical extracts do not account for bioavailability and metabolism. Therefore, an *in vivo* study was undertaken in normal mice to define the potential transcriptional activity of EGb 761 [7]. Several reports describe *in vivo* effects of orally administered EGb 761 in rodents [19,20] and humans [21,22].

The cortex and hippocampus were chosen as the target sites for gene expression analysis of ~12,000 mouse genes (affymetrix and murine genome Mu74Av2). Mice were fed a diet supplemented with 1 mg of EGb 761 per day [7]. Analysis

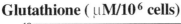

Glutathione ($\mu M/10^6$ cells)

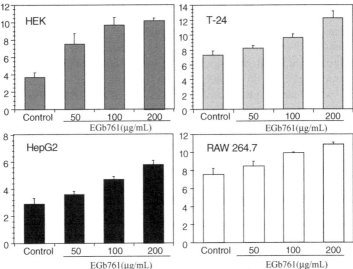

FIGURE 17.3 Glutathione levels are elevated in four different cell types after being treated in culture with the extract from *Gingko biloba* leaves (EGb 761). Data shown for the concentration-dependent increase of glutathione in HEK Human, keratinocytes, T-24 human bladder cells, HepG2 liver cells, and RAW murine macrophages. (Data from References 13 and 14.)

of the expression of mouse genes showed that the expression of 43 genes and 13 genes were upregulated at least twofold in the cortex and hippocampus, respectively. These observations provide definitive molecular evidence for the presence of neuroactive substances derived from EGb 761 in the brain.

In the hippocampus and cortex of mice supplemented with EGb 761, the gene responses showed significant increase in transcripts that are known to encode brain proteins vital in neuronal and synaptic plasticity [7].

17.7   RESPONSES OF HUMAN SKIN KERATINOCYTE GENES TO PYCNOGENOL [23]

Pine bark has been used for centuries as an herbal remedy. During the 15th century, pine decoctions were used for wound healing, as noted by Minner in 1497 in *Thesaurus Medicaminum*. Maritime North American Indians were the first reported to have used pine bark (decoction of white pine bark) to treat Jacques Cartier's crew from scurvy during the winter of 1535. Maritime Indians used various barks to treat wounds and skin sores [24]. The French maritime pine (*Pinus maritime*) bark extract (Pycnogenol) is a mixture particularly rich in oligomeric procyanidins and other bioflavonoids such as taxifolin, catechin, and epicatechin. The French maritime pine grows on the weather beaten sand dunes of the Bay of Biscay in southwestern France. Pine bark extracts have become popular as dietary supplements. Besides their antioxidant functions, their other biological properties are now being characterized.

Genes are the Rosetta stone of human health and disease. Thus, gene expression analysis or genomics studies using recently developed complementary deoxyribonucleic acid (cDNA) arrays help in identifying markers of disease, therapeutic targets, and potential pharmacological activities [25]. Using this approach, the effects of pycnogenol on the gene expression profile of the human keratinocyte (HaCat) cell line were investigated.

Numerous flavonoids, from kinases to transcription factors have been reported to affect cellular signaling processes. In fact, oligomeric procyanidins from pine bark were found to prevent the activation of the proinflammatory transcription factor NF-κB on ultraviolet (UV) exposure [26]. The same procyanadins are also potent modulators of nitric oxide metabolism [27,28].

To understand the protective mechanism afforded by Pycnogenol-supplemented cells, their basal gene expression profile was determined and compared to that of nonsupplemented cells [29]. As expected, only a small proportion (83 genes) of the 588 genes analyzed were detected in either group. However, of these 83 genes, 39 genes showed an expression that was significantly (more than twofold) increased or decreased (Table 17.1).

Interestingly, a group of overexpressed genes is involved in stress response. Thioredoxin-peroxidase-2 plays an important role in eliminating peroxides generated during cellular metabolism and signaling cascades. This effect should be related to the increase of glutathione level in keratinocytes (Figure 17.3) by Pycnogenol observed by Rimbach and colleagues [30]. The ultraviolet (UV) excision repair protein (HHR23B) is involved in nucleotide excision repair of DNA damage. Finally, the heat shock protein HSP70 is generally expressed in response to stress. However, antioxidants such as curcumin have been observed to induce its expression, thereby increasing cell resistance to stress [31]. The expression of the inhibitor for helix-loop-helix protein ED-3 is known to be dysregulated in keratinocyte cell lines such as HaCat [32]. The increased dynein-light-chain-1 (dlc1) expression points to the regulation of intracellular trafficking through cytoskeleton interaction.

17.8 CONCLUSION

In the future, the use of genomic and proteonomic techniques offer attractive opportunities to explore and unravel the molecular basis of complex mixtures from natural sources in regard to their action on human health.

Studies using food factors, purified natural antioxidant substances, botanical or herbal extracts, and traditional medicines containing complex mixtures of substances can now be analyzed without resorting, at the outset, to fractionate

TABLE 17.1
Genes Overexpressed or Downregulated in Pycnogenol-Treated Human Keratinocytes

GeneBank	Genes Overexpressed in PBE-Supplemented Cells	P/C
X69111	Inhibitor for helix-loo-helix protein D-3	10.34
X67643	Heat shock 70-kD protein 1	5.70
M11886	HLA class 1 histocompatibilit antigen C-4 alpha chain	4.92
D21090	UV excision repair protein HHR 236	4.69
U32944	Cytoplasmic dynein light chain 1	4.56
X67951	Thioredoxin peroxidase 2	3.93
M36429	Transducin beta-2	3.68
X00351	Beta-actin	2.00
X53587	Beta4-integrin	1.70
K00558	Alpha-tubulin	1.60
	Housekeeping Genes	
X56932	23 kDa highly basic protein	1.10
	Genes Downregulated in Control Cells	
M22489	Bon inorphogenic protein 2A	0.08
X01060	CD71, transferin receptor protein	0.07
X59798	Cyclin D1 (G/S specific)	0.07
D13866	Alpha-catenin	0.07
M74088	Adenomatous polyposis coli protein	0.05
X06234	Calgrenulin A	0.04
X06233	Calgranulin B	0.04
M28372	Cellular nucleic acid binding protein	0.04

Note: P/C: ratio of relative abundancy of a given mRNA from Pycnogenol-supplemented cells or its equivalent from control cells: $C/P = 1/CP^{-1}$.

and determinate molecular structures. Natural products' chemistry can now make use of the combined tools of functional genomics and proteonomics to investigate the molecular basis of lifestyle-related diseases, behavior, and cognitive function. Future studies investigating the effects of complex botanical extracts and preparations will provide guidance and direction toward identifying the nature of the active components, thus complementing natural product chemistry analysis.

It is essential for optimal human health that critical and objective analysis of the *in vivo* effects of dietary factors, antioxidant supplements, herbal and traditional medical preparations and, indeed, pharmacological interventions be investigated. With modern techniques using systems cells, animal models, and human studies as models, it is now possible to obtain global insight of their effects on the response of the entire human genome. In this way, we can establish the safety and efficiency of their use for human consumption.

REFERENCES

1. Packer L and Yodoi J, Eds., *Redox Regulation of Cell Signaling and its Clinical Application*. The Oxidative Stress and Disease Series, Eds., L. Packer and E. Cadenas, 1999; Marcel Dekker: New York.
2. Sen CK, Sies H, Baeurerle P, Eds., *Antioxidant and Redox Regulation of Genes*, 1999; Academic Press.
3. Cadenas E and Packer L. *Handbook of Antioxidants*. The Oxidative Stress and Disease Series, Eds., L. Packer and E. Cadenas, 2001; Marcel Dekker: New York.
4. Packer L, Traber M, Kraemer K, Frei B. *The Antioxidant Vitamins C and E*, 2002; AOCS Press: Champaign, IL.
5. Rice-Evans CA and Packer L, Eds., *Flavonoids in Health and Disease,* 2nd ed. Revised and Expanded. The Oxidative Stress and Disease Series, Eds., L. Packer and E. Cadenas, 2003; Marcel Dekker: New York, pp.1–467.
6. Azzi A, Davies K, Kelly F. Free radical biology — terminology and critical thinking. *FEBS Lett*, 2003; (558): pp. 3–6.
7. Watanabe CM, Wolffram S, Ader P, Rimbach G, Packer L, Maguire JJ, Schultz PG, Gohil K. The *in vivo* neuromodulatory effects of the herbal medicine ginkgo biloba. *Proc Natl Acad Sci USA*, 2001; 98(12): pp. 6577–6580.
8. Shimoi K, Mochizuki M, Tomita I, Kaji K, Kuruto R, Nozawa R, Kumazawa S, Terao J, Nakayama T. Vascular permeability and functional activity of quercetin conjugates. In *1st International Conference on Polyphenols and Health*, 2003; Vichy: France.
9. Sies H. Biochemistry of oxidative stress. *Angew Chem Int,* ed., 1986; 25: pp. 1058–1071.
10. Sies H. Strategies of antioxidant defense. *Eur J Biochem*, 1993; 215: pp. 213–219.
11. Packer L. Vitamin E is nature's master antioxidant. *Sci Am, science and medicine*, 1994; 1: pp. 54–63.
12. Talalay P. Chemoprotection against cancer by induction of phase 2 enzymes. *Biofactors*, 2000; 12(1–4): pp. 5–11.
13. Rimbach G, Gohil K, Matsugo S, Moini H, Saliou C, Virgili F, Weber S, Packer L. Induction of glutathione synthesis in human keratinocytes in ginkgo biloba. *Biofactors*, 2001; 15: pp. 39–52.
14. Rimbach G, Wolffram S, Watanabe CG, Packer L, Gohil K. Effect of ginkgo biloba (EGb 761) on differential gene expression. *Pharmacopsychiatry*, 2003; 36: pp. S95–S99.
15. Gohil K, Moy R, Farzin S, Maguire JJ, Packer L. mRNA Expression profile of a human cancer cell line in response to ginkgo biloba extract: induction of antioxidant response and the golgi system. *Free Radic Res*, 2000; 33: pp. 831–849.
16. Gohil K, Packer L. Ginkgo biloba extract and gene expression, in *Micro Nutrients and Health: Molecular Biological Mechanisms*, Eds., K. Nesaretnam and L. Packer, 2001; AOCS Press: Champaign, IL. pp. 217–224.
17. Gohil K, Packer L. Bioflavonoid-rich botanical extracts show antioxidant and gene regulatory activity. *Ann NY Acad Sci*, 2002; 957: pp. 1–8.
18. Gohil K, Packer L. Global gene expression analysis identifies cell and tissue specific actions of ginkgo biloba extract, EGb 761. *Cell Mol Biol*, 2002; 48: pp. 625–531.

19. Hoyer S, Lannert H, Noldner M, Chatterjee S. Damaged neurol energy metabolism and behavior are improved by ginkgo biloba extract (EGb 761). *J Neural Transm*, 1999; 106: pp. 1171–1188.

20. Stoll S, Scheuer K, Pohl O, Muller W. Ginkgo biloba extract (EGb 761) independently improves changes in passive avoidance learning and brain membrane fluidity in the aging mouse. *Pharmacopsychiatry*, 1996; 29: pp. 144–149.

21. Cesarani A, Meloni F, Alpini D, Barozzi S, Verdorio L, Boscani P. *Ginko biloba* (EGb 761) in the treatment of equilibrium disorders. *Adv Ther*, 1998; 15: pp. 291–304.

22. Itil T, Eralp E, Tsambis E, Itil K, Stein U. Central nervous system effects of ginkgo biloba, a plant extract. *Am J Ther*, 1996; 3: pp. 63–73.

23. Rihn B, Saliou C. Gene modulation of HaCaT cells induced by pine bark extract, in *Flavonoids in health and disease*, Eds., C. Rice-Evans and L. Packer, 2003; Marcel Dekker: New York.

24. Chandler F, Freeman L, Hooper S. Herbal remedies of the maritime Indians. *J Ethnopharmacol*, 1979; 1: pp. 49–68.

25. Gerhold D, Rushmore T, Caskey C. DNA chips: promising toys have become powerful tools. *Trends Biochem Sci*, 1999; 24: pp. 168–173.

26. Saliou C, Rimbach G, Moini H, McLaughlin L, Hosseini S, Lee J, Watson RR, Packer L. Solar ultraviolet-induced erythema in human skin and nuclear factor-kappa-B-dependent gene expression in keratinocytes are modulated by a French maritime pine bark extract. *Free Radic Biol Med*, 2001; 30(2): pp. 154–160.

27. Packer L, Rimbach G, Virgili F. Antioxidant activity and biologic properties of a procyanidin-rich extract from pine (*Pinus maritima*) bark, pycnogenol. *Free Radic Biol Med*, 1999; 27: pp. 704–724.

28. Park YC, Rimbach G, Saliou C, Valacchi G, Packer L. Activity of monomeric, dimeric, and trimeric flavonoids on NO production, TNF-alpha secretion, and NF-kappaB-dependent gene expression in RAW 264.7 macrophages. *FEBS Lett*, 2000; 465(2–3): pp. 93–97.

29. Rihn B, Saliou C, Bottin MC, Keith G, Packer L. From ancient remedies to modern therapeutics: pine bark uses in skin disorders revisited. *Phytother Res*, 2001; 15(1): pp. 76–78.

30. Rimbach G, Virgili F, Park YC, L. P. Effect of procyanidins from *Pinus maritima* on glutathione levels in endothelial cells challenged by 3-morpholinosydnonimine or activated macrophages. *Redox Rep*, 1999; 4(4): pp. 171–177.

31. Sood A, Mathew R, Trachtman H. Cytoprotective effect of curcumin in human proximal tubule epithelial cells exposed to shiga toxin. *Biochem Biophys Res Commun*, 2001; 283: pp. 36–41.

32. Langlands K, Down G, Kealey T. ID proteins are dynamically expressed in normal epidermis and dysregulated in squamous cell carcinoma. *Cancer Res*, 2000; 60: pp. 5929–5933.

18 New Horizons in Vitamin E Research

Etsuo Niki
National Institute of Advanced Industrial Science and
Technology, Ikeda, Japan

CONTENTS

18.1 INTRODUCTION

Vitamin E was discovered in 1922 as an essential dietary factor required for reproduction. The role and action of vitamin E as a radical-scavenging antioxidant have been the subject of extensive studies and is well documented. New findings, observations, and features are emerging even today. In this brief overview, several issues will be discussed.

18.2 UPTAKE, TRANSFER, DISTRIBUTION, AND METABOLISM (1–3)

Vitamin E is absorbed together with lipids in the intestine, packed into chylomicrons, and transported to the liver. This process is similar for all eight forms of vitamin E, but hepatic α-tocopherol transfer protein (α-TTP) not only specifically sorts out the α-form of all tocopherols (T) and tocotrienols (T3) but also has a preference for 2R-stereoisomers and tocopherols over 2S-stereoisoforms and tocotrienols. α-TTP facilitates RRR-α-T incorporation into very low density lipoproteins, which deliver it to peripheral cells. Thus, RRR-α-T is the predominant form of vitamin E found in plasma and all tissues except the liver, despite the high intake of γ-T. The plasma levels of vitamin E after supplementation cease to increase at around 80 μM despite increasing dosages of up to 800 mg per day.

217

It has been also found that free and esterified α-tocopherol have the same bioavailability (4).

Furthermore, a family of cellular tocopherol associated proteins (TAPs) has been identified (5). It is suggested that TAP functions in intracellular tocopherol traffic. It has been also reported that TAP is a cytosolic squalene transfer protein and enhances cholesterol biosynthesis (6).

Most of the ingested β-, γ-, and δ-T and T3 are either metabolized quickly and secreted into bile or are not taken up and are excreted in the feces (7). 2(2'-Carboxyethyl)-6-hydroxychroman (CEHC) has received much attention recently as a metabolite of vitamin E and its function as an endogenous natriuretic factor. Dietary supplements can have adverse effects. Unlike other fat-soluble vitamins, vitamin E is not accumulated in the liver to toxic levels, suggesting efficient metabolism and excretion (8).

18.3   ROLE AND ACTION OF VITAMIN E ISOFORMS

Vitamin E has eight major isoforms, that is, α-, β-, γ-, and δ-T and T3. As mentioned earlier, the specific α-TTP mediates the selective transfer of α-TTP into lipoproteins, and accordingly α-TTP is found most prevalently in plasma and tissues. The specific function and role of other isoforms of vitamin E have been recognized recently. Among others, γT has long been considered to be of minor importance compared to αT, but it has been found that γT may have certain effects that αT does not (9). These functions include the control of nitric-oxide-related toxicity, inhibitory effects on platelet aggregation and thrombogenesis, prostate cancer cell proliferation, and cyclo-oxygenase activity. It has been also proposed that T3 may have a greater effect against some specific oxidative stress than the corresponding T. There are some conflicting reports on the relative antioxidant activities of T and T3. This is partly because the antioxidant activities have been assessed under different conditions and for different activities. It was confirmed recently that T3 and the corresponding T have the same chemical reactivities toward free radicals (10). It has been also found that T3 is more readily incorporated into cultured cells than the corresponding T (11–13), but T3 is more rapidly metabolized *in vivo* (1). Thus, it is understandable that the apparent antioxidant activities depend on the substrates, methods, and experimental conditions. For example, the relative antioxidant activities of T and T3 vary markedly when assessed using the concentrations added to the cell culture medium and using those taken into the cells.

18.4   TOCOPHEROL-MEDIATED PEROXIDATION (TMP)

In 1992, Bowry, Ingold, and Stocker (14) reported that α-T might act, under certain circumstances, as a pro-oxidant against LDL oxidation, which is accepted as an important initial event in the pathogenesis of atherosclerosis. They interpreted this pro-oxidant action of αT, named tocopherol-mediated peroxidation (TMP), using phase-transfer and chain-transfer mechanisms. αT carries aqueous radicals into low-density lipoprotein (LDL) particles and propagates chain oxidation through the hydrogen atom abstraction from lipids by αT radical. Stocker and his colleagues have claimed that this effect is important in the LDL oxidation induced by various oxidants.

That αT acts as a pro-oxidant under certain circumstances has been known for years, and it has been also known that vitamin C inhibits such a pro-oxidant action of αT. In fact, although the pro-oxidant action of αT can be observed against isolated LDL oxidation or oxidation of plasma after removal of vitamin C; such an action has not been observed as expected in the presence of vitamin C. Therefore, the pro-oxidant action of αT by TMP should not be important *in vivo*.

18.5 NONANTIOXIDANT FUNCTION

Another interesting and important feature of vitamin E is its function as a cellular signaling molecule. It has been shown that αT inhibits cell proliferation, platelet aggregation, monocyte adhesion, and the oxygen burst in neutrophils; these effects were due to the inhibition of protein kinase C (15). Interestingly, βT *per se* is ineffective, but it prevents αT from the inhibition of PKC, suggesting that these functions are independent of antioxidant properties. The effects of αT on various cellular reactions and growth inhibition on different cell lines are summarized in Table 18.1 and Table 18.2, respectively, of Reference 15. The physiological significance of these functions should be elucidated in the future studies.

18.6 EFFECTS AGAINST ATHEROSCLEROSIS

The oxidation hypothesis that the oxidative modification of LDL plays a pivotal role in the progression of atherosclerosis has been widely accepted. In fact, numerous *in vitro* and *in vivo* studies have shown that oxidized LDL is uptaken by macrophages through scavenger receptors, causing cholesterol accumulation and foam cell formation in the subendothelial space. Furthermore, oxidized LDL has various proatherogenic effects such as expression of various cytokines, adhesion molecules and chemokines, and stimulation of smooth-muscle cell proliferation. This oxidation hypothesis implies that antioxidants should be effective in protecting against atherosclerosis. The results of many studies both *in vitro* and *in vivo* support the beneficial role of antioxidants, but in some animal models the negative effects of antioxidants have been reported. Furthermore, large-scale, double-blind intervention studies have not always shown consistent efficacy of antioxidants. These inconsistent results have cast doubts on the oxidation hypothesis.

However, such inconsistent results may not be surprising, because LDL oxidation *in vivo* must be induced by several oxidants and proceed by several mechanisms and the efficacy of antioxidants depends on the type of oxidants and

TABLE 18.1
The Growth Inhibitory Effect of α-Tocopherol on Different Cell Lines

Reaction	Proposed Mechanism[a]
Inhibition of cell proliferation	NA
Inhibition of platelet adhesion and aggregation	NA/ND/A
Inhibition of cell adhesion	NA/ND/A
Inhibition of ROS[b] in monocytes and neutrophils	NA/A
Inhibition of PKC	NA/A
Activation of PP$_2$A	NA
Inhibition of 5-lipoxygenase	NA/A
Activation of diacylglycerol kinase	NA/A
Inhibition of α-tropomyosin expression	NA
Inhibition of liver collagen α1 expression	ND
Inhibition of collagenase MMP1 expression	NA
Modulation of α-TTP expression	NA
Inhibition of scavenger receptor SR-A	NA
Inhibition of scavenger receptor CD36	NA
Inhibition of ICAM-1 and VCAM-1 expression	ND

[a] A, antioxidant, NA, nonantioxidant; ND, not discussed.

TABLE 18.2
Effect of RRR-α-Tocopherol and Their Supposed Molecular Mechanisms

Sensitive Lines	Insensitive Lines	Tissue and Origin
A10		Rat aorta smooth muscle
A7r5′		Rat aorta smooth muscle
T/G		Human aorta smooth muscle
NB2A		Mouse neuroblastoma
Balb/3T3		Mouse fibroblast
Human fibroblast		Primary cell lines
DU-145, PC-3		Human prostate cancer
	Cells	
LNCaP		Human Prostate Cancer (androgen sensitive)
Human leukaemia		U937
Mouse fibroblast		Balb/c-3T3
Glioma		C6
	P388-Dl	Mouse monocyte macrophage
	LR73	Chinese hamster ovary
	Saos-2	Human osteosarcoma
	Human hepatocarcinoma	HepG2
	Human colon adenocarcinoma	CaCo2

mechanisms (16). Vitamin E, for example, which is the antioxidant most frequently used to verify the oxidation hypothesis, can function as an antioxidant against free radical-mediated oxidation, whereas it is not capable of inhibiting oxidation by nonradical mechanisms.

It has been shown that LDL oxidation is induced by various oxidants and by different mechanisms. They can be characterized by two distinct features — radical and nonradical oxidants and mechanisms. Different types of free radicals generated from cells are capable of inducing chain oxidation of polyunsaturated lipids to give modified LDL. On the other hand, oxidants such as lipoxygenase and hypochlorite oxidize LDL, lipids, and apoB by nonradical mechanism. It is known that aldehydes modifies apoB by nonradical mechanism. Aldehydes may also be generated by radical and nonradical mechanisms. Most likely, the relative importance of radical and nonradical oxidation may vary with occasion and local areas.

Various antioxidants function *in vivo* on different occasion by different mechanisms. As stated earlier, the radical-scavenging antioxidants such as vitamin E are effective against free-radical-mediated oxidation but not against nonradical oxidation. For example, vitamin E is not capable of inhibiting the oxidation of amino groups by hypochlorite to give chloramines, which break down to aldehydes. Vitamin E does not inhibit 15-cipoxygenase-induced LDL oxidation. The lipid peroxidation of cholesteryl esters and phospholipids in LDL particles by 15-cipoxygenase in plasma does not always proceed by enzymatic, specific mechanisms and some oxidation may proceed by free radical mechanisms. However, vitamin E cannot inhibit much of LDL oxidation by 15-lipo-oxygenase (18).

It is important to understand that vitamin E and other radical-scavenging antioxidants are not effective in inhibiting some of the LDL oxidation and that the rates of inhibitable oxidation and hence the efficacy of antioxidants as well may vary with subjects and occasions.

The antioxidants that suppress LDL oxidative modification should be effective in inhibiting or at least delaying atherosclerosis. However, because atherosclerosis is a complex disease that develops slowly over decades and that the oxidative modification may be induced by various oxidants and proceed by various mechanisms *in vivo*, antioxidants with different functions may be required in sufficient quantity and quality to inhibit the progression of atherosclerosis completely. This may indicate that any particular single antioxidant may not be good enough to suppress atherosclerosis. Thus, vitamin E, in spite of its potent radical-scavenging activity and also its antiatherogenic properties not related to antioxidant function (19), may not be able to inhibit atherosclerosis and related diseases completely. The efficacy of vitamin E supplementation has been argued for several years. It may depend on the mechanisms of pathogenesis and also on the stage of disease, diet, and lifestyle. The difficulty in obtaining unequivocal results in clinical trials of antioixidant supplementation involving a large number of subjects is easily understood. Although the pro-oxidant action of vitamin E may not be completely ruled out, in view of the very low risk of reasonable supplementation with vitamin E some supplementation especially from the early stage may be prudent (20).

REFERENCES

1. Brigelius-Flohe R, Traber MG. Vitamin E: function and metabolism. *FASEB J* 1999; 13: 1145–1155.
2. Hosomi A, Arita M, Sato Y, Kiyose C, Ueda T, Igarashi O, Arai H, Inoue K. Affinity for alpha-tocopherol transfer protein as a determination of the biological activities of vitamin E analogs. *FEBS Lett* 1997; 409: 105–108.
3. Kiyose C, Muramatsu R, Kameyama Y, Ueda T, Igarashi O. Biodiscrimination of alpha-tocopherol stereoisomers in humans after oral administration. *Am J Clin Nutr* 1997; 65: 785–789.
4. Burton GW, Ingold KU, Foster DO, Cheng SC, Webb A, Hughes L, Lusztyk E. Comparison of free α-tocopherol and α-tocopheryl acetate as sources of vitamin E in rats and humans. *Lipids* 1988; 23: 834–840.
5. Stocker A, Azzi A. Tocopherol-binding proteins: their function and physiological significance. *Antioxid Redox Signal* 2000; 2: 397–404.
6. Shibata N, Arita M, Misaki Y, Dohmae N, Takio K, Ono T, Inoue K, Arai H. *Proc Natl Acad Sci USA* 2001; 98: 2244–2249.
7. Drevon CA. Absorption, transport and metabolism of vitamin E. *Free Radic Res Commun* 1991; 14: 229–246.
8. Traber MG. Vitamin E, nuclear receptors and xenobiotic metabolism. *Arch Biochem Biophys* 2004; 423: 6–11.
9. Jiang Q, Christen S, Shigenaga MK, Ames BN. γ-Tocopherol, the major form of vitamin E in the U.S. diet, deserves more attention. *Am J Clin Nutr* 2001; 74: 714–722.
10. Yoshida Y, Niki E, Noguch N. Comparative study on the action of tocopherols and tocotrienols as antioxidant: chemical and physical effects. *Chem Phys Lipids* 2003; 123: 63–75.
11. Sen CK, Khanna S, Roy S, Packer L. Molecular basis of vitamin E action: tocotrienol potently inhibits glutamate-induced pp60[s-Src] kinase activation and death of HT4 neuronal cells. *J Biol Chem* 2000; 275: 13049–13055.
12. Noguchi N, Hanyu R, Nonaka A, Okimoto Y, Kodama T. Inhibition of THP-1 cell adhesion to endothelial cells by α-tocopherol and α-tocotrienol is dependent on intracellular concentration of the antioxidants. *Free Radic Biol Med* 2003; 34: 1614–1620.
13. Saito Y, Yoshida Y, Akazawa T, Takahashi K, Niki E. Cell death caused by selenium deficiency and protective effect of antioxidants. *J Biol Chem* 2003; 278: 39428–39434.
14. Bowry VW, Ingold KU, Stocker R. Vitamin E in human low-density lipoprotein: when and how this antioxidant becomes a pro-oxidant. *Biochem J* 1992; 288: 341–344.
15. Azzi A, Ricciarelli R, Zingg J-M. Non-antioxidant molecular functions of α-tocopherol (vitamin E). *FEBS Lett* 2002; 519: 8–10.
16. Niki E. Antioxidants and atherosclerosis. *Biochem Soc Trans* 2004; 32: 156–159.
17. Hazell LJ, Stocker R. α-Tocopherol does not inhibit hypochlorite-induced oxidation of apolipoprotein B-100 of low-density lipoprotein. *FEBS Lett* 1997; 414: 541–544.
18. Noguchi N, Yamashita H, Hamahara J, Nakamura A, Kühn H, Niki E. The specificity of lipoxygenase-catalyzed lipid peroxidation and the effects of radical-scavenging antioxidants. *Biol Chem* 2002; 383: 619–626.

19. Azzi A, Breyer I, Feher M, Ricciarelli R, Stocker A, Zimmer S, Zingg JM. Nonantioxidant functions of α-tocopherol in smooth muscle cells. *J Nutr* 2001; 131: 378S–381S.

20. Pryor WA. Vitamin E and heart diseases: basic science to chemical intervention trials. *Free Radic Biol Med* 2000; 28: 141–164.

19 Gastrointestinal Inflammatory Diseases: Role of Chemokine

Toshikazu Yoshikawa and Yuji Naito
Kyoto Prefectural University of Medicine, Kyoto, Japan

CONTENTS

19.1 INTRODUCTION

Esophagogastrointestinal inflammation is a highly complex biochemical protective response to cellular or tissue injury. When this process occurs in an uncontrolled manner, the result is excessive cellular or tissue damage, which results in chronic inflammation and destruction of normal tissue. Current evidence suggests that *Helicobacter pylori* infection and nonsteroidal antiinflammatory drug (NSAID) ingestion are major causative factors in the pathogenesis of gastric mucosal injury in humans, and chemical agents such as gastric acid, bile acids, and pancreatic protease also cause esophageal inflammation. However, the cause of inflammatory bowel disease is unknown. In response to *H. pylori* infection, NSAID, or chemical agents, neutrophils are recruited to the site of inflammation and generate reactive oxygen and nitrogen species and proteases. Extravascularly migrated neutrophils infiltrate the region around bacteria and target cells, depending on the concentration of the chemoattractants, including interleukin (IL)-8, and take actions advantageous to the body, such as killing bacteria and cancer cells; they may also injure normal cells and tissue. It has been shown that the interaction between leukocytes and vascular endothelial cells is regulated by various cell adhesion molecules and that this interaction is directly or indirectly modified by many factors including inflammatory chemokines and reactive oxygen species (ROS). In addition to oxidative stress induced by ROS, our recent findings indicate that gas-like mediators play a crucial role in the regulation of gastrointestinal inflammation. The inhibition of inducible nitric oxide synthase attenuates esophagogastrointestinal inflammation. This paper describes the potential role of chemokine and activated neutrophils in esophagogastrointestinal inflammation induced by chemical agents, *H. pylori*, NSAID, and inflammatory bowel disease.

19.2 INFLAMMATORY RESPONSE IN GASTROESOPHAGEAL REFLUX DISEASE (GERD)

The pathophysiology of GERD involves the contact of esophageal epithelium with gastric/duodenal juice in the refluxate. Recently, several studies have shown that mucosal immune and inflammatory responses, characterized by specific cytokine and chemokine profiles, may determine the diversity of esophageal phenotypes of GERD. Fitzgerald et al. (1) first reported that reflux esophagitis is characterized by an acute inflammatory response with significantly increased levels of the proinflammatory cytokines (IL-1β, IL-8, and interferon-γ) compared with noninflamed squamous esophagus. Isomoto et al. (2) have also demonstrated that the presence of intraepithelial neutrophils and eosinophils, which also indicate reflux esophagitis, is associated with high levels of IL-8 and regulated on activation normal T-cell expressed and presumably secreted (RANTES), respectively. Furthermore, they have shown that the IL-8 levels are significantly decreased after proton pump inhibitor treatment. Their data indicate that chemokine production locally in the esophageal mucosa may be involved in the development and progression of reflux esophagitis (Figure 19.1). The importance of gastric acid in the development of esophageal inflammation is also supported by a recent study using an experimental esophagitis model in rats (3). Hamaguchi

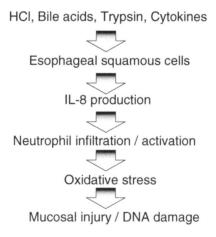

FIGURE 19.1 Chemokine, neutrophil activation, and inflammation in gastroesophageal reflux disease.

et al. showed that (3) treatment with rabeprazole, a proton pump inhibitor, almost completely inhibits development of chronic acid reflux esophagitis and significantly decreases expression of growth related oncogene (GRO)/cytokine-induced neutrophil chemoattractant-2α (CINC-2α), a rat IL-8-like chemokine, in esophageal tissue, compared with control. We have recently investigated the relationship between the IL-8/monocyte chemoattractant protein 1 (MCP-1) mRNA expression and endoscopic grading of reflux esophagitis according to the Los Angeles classification (4). The expression of IL-8 mRNA determined by the real-time PCR correlated with the endoscopic severity of GERD and increased in endoscopic-negative GERD compared to normal subjects. There is no correlation between the MCP-1 mRNA expression and endoscopic severity or between severity of subjective symptoms (QUEST score) and endoscopic grading. Kanazawa et al. (5) also reported higher expression levels of IL-8 mRNA in esophageal mucosa of patients with nonerosive reflux disease (NERD) than those in asymptomatic controls, suggesting that IL-8 is implicated in the pathogenesis of NERD.

Judging from immunohistochemical staining (1), there was some esophageal epithelial expression of IL-8. In order to confirm the epithelial cell layer as a potential source of cytokines, we have determined whether cultured human esophageal epithelial cells (HEEC) produce IL-8, and to identify molecular mechanism involved in IL-8 production (6), HEEC were isolated from normal esophageal mucosa by Shimada et al. (7) (Kyoto University). Stimulation of HEEC with cholic acid (CA) or taurochenodeoxycholic acid (TCDA) resulted in IL-8 production; CA, especially, induced rapid and enhanced expression of IL-8 mRNA. CA also induced p38 MAPK phosphorylation, as determined by Western blotting, and FR167653, a specific inhibitor of p38 MAPK, significantly inhibited IL-8 production elicited by CA. These results indicate that bile acids induced IL-8 production from HEEC via p38 MAPK pathways, which may be involved in the pathogenesis of reflux esophagitis. In addition to proinflammatory properties, IL-8 is known to be linked to angiogenesis initiation and metastasis and is capable of suppressing apoptosis. Further examination will be needed to clarify the role of IL-8 expression of esophageal squamous cells in the progression of GERD or in the development of Barrett's esophagus.

Recently, we investigated whether blockade of pancreatic protease could attenuate the severity of reflux esophagitis induced by the gastroduodenal juice (8). With male Wistar rats, esophago-gastroduodenal anastomosis is accomplished by anastomosing the jejunum to the gastroesophageal junction under diethyl-ether inhalation anesthesia. Four and eight weeks after surgery, gastroduodenal reflux led to esophageal erosions and ulcer formation, and marked thickening of the esophageal wall. Histological study showed thickness of esophageal mucosa, hyperplasia of the epidermis and basal cells, ulcer formation, and marked infiltration of inflammatory cells. Because inducible forms of cyclooxygenase and nitric oxide synthase, COX-2 and iNOS, have been implicated in the pathogenesis of esophageal inflammation, the expression of these genes was evaluated. iNOS/COX-2 mRNA and protein expression was markedly enhanced in the esophagus 4 and 8 weeks after surgery. Treatment with camostat mesilate, an oral

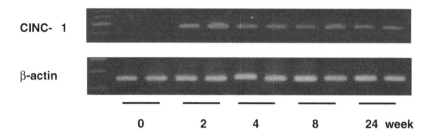

FIGURE 19.2 Cytokine-induced neutrophil attractant-1 (CINC-1) mRNA expression in a chronic esophagitis model induced by gastroduodenal reflux in rats.

protease inhibitor, significantly inhibited both ulcerative changes and mucosal hyperplasia. The enhanced expression of COX-2 and iNOS was also markedly inhibited in the camostat mesilate-treated group. In addition to these genes, the expression of cytokine-induced neutrophil attractant-1 (CINC-1), a rat IL-8-like chemokine, was also enhanced in this esophagitis model (Figure 19.2), and this enhancement was markedly inhibited by chronic treatment with camostat mesilate. These results indicate that trypsin, mainly inhibited by camostat mesilate, plays an important role in mucosal damage and in the expression of inflammation-associated genes, including chemokines, in chronic esophagitis induced by gastroduodenal reflux in rats.

19.3   *H. PYLORI* INFECTION AND NSAID-INDUCED GASTRIC MUCOSAL INJURY

Both *H. pylori* and NSAID use are well-established risk factors for gastrointestinal mucosal injury. Recent experimental studies, including those from our laboratory, have indicated that neutrophil adherence to the endothelium via various adhesion molecules are involved in the development of gastric mucosal injury induced by *H. pylori* infection or NSAID use (9–12). We previously reported that *H. pylori* and NSAIDs cause neutrophils to express adhesion molecules followed by accumulation of neutrophils in the gastric mucosa (11,12). Activated neutrophils have been suggested as injuring endothelial and epithelial cells by producing ROS and proteases (13,14). In a randomized, controlled trial, Taha et al. (15) demonstrated that gastric neutrophils associated with *H. pylori* infection increased the incidence of ulceration in long-term NSAID users. These findings suggest that *H. pylori* and NSAIDs can elicit an acute inflammatory response in the gastric mucosa, leading to neutrophil-mediated tissue injury.

In 1996, Hirayama and colleagues (16,17) described a model in which Mongolian gerbils infected with *H. pylori* developed pathological changes in the stomach that mimicked those seen in humans who harbor the bacteria. These changes included a high incidence of gastritis after 6 weeks of infection and gastric ulceration after 6 months of infection. Using these animals, we recently found that *H. pylori* infection potentiates aspirin-induced gastric mucosal injury

(18). Three weeks after inoculation with *H. pylori*, aspirin (400 mg/kg) suspended in 0.8 ml of 0.25% carboxymethylcellulose (CMC) and 0.1 N HCl were administered orally to gerbils fasted for 18 h. Hemorrhagic erosions in the gastric mucosa caused by aspirin were much more severe in gerbils with *H. pylori* infection than in uninfected gerbils. Whereas minimal neutrophil infiltration resulted when aspirin alone was administered, neutrophil accumulation was more prominent when *H. pylori* infection was also present. Furthermore, aggravation of gastric mucosal lesions induced by the combination of aspirin and *H. pylori* infection was significantly inhibited in neutrophil-depleted gerbils. These findings indicate that the effects of aspirin on the gastric mucosa may be potentiated by *H. pylori* infection via neutrophil-dependent mechanisms.

Accumulation of neutrophils in *H. pylori*-infected human gastric mucosa is related to increased concentrations of IL-8 released by gastric epithelial cells (19,20). In our study, we measured KC, a mouse IL-8-like neutrophil chemoattractant in gastric mucosa. KC is considered to be both a potent chemoattractant and an upregulator of CD11b/CD18 cell surface expression in rodent neutrophils. We found that gastric mucosal KC content in gerbils exposed to *H. pylori* alone was somewhat greater than in mucosa exposed to aspirin alone. Furthermore, in gerbils treated with aspirin after inoculation with *H. pylori*, the KC content in gastric mucosa was significantly greater than in gastric mucosa of gerbils that received only aspirin. These findings suggest that increased KC content may be involved in accumulation of neutrophils in the gastric mucosa, and that administration of aspirin to gerbils three weeks after *H. pylori* inoculation produced severe gastric mucosal injury via marked infiltration of neutrophils (Figure 19.3).

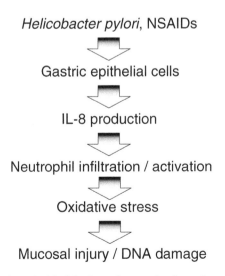

FIGURE 19.3 Role of interleukin 8 in the pathogenesis of gastric mucosal injury induced by *Helicobacter pylori* and nonsteroidal antiinflammatory drugs (NSAID).

19.4 INDUCIBLE NITRIC OXIDE SYNTHASE IN INTESTINAL INFLAMMATION

It has been demonstrated that nitric oxide (NO) production as well as the expression of inducible NO synthase (iNOS) in the intestinal mucosa appear to be enhanced in active human inflammatory bowel disease (21–24), in gut inflammation induced by exogenous agents (25–27), or in the spontaneous colitis observed in genetically engineered rodents (28,29). Immunohistochemical studies with an iNOS antibody has shown that iNOS is expressed intensively in the inflamed colonic epithelium of patients with ulcerative colitis and Crohn's disease by polymorphonuclear leukocytes that have migrated into the colonic lumen and formed crypt abscesses, and in a limited number of mononuclear cells of the inflamed lamina propria (22,24). Thus, NO synthesis may be augmented in inflammatory bowel disease and, therefore, may influence the disease process. One mechanism for NO-induced cytotoxicity operates through the interaction of NO with superoxide to produce peroxynitrite, an oxidizing agent that initiates lipid peroxidation, sulfhydryl oxidation of proteins, and nitration of aromatic amino acids (Figure 19.4). Nitrotyrosine immunoreactivity, an index of nitrosative stress, has been observed not only in the lamina propria but also in the surface epithelium of colitis models (30,31) and in patients with inflammatory bowel disease (22,24).

We have clearly demonstrated that NO is overproduced, accompanied by the increase in iNOS mRNA expression, during the development of dextran sulfate sodium (DSS)-induced colitis (27). A DSS-colitis model used in the present study has been proven to have histological relevance to human ulcerative colitis, including focal crypt lesions, mucosal and submucosal inflammation, and granulocyte infiltration (32). We have also reported that treatment with the potent iNOS inhibitor ONO-1714 ameliorated colonic mucosal injury induced by DSS in mice (27). ONO-1714, a novel cyclic amidine analogue, inhibits human iNOS with a Ki of 1.88 nM and, rodent iNOS with similar potency *in vitro* (33). When the inhibitory activity of ONO-1714 (IC50 of 4.0 nM) is compared for mouse iNOS,

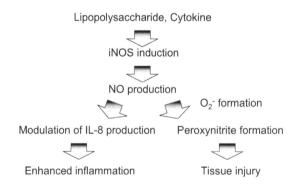

FIGURE 19.4 Role of nitric oxide (NO) produced by inducible NO synthase (iNOS) in intestinal inflammation.

it is found to be 875-fold and 4900-fold more potent than L-NMMA and aminoguanidine, respectively (34). In terms of human iNOS selectivity, ONO-1714 is approximately 34- and 2-fold more selective than L-NMMA and aminoguanidine, respectively. Therefore, ONO-1714 represents a potentially beneficial tool for clarifying the role of iNOS in disease status and may have considerable therapeutic potential as a novel drug.

One mechanism by which the overproduction of NO may promote inflammatory injury is via its ability to mediate chemotaxis of both granulocytes and monocytes. A previous report suggests that NO may enhance IL-8 release by cultured human endothelial cell line, indicating that NO may enhance neutrophil chemotaxis via an IL-8-dependent mechanism (35). Inhibition of NOS activity has been shown to inhibit chemotaxis by neutrophils and monocytes *in vitro*. In our data, the increase in colonic myeloperoxidase (MPO) activity, an indicator of neutrophil infiltration, was significantly inhibited by treatment with ONO-1714. This is consistent with the results of a study by Hogaboam (36) in which colonic MPO was reduced in trinitrobenzenesulfonic acid sodium (TNBS)-colitis rats orally treated with L-NAME, but contrasts with the results obtained by Southey (37). In the latter study, an iNOS inhibitor, S-(2-aminoethyl)isothiouromium bromide (ITU), did not affect the elevated colonic MPO activity in these rats. In addition, a recent study using mice lacking a functional gene for iNOS (iNOS$^{-/-}$) demonstrated that similar high concentrations of MPO activity were found in TNBS-treated iNOS$^{-/-}$ mice as in iNOS$^{+/+}$ mice (31). The reason for iNOS inhibition showing opposing effects in TNBS- and DSS-induced colitis models is not clear. Further studies in DSS-induced colitis in knockout animals or TNBS-induced colitis treated with a selective iNOS inhibitor ONO-1714 will help resolve these issues.

ACKNOWLEDGMENTS

This work was supported by grants-in-aid for scientific research (14570493 YN and 15390178 TY) from the Ministry of Education, Culture, Sports, and Science and Technology of Japan and grants from Bio-Oriented Technology Research Advancement Institution and the Ministry of Agriculture, Forestry, and Fisheries of Japan.

REFERENCES

1. Fitzgerald RC, Onwuegbusi BA, Bajaj-Elliott M, Saeed IT, Burnham WR, Farthing MJ. Diversity in the oesophageal phenotypic response to gastro-oesophageal reflux: immunological determinants. *Gut* 2002; 50: 451–459.
2. Isomoto H, Wang A, Mizuta Y, Akazawa Y, Ohba K, Omagari K, Miyazaki M, Murase K, Hayashi T, Inoue K, Murata I, Kohno S. Elevated levels of chemokines in esophageal mucosa of patients with reflux esophagitis. *Am J Gastroenterol* 2003; 98: 551–556.

3. Hamaguchi M, Fujiwara Y, Takashima T, Hayakawa T, Sasaki E, Shiba M, Watanabe T, Tominaga K, Oshitani N, Matsumoto T, Higuchi K, Arakawa T. Increased expression of cytokines and adhesion molecules in rat chronic esophagitis. *Digestion* 2003; 68: 189–197.

4. Uchiyama K, Yoshida N, Higashihara H, Immamoto E, Tomatsuri N, Ishikawa T, Ueda M, Takagi T, Naito Y, Yoshikawa T. Transition of chemokine expression in esophageal mucosa after the treatment of GERD patients with proton pump inhibitors. *Gastroenterology* 2003; 124: A-409.

5. Kanazawa Y, Isomoto H, Wen CY, Wang AP, Saenko VA, Ohtsuru A, Takeshima F, Omagari K, Mizuta Y, Murata I, Yamashita S, Kohno S. Impact of endoscopically minimal involvement on IL-8 mRNA expression in esophageal mucosa of patients with non-erosive reflux disease. *World J Gastroenterol* 2003; 9: 2801–2804.

6. Imamoto E, Yoshida N, Uchiyama K, Higashihara H, Tomatsuri N, Ueda M, Takagi T, Naito Y, Yoshikawa T. Effects of bile acids and acidic exposure on IL-8 expression in human esophageal epithelial cells. *Gastroenterology* 2003; 124: A-409.

7. Kawabe A, Shimada Y, Soma T, Maeda M, Itami A, Kaganoi J, Kiyono T, Imamura M. Production of prostaglandin E2 via bile acid is enhanced by trypsin and acid in normal human esophageal epithelial cells. *Life Sci* 2004; 75: 21–34.

8. Kuroda M, Naito Y, Takagi T, Uchiyama K, Isozaki Y, Katada K, Sakamoto N, Shimozawa M, Yoshida N, Yoshikawa T. Gastroduodenal reflux induces cyclooxygenase-2 and inducible nitric oxide synthase in the esophageal mucosa in rats: evidence for involvement of pancreatic proteases. *Gastroenterology* 2003; 124: A-230.

9. Wallace JL, Arfors KE, McKnight GW. A monoclonal antibody against the CD18 leukocyte adhesion molecule prevents indomethacin-induced gastric damage in the rabbit. *Gastroenterology* 1991; 100: 878–883.

10. Wallace JL, McKnight W, Miyasaka M, Tamatani J, Paulson J, Anderson DC, Granger DN, Kubes P. Role of endothelial adhesion molecules in NSAID-induced gastric mucosal injury. *Am J Physiol* 1993; 256: G993–G998.

11. Yoshida N, Takemura T, Granger DN, Anderson DC, Wolf RE, McIntire LV, Kvietys PR. Molecular determinations of aspirin-induced neutrophil adherence to endothelial cells. *Gastroenterology* 1993; 105: 715–724.

12. Yoshida N, Granger DN, Evans Jr DJ, Evans DG, Graham DY, Anderson DC, Wolf RE, Kvietys PR. Mechanisms involved in *Helicobacter pylori*-induced inflammation. *Gastroenterology* 1993; 105: 1431–1440.

13. Yoshida N, Cepinskas G, Granger DN, Anderson DC, Wolf RE, Kvietys PR. Aspirin-induced, neutrophil-mediated injury to vascular endothelium. *Inflammation* 1995; 19: 297–312.

14. Takemura T, Granger DN, Evans Jr. DJ, Evans DG, Graham DY, Anderson DC, Wolf RE, Cepinskas G, Kvietys PR. Extract of Helicobacter pylori induces neutrophils to injure endothelial cells and contains antielastase activity. *Gastroenterology* 1996; 110: 21–29.

15. Taha AS, Dahill S, Morran C, Hudson N, Hawkey CJ, Lee FD, Sturrock RD, Russell RI. Neutrophils, Helicobacter pylori, and nonsteroidal anti-inflammatory drug ulcers. *Gastroenterology* 1999; 116: 254–258.

16. Hirayama F, Takagi S, Yokoyama Y, Iwao E, Ikeda Y. Establishment of gastric Helicobacter pylori infection in Mongolian gerbils. *J Gastroenterol* 1996; 31 Suppl. 9: 24–28.

17. Hirayama F, Takagi S, Iwao E, Yokoyama Y, Haga K, Hanada S. Development of poorly differentiated adenocarcinoma and carcinoid due to long-term Helicobacter pylori colonization in Mongolian gerbils. *J Gastroenterol* 1999; 34: 450–454.

18. Yoshida N, Sugimoto N, Hirayama F, Nakamura Y, Ichikawa H, Naito Y, Yoshikawa T. Helicobacter pylori infection potentiates aspirin induced gastric mucosal injury in Mongolian gerbils. *Gut* 2002; 50: 594–598.

19. Crabtree JE, Covacci A, Farmery SM, Xiang Z, Tompkins DS, Perry S, Lindley IJ, Rappuoli R. Helicobacter pylori induced interleukin-8 expression in gastric epithelial cells is associated with CagA positive phenotype. *J Clin Pathol* 1995; 48: 41–45.

20. Crabtree JE, Lindley IJ. Mucosal interleukin-8 and Helicobacter pylori-associated gastroduodenal disease. *Eur J Gastroenterol Hepatol* 1994.

21. Dijkstra G, Moshage H, van Dullemen HM, de Jager-Krikken A, Tiebosch AT, Kleibeuker JH, Jansen PL, van Goor H. Expression of nitric oxide synthases and formation of nitrotyrosine and reactive oxygen species in inflammatory bowel disease. *J Pathol* 1998; 186: 416–421.

22. Kimura H, Hokari R, Miura S, Shigematsu T, Hirokawa M, Akiba Y, Kurose I, Higuchi H, Fujimori H, Tsuzuki Y, Serizawa H, Ishii H. Increased expression of an inducible isoform of nitric oxide synthase and the formation of peroxynitrite in colonic mucosa of patients with active ulcerative colitis. *Gut* 1998; 42: 180–187.

23. Rachmilewitz D, Eliakim R, Ackerman Z, Karmeli F. Direct determination of colonic nitric oxide level — a sensitive marker of disease activity in ulcerative colitis. *Am J Gastroenterol* 1998; 93: 409–412.

24. Singer I, Kawka DW, Scott S, Weidner JR, Mumford RA, Riehl TE, Stenson WF. Expression of inducible nitric oxide synthase and nitrotyrosine in colonic epithelium in inflammatory bowel disease. *Gastroenterology* 1996; 111: 871–885.

25. Kankuri E, Asmawi MZ, Korpela R, Vapaatalo H, Moilanen E. Induction of iNOS in a rat model of acute colitis. *Inflammation* 1999; 23: 141–152.

26. Miampamba M, Sharkey KA. Temporal distribution of neuronal and inducible nitric oxide synthase and nitrotyrosine during colitis in rats. *Neurogastroenterol Motil* 1999; 11: 193–206.

27. Naito Y, Takagi T, Ishikawa T, Handa O, Matsumoto N, Yagi N, Matsuyama K, Yoshida N, Yoshikawa T. The inducible nitric oxide synthase inhibitor ONO-1714 blunts dextran sulfate sodium colitis in mice. *Eur J Pharmacol* 2001; 412: 91–99.

28. Aiko S, Grisham MB. Spontaneous intestinal inflammation and nitric oxide metabolism in HLA-B27 transgenic rats. *Gastroenterology* 1995; 109: 142–150.

29. Harren M, Schonfelder G, Paul M, Horak I, Riecken EO, Wiedenmann B, John M. High expression of inducible nitric oxide synthase correlates with intestinal inflammation of interleukin-2-deficient mice. *Ann N Y Acad Sci* 1998; 859: 210–215.

30. Zingarelli B, Cuzzocrea S, Szabo C, Salzman AL. Mercaptoethylguanidine, a combined inhibitor of nitric oxide synthase and peroxynitrite scavenger, reduces trinitrobenzene sulfonic acid-induced colonic damage in rats. *J Pharmacol Exp Ther* 1998; 287: 1048–1055.

31. Zingarelli B, Szabo C, Salzman AL. Reduced oxidative and nitrosative damage in murine experimental colitis in the absence of inducible nitric oxide synthase. *Gut* 1999; 45: 199–209.

32. Okayasu I, Hatakeyama S, Yamada M, Ohkusa T, Inagaki Y, Nakaya R. A novel method in the induction of reliable experimental acute and chronic ulcerative colitis in mice. *Gastroenterology* 1990; 98: 694–702.

33. Naka M, Nanbu T, Kobayashi K, Kamanaka Y, Komeno M, Yanase R, Fukutomi T, Fujimura S, Seo HG, Fujiwara N, Ohuchida S, Suzuki K, Kondo K, Taniguchi N. A potent inhibitor of inducible nitric oxide synthase, ONO-1714, a cyclic amidine derivative. *Biochem Biophys Res Commun* 2000; 270: 663–667.

34. Naka K, Yokozaki H, Yasui W, Tahara H, Tahara E, Tahara E. Effect of antisense human telomerase RNA transfection on the growth of human gastric cancer cell lines. *Biochem Biophys Res Commun* 1999; 255: 753–758.

35. Villarete LH, Remick DG. Nitric oxide regulation of IL-8 expression in human endothelial cells. *Biochem Biophys Res Commun* 1995; 211: 671–676.

36. Hogaboam CM, Jacobson K, Collins SM, Blennerhassett MG. The selective beneficial effects of nitric oxide inhibition in experimental colitis. *Am J Physiol* 1995; 268: G673–684.

37. Southey A, Tanaka S, Murakami T, Miyoshi H, Ishizuka T, Sugiura M, Kawashima K, Sugita T. Pathophysiological role of nitric oxide in rat experimental colitis. *Int J Immunopharmacol* 1997; 19: 669–676.

20 Oxidative Stress Involvement in Diabetic Nephropathy and Its Prevention by Astaxanthin

Yuji Naito, Kazuhiko Uchiyama, and Toshikazu Yoshikawa
Kyoto Prefectural University of Medicine, Kyoto, Japan

CONTENTS

20.1 INTRODUCTION

Diabetic nephropathy is characterized by the enlargement of glomerular mesangium because of the accumulation of extracellular matrix proteins, and is a leading cause of end-stage renal disease (1,2). Clinical studies in subjects with type 1 and type 2 diabetes clearly link hyperglycemia to vascular complications, including diabetic nephropathy (3,4). Hyperglycemia is responsible for the development and progression of diabetic nephropathy through metabolic derangements, including increased oxidative stress, renal polyol formation, activation of protein kinase C-mitogen-activated protein kinases, and accumulation of advanced glycation end products, as well as such hemodynamic factors as systemic hypertension and increased intraglomerular pressure (5). However, the definite mechanism of mesangial cell activation by high ambient glucose has not been investigated.

20.2 OXIDATIVE STRESS IN DIABETIC NEPHROPATHY

We postulated that increased oxidative stress due to high glucose levels is important in the pathogenesis of diabetic nephropathy. Studies that have used natural and synthetic antioxidants have provided convincing evidence that glomerular hypertrophy and accumulation of collagen and transforming growth factor (TGF)-β by high glucose is largely mediated by reactive oxygen species (ROS) (6–9). Under diabetic conditions, ROS are produced by the nonenzymatic glycation reaction of proteins, mitochondria, and protein kinase C-dependent activation of NAD(P)H oxidase in mesangial cells, infiltrated inflammatory cells, and endothelial cells (10,11). In addition, the persistence of hyperglycemia has been reported to increase the production of ROS through glucose auto-oxidation, abnormal metabolism of prostaglandins, and high polyol pathway flux. A recent study using a suppression-subtractive hybridization has demonstrated that high glucose induces actin cytoskeleton regulatory genes in mesangial cells and that the induction is dependent on mitochondria-induced ROS and is independent of protein kinase C and TGF-β (12).

The oxidation products of biological components are generally accepted as markers of oxidative stress, and these products have been measured in many clinical specimens, including lipid peroxides, malondialdehyde, and 4-hydroxy-2-nonenal (all of which are indices of lipid peroxidation); isoprostane (a product of oxidation of arachidonic acid); 8-hydroxydeoxyguanosine (8-OHdG) and thymine glycol (both are markers of DNA oxidation); and protein carbonyl, hydroxyleucine, hydrovaline, and nitrotyrosine (products of oxidation of protein and amino acids) (13). In 1994, Ha et al. (14) first demonstrated that the formation of 8-OHdG is closely related to the process of diabetic nephropathy in experimental diabetic rodents. They showed that the 8-OHdG levels in both rat renal cortex and papilla were significantly increased compared to those of controls after streptozotocin (STZ) administration and that daily injection of insulin after STZ treatment significantly reduced both urinary albumin excretion and 8-OHdG formation, which suggests that these are associated with the diabetic state induced by STZ rather than a direct nephrotoxic effect of the drug (14). Recent clinical evaluation indicates that 8-OHdG in urine is a useful clinical marker not only for detecting micro- and macrovascular complications (15) but also for predicting the development of diabetic nephropathy in diabetic patients (16). Hinokio et al. (17) have reported a direct association between oxidative DNA damage and the complication of diabetes by measuring the 8-OHdG levels in the urine and blood mononuclear leukocytes of patients with type 2 diabetes. These data, therefore, indicate that the agent that has antioxidative actions is one of the factors that reduce glucose toxicity and prevent diabetic nephropathy.

20.3 PREVENTION OF DIABETIC NEPHROPATHY
BY ANTIOXIDANTS

The protective effect that antioxidants have on certain aspects of nephropathy in diabetic animals has been reported. Recently, it was reported that antioxidant treatment with vitamin E, probucol, alpha-lipoic acid, or taurine normalized not only diabetes-induced renal dysfunction such as albuminuria and glomerular hypertension but also glomerular pathologies (7). Ueno et al. (18) have also reported that dietary glutathione can exert beneficial effects on diabetic complications in STZ-induced diabetic rats. The limited data available suggest that antioxidants help protect against diabetic nephropathy in humans, and a combined daily treatment with 680 IU vitamin E and 1250 mg vitamin C for 4 weeks reduced the urinary albumin excretion rate by 19% in a crossover study of 30 patients with type 2 diabetes (19). Kaneto et al. (9) also demonstrated that antioxidant treatment can exert beneficial effects in db/db mice with preservation of *in vivo* β-cell function. The db/db mouse exhibits clinical and histological features of diabetic nephropathy that track the human disease. This animal exhibits hyperglycemia and renal insufficiency by the age of 16 weeks. The kidneys show characteristic histological lesions of diabetic nephropathy, including mesangial matrix expansion and glomerular basement membrane thickening (20). However, the effects of astaxanthin on oxidative damage and the development of diabetic renal complication have not been investigated.

20.4 PREVENTION OF DIABETIC NEPHROPATHY BY ASTAXANTHIN

Astaxanthin, which is a common pigment in algae, fish, and birds, is a carotenoid that has many highly potent pharmacological effects, such as antioxidative activity (21–23), immunomodulating actions (24,25), anticancer activity (26), and anti-inflammation action (27,28). Astaxanthin is reported to be more effective than other antioxidants such as vitamin E and β-carotene in preventing lipid peroxidation in solution and various biomembrane systems (21,22). Goto et al. (29) reported that the efficient antioxidant activity of astaxanthin could be due to the unique structure of its terminal ring moiety. Astaxanthin traps radicals not only at the conjugated polyene chain but also in its terminal ring moiety, in which the hydrogen atom at the C3 methine is suggested to be a radical-trapping site. Furthermore, our recent data demonstrated that esterified astaxanthin was efficiently absorbed and transported into the kidney (30). Esterified astaxanthin is known to be more effectively absorbed from the small intestine compared to nonesterified astaxanthin, and is mainly metabolized in the liver.

Recently, astaxanthin has been made available for *in vivo* study because astaxanthin esters biosynthesized in the unicellular microalga *Haematococcus pluvialis* have been obtained from the culture of the algae (Figure 20.1) (23,31). A randomized clinical trial revealed that 6 mg of astaxanthin per day from a *Haematococcus pluvialis* algal extract can be safely consumed by healthy adults (32). We previously reported the potential usefulness of astaxanthin treatment for reducing glucose toxicity using db/db mice, a rodent model of type 2 diabetes. The function of islet cells to secrete insulin, as determined by the intraperitoneal glucose tolerance test, was preserved in the astaxanthin-treated group, although

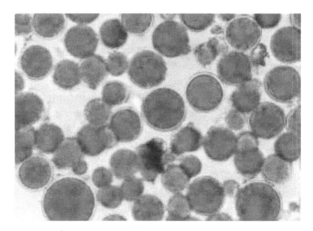

FIGURE 20.1 Astaxanthin esters biosynthesized in the unicellular microalga *Haematococcus pluvialis.*

a histologic study of the pancreas revealed no significant differences in the β-cell mass between astaxanthin-treated and astaxanthin-untreated db/db mice (33).

Recently, we examined whether chronic administration of astaxanthin could prevent diabetic nephropathy in db/db mice (30). The mice were divided into three groups as follows: nondiabetic db/m, diabetic db/db, and diabetic db/db treated with astaxanthin. The diet for the astaxanthin supplementation group was prepared by mixing CE-2 powder with astaxanthin (Fuji Chem. Industry Co., Ltd., Toyama, Japan) at 0.02%. Throughout the experimental periods (6–18 weeks of age), blood glucose levels were consistently higher in db/db mice than in db/m mice. At 18 weeks of age, the blood glucose levels of mice in the astaxanthin-treated group (338.0 ± 43.8 mg/dl) were lower than those of mice in the nontreated db/db group (417.6 ± 13.7 mg/dl), but the levels continued to be high (more than 300 mg/kg) throughout the experiment period compared with those of db/m mice (111.1 ± 3.4 mg/dl at 18 weeks of age). The mean body weight of db/db mice at 18 weeks of age (47.6 ± 0.6 g) was greater than that of db/m mice (29.5 ± 1.1 g). There were no differences in body weight and food intake between astaxanthin-treated and nontreated db/db mice. Albumin levels in urine of db/db mice significantly increased compared with those of db/m mice at 18 weeks of age (Figure 20.2). The increase in urinary albumin at 18 weeks of age was significantly inhibited by chronic treatment with astaxanthin. The glomerular histology in db/db mice showed accelerated mesangial expansion characterized by an increase in PAS-positive mesangial matrix area relative to that observed in db/m mice at 18 weeks of age. On the other hand, therapy with astaxanthin for 12 weeks partially reversed the mesangial matrix accumulation that had been established by 18 weeks of age. The relative mesangial area calculated by the ratio of the mesangial area to the total glomerular area was increased by 250% in db/db mice as compared to db/m mice. Administration of astaxanthin significantly ameliorated the increase in the relative mesangial area in db/db mice.

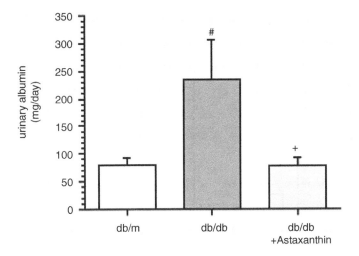

FIGURE 20.2 Effect of astaxanthin on the urinary albumin excretion rate of diabetic db/db mice. A 24-h urine sample from each mouse was collected in metabolic cages 12 weeks after the start of this experiment. Urine samples were processed to measure urinary albumin concentration using a competitive ELISA. Data shown are the mean ± SE from 5 mice. $^#p < .01$ vs. db/m mice and $^+p < .05$ vs. db/db diabetic mice. (Data from Naito Y, Uchiyama K, Aoi W, Hasegawa G, Nakamuara N, Yoshida N, Maoka T, Takahashi J, Yoshikawa T. Prevention of diabetic nephropathy by treatment with astaxanthin in diabetic db/db mice. *BioFactors* 2004; 20: 49–59.)

In nontreated db/db mice, the increase in urinary 8-OHdG levels was significantly enhanced at 18 weeks of age, but the enhancement was significantly inhibited by the treatment with astaxanthin (Figure 20.3). To evaluate the damage caused by oxidative stress, we performed 8-OHdG immunostaining in the kidneys of mice in each group. The results revealed that 8-OHdG immunoreactive cells in the glomeruli of nontreated db/db mice were more numerous than those in astaxanthin-treated db/db mice. The number of 8-OHdG-positive cells in db/db mice was significantly increased compared with that of db/m mice. The increase in the number of 8-OHdG-positive cells was significantly inhibited by the treatment with astaxanthin. These results indicate that the urinary level of 8-OHdG reflects renal oxidative damage in diabetes mellitus and might be the biomarker of diabetic nephropathy. The administration of dietary astaxanthin improved renal dysfunction in diabetic mice through its antioxidant function, and the urinary and tissue 8-OHdG data support these results. The diabetes-induced changes observed in the present study were markedly prevented by the potent antioxidant astaxanthine. This is quite consistent with the effects of this compound in other nondiabetic models of oxidative stress (34). In conclusion, our present results reveal, for the first time, that astaxanthin can exert beneficial effects on renal mesangial cells in diabetic db/db mice. Thus, use of astaxanthin might be a novel approach for the prevention of diabetes nephropathy.

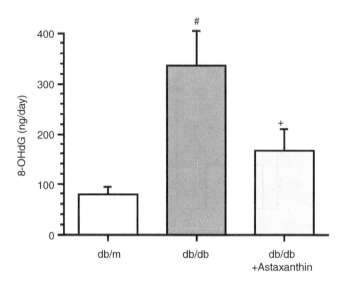

FIGURE 20.3 Effect of astaxanthin on the urinary 8-hydroxydeoxyguanosine (8-OHdG) excretion rate of db/db mice. Data shown are the mean ± SE from 5 mice. $^\#p < .01$ vs. db/m mice and $^+p < .05$ vs. db/db diabetic mice.

ACKNOWLEDGMENTS

This work was supported by the grant-in-aid for Scientific Research (14570493 YN and 15390178 TY) from the Ministry of Education, Culture, Sports, Science, and Technology of Japan and by grants from the Bio-Oriented Technology Research Advancement Institution and the Ministry of Agriculture, Forestry and Fisheries of Japan.

REFERENCES

1. Ziyadeh FN. The extracellular matrix in diabetic nephropathy. *Am J Kidney Dis* 1993; 22: 736–744.
2. Zhu D, Kim Y, Steffes MW, Groppoli TJ, Butkowski RJ, Mauer SM. Glomerular distribution of type IV collagen in diabetes by high resolution quantitative immunochemistry. *Kidney Int* 1994; 45: 425–433.
3. Group TDCaCTR. The effect of intensive treatment of diabetes on the development and progression of long-term complications in insulin-dependent diabetes mellitus. *N Engl J Med* 1993; 329: 977–986.
4. Group UPDSU. Intensive blood-glucose control with sulphonylureas or insulin compared with conventional treatment and risk of complications in patients with type 2 diabetes (UKPDS33). *Lancet* 1998; 352: 837–853.
5. Kikkawa R, Koya D, Haneda M. Progression of diabetic nephropathy. *Am J Kidney Dis* 2003; 41: S19–21.

6. Inoguchi T, Tsubouchi H, Etoh T, Kakimoto M, Sonta T, Utsumi H, Sumimoto H, Yu HY, Sonoda N, Inuo M, Sato N, Sekiguchi N, Kobayashi K, Nawata H. A Possible Target of Antioxidative Therapy for Diabetic Vascular Complications-Vascular NAD(P)H Oxidase. *Curr Med Chem* 2003; 10: 1759–1764.

7. Koya D, Hayashi K, Kitada M, Kashiwagi A, Kikkawa R, Haneda M. Effects of antioxidants in diabetes-induced oxidative stress in the glomeruli of diabetic rats. *J Am Soc Nephrol* 2003; 14 Suppl. 3: S250–253.

8. Maritim AC, Sanders RA, Watkins JB. Effects of alpha-lipoic acid on biomarkers of oxidative stress in streptozotocin-induced diabetic rats. *J Nutr Biochem* 2003; 14: 288–294.

9. Kaneto H, Kajimoto Y, Miyagawa J, Matsuoka T, Fujitani Y, Umayahara Y, Hanafusa T, Matsuzawa Y, Yamasaki Y, Hori M. Beneficial effects of antioxidants in diabetes: possible protection of pancreatic beta-cells against glucose toxicity. *Diabetes* 1999; 48: 2398–2406.

10. Catherwood MA, Powell LA, Anderson P, McMaster D, Sharpe PC, Trimble ER. Glucose-induced oxidative stress in mesangial cells. *Kidney Int* 2002; 61: 599–608.

11. Li JM, Shah AM. ROS Generation by Nonphagocytic NADPH Oxidase: Potential Relevance in Diabetic Nephropathy. *J Am Soc Nephrol* 2003; 14 (Suppl. 3): S221–226.

12. Clarkson MR, Murphy M, Gupta S, Lambe T, Mackenzie HS, Godson C, Martin F, Brady HR. High glucose-altered gene expression in mesangial cells. Actin-regulatory protein gene expression is triggered by oxidative stress and cytoskeletal disassembly. *J Biol Chem* 2002; 277: 9707–9712.

13. Yoshikawa T, Naito Y. What is oxidative stress? *JMAJ* 2002; 45: 271–276.

14. Ha H, Kim C, Son Y, Chung MH, Kim KH. DNA damage in the kidneys of diabetic rats exhibiting microalbuminuria. *Free Radic Biol Med* 1994; 16: 271–274.

15. Nishikawa T, Sasahara T, Kiritoshi S, Sonoda K, Senokuchi T, Matsuo T, Kukidome D, Wake N, Matsumura T, Miyamura N, Sakakida M, Kishikawa H, Araki E. Evaluation of urinary 8-hydroxydeoxy-guanosine as a novel biomarker of macrovascular complications in type 2 diabetes. *Diabetes Care* 2003; 26: 1507–1512.

16. Hinokio Y, Suzuki S, Hirai M, Suzuki C, Suzuki M, Toyota T. Urinary excretion of 8-oxo-7, 8-dihydro-2′-deoxyguanosine as a predictor of the development of diabetic nephropathy. *Diabetologia* 2002; 45: 877–882.

17. Hinokio Y, Suzuki S, Hirai M, Chiba M, Hirai A, Toyota T. Oxidative DNA damage in diabetes mellitus: its association with diabetic complications. *Diabetologia* 1999; 42: 995–998.

18. Ueno Y, Kizaki M, Nakagiri R, Kamiya T, Sumi H, Osawa T. Dietary glutathione protects rats from diabetic nephropathy and neuropathy. *J Nutr* 2002; 132: 897–900.

19. Graede P, Poulsen HE, Parving HH, Pedersen O. Double-blind, randomized study of the effect of combined treatment with vitamin C and E on albuminuria in type 2 diabetic patients. *Diabet Med* 2001; 18: 756–760.

20. Sharma K, McCue P, Dunn SR. Diabetic kidney disease in the db/db mouse. *Am J Physiol Renal Physiol* 2003; 284: F1138–1144.

21. Naguib YM. Antioxidant activities of astaxanthin and related carotenoids. *J Agric Food Chem* 2000; 48: 1150–1154.

22. Fukuhara K, Inokami Y, Tokumura A, Terao J, Suzuki A. Rate conctants for quenching singlet oxygen and activities for inhibiting lipid peroxidation of carotenoids and alpha-tocopherol in liposomes. *Lipids* 1998; 33: 751–756.

23. Kobayashi M. *In vivo* antioxidant role of astaxanthin under oxidative stress in the green alga Haematococcus pluvialis. *Appl Microbiol Biotechnol* 2000; 54: 550–555.

24. Jyonouchi H, Zhang L, Gross M, Tomita Y. Immunomodulating actions of carotenoids: enhancement of in vivo and in vitro antibody production to T-dependent antigens. *Nutr Cancer* 1994; 21: 47–58.

25. Kurihara H, Koda H, Asami S, Kiso Y, Tanaka T. Contribution of the antioxidative property of astaxanthin to its protective effect on the promotion of cancer metastasis in mice treated with restraint stress. *Life Sci* 2002; 70: 2509–2520.

26. Chew BP, Park JS, Wong MW, Wong TS. A comparison of the anticancer activitaies of dietary bata-carotene, canthaxanthin and astaxanthin in mice *in vivo*. *Anticancer Res* 1999; 19: 1849–1853.

27. Bennedsen M, Wang X, Willen R, Wadstroem T, Andersen LP. Treatment of H. pylori infected mice with antioxidant astaxanthin reduces gastric inflammation, bacterial load and modulated cytokine release by splenocytes. *Immun Letters* 1999; 70: 185–189.

28. Ohgami K, Shiratori K, Kotake S, Nishida T, Mizuki N, Yazawa K, Ohno S. Effects of astaxanthin on lipopolysaccharide-induced inflammation *in vitro* and *in vivo*. *Invest Ophthalmol Vis Sci* 2003; 44: 2694–2701.

29. Goto S, Kogure K, Abe K, Kimata Y, Kitahama K, Yamashita E, Terada H. Efficient radical trapping at the surface and inside the phospholipid membrane is responsible for highly potent antiperoxidative activity of the carotenoid astaxanthin. *Biochim Biophys Acta* 2001; 1512: 251–258.

30. Naito Y, Uchiyama K, Aoi W, Hasegawa G, Nakamuara N, Yoshida N, Maoka T, Takahashi J, Yoshikawa T. Prevention of diabetic nephropathy by treatment with astaxanthin in diabetic db/db mice. *BioFactors* 2004; 20: 49–59.

31. Steinbrenner J, Linden H. Light induction of carotenoid biosynthesis genes in the green alga Haematococcus pluvialis: regulation by photosynthetic redox control. *Plant Mol Biol* 2003; 52: 343–356.

32. Spiller GA, Dewell A. Safety of an astaxanthin-rich *Haematococcus pluvialis* algal extract: a randomized clinical trial. *J Med Food* 2003; 6: 51–56.

33. Uchiyama K, Naito Y, Hasegawa G, Nakamura N, Takahashi J, Yoshikawa T. Astaxanthin protects beta-cells against glucose toxicity in diabetic db/db mice. *Redox Rep* 2002; 7: 290–293.

34. Aoi W, Naito Y, Sakuma K, Kuchide M, Tokuda H, Maoka T, Toyokuni S, Oka S, Yasuhara M, Yoshikawa T. Astaxanthin limits exercise-induced skeletal and cardiac muscle damage in mice. *Antioxid Redox Signal* 2003; 5: 139–144.

21 Preventive and Therapeutic Effects of Plant Polyphenols through Suppression of Nuclear Factor-Kappa B

Navindra P. Seeram
UCLA, Los Angeles, California

Haruyo Ichikawa, Shishir Shishodia, and Bharat B. Aggarwal
University of Texas, M.D. Anderson Cancer Center, Houston, Texas

CONTENTS

21.1 OVERVIEW

Nuclear transcription factor-κB (NF-κB) regulates the expression of over 200 different genes. The activation of NF-κB has now been linked with a variety of inflammatory diseases, including cancer, atherosclerosis, myocardial infarction, diabetes, allergy, asthma, arthritis, Crohn's disease, multiple sclerosis, Alzheimer's disease, osteoporosis, psoriasis, septic shock, and AIDS. There is much evidence suggesting that phytochemicals can inhibit the pathways that lead to the activation of this transcription factor and have the potential to prevent and treat the diseases mentioned. These phytochemicals are derived from plants such as turmeric, red pepper, cloves, ginger, cumin, anise and fennel, rosemary, garlic, green tea, basil, cauliflower, cabbage, artichoke, lemon, and pomegranate.

21.2 INTRODUCTION

Plant extracts and natural compounds purified from plants have been used by humans for many centuries for the treatment and alleviation of a variety of inflammation-related diseases, including cancer. Eastern medicine, viz., traditional Chinese medicine (TCM) and the Indian ayurvedic system of medicine, continue to prescribe complex mixtures of herbs and herbal extracts for the treatment of cancer. Recent research has shown that a mechanism-based approach that targets the means by which cancer cells prosper has significant advantages over the current methods of cancer treatment, including chemotherapy, with their attendant adverse effects. The regulation of the cell cycle (cell survival, proliferation, and death) requires the integration of a myriad of cell-signaling factors, including those that direct the transcription of genes coding for integral cell proteins. Nuclear factor (NF)-κB is a transcription factor that regulates the expression of genes involved in cancer and other diseases.

21.3 NF-κB AND DISEASE

NF-κB, discovered by David Baltimore in 1986, is a ubiquitous factor that resides in the cytoplasm in an inactive state. When activated, it is translocated to the nucleus and induces gene transcription. NF-κB is activated by free radicals, inflammatory stimuli, carcinogens, tumor promoters, endotoxins, gamma radiation, UV light, and x-rays. On activation, it induces the expression of more than 200 genes, and these genes have been shown to suppress apoptosis and induce cellular transformation, proliferation, invasion, metastasis, chemoresistance, radioresistance, and inflammation.[1-3] The activated form of NF-κB has been found to mediate cancer,[1,4,5] atherosclerosis,[6] myocardial infarction,[7] diabetes,[8] allergy,[9,10] asthma,[11] arthritis,[12] Crohn's disease,[13] multiple sclerosis,[14] Alzheimer's disease,[15,16] osteoporosis, psoriasis, septic shock, AIDS, and other inflammatory diseases[17-19] (Figure 21.1). That NF-κB has been linked to wide variety of diseases is not too surprising because most diseases are caused by dysregulated inflammatory mechanisms.[20] Thus, agents that can suppress NF-κB activation can, in principle, either prevent, delay the onset, or treat NF-κB -linked diseases.

Ever since research has shown that there is an intrinsic link between inflammation and various diseases, it has become obvious that inhibition of NF-κB activity is desirable in the treatment of not only inflammation but also the disease itself. For instance, aberrant NF-κB activation is a known factor in oncogenesis, tumor growth, and metastasis, and specific constitutive activation of NF-κB has been identified in a number of cancers including, breast, ovarian, colon, and prostate cancer and Hodgkin's lymphoma (for references see Reference 1). Hence,

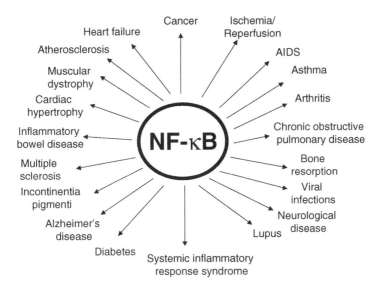

FIGURE 21.1 NF-κB-linked diseases.

this transcription factor is an important target in the prevention of diseases and their treatment in humans.

21.4　NF-κB AS A THERAPEUTIC TARGET

NF-κB represents an important and very attractive therapeutic target for plant-derived polyphenols. Much attention has been paid in the last decade to the identification of compounds that selectively interfere with this pathway. More recently, a great number of plant-derived natural products have been evaluated as possible inhibitors of the NF-κB pathway (Table 21.1, Figure 21.2). This chapter focuses on plant extracts, plant isolates, and distinct classes of plant-derived compounds that form part of this group. It is noteworthy that there are also reports of synthetic compounds and compounds from nonplant sources (e.g., caffeic acid phenethyl ester [CAPE] identified from honey bee propolis) that are known to block the activity of NF-κB.[21]

NF-κB plays a central role in inflammation, and research has made it clear that most diseases are linked to inflammation. Because NF-κB can also regulate the expression of many key genes involved in a variety of human cancers, it represents a relevant and promising target for new therapeutic agents. Many pharmaceutical and biotechnology companies have drug discovery programs that target NF-κB and have been investing heavily in the search for proteins that regulate this transcription factor. However, due to the ubiquitous nature of NF-κB, many of these drugs may exhibit undesirable side effects.

The mechanism-based approach to combat diseases from different angles with combinations of naturally derived compounds has a distinct advantage in

TABLE 21.1

Natural Products from Plants that Exhibit Chemopreventive and Therapeutic Activities against Cancer

Compound	Source	Botanical Name	Structure	↓NF-κB	P	T	Ref.
Polyphenol							
Amentoflavone (biapigenin)*	Gingko	*Gingko biloba; Cypress* spp.; *Galeobdolon chinense; Garcinia intermedia; Selaginella* spp.; *Biophytum sensitivum*		+	−	−	22,23
Apigenin*	Constituents of fruits and vegetables	*Scutellaria* spp. (incl. in Chinese herbal mixture, PCSPES, *Huang-Qi; Qingkailing; Shuanghuanglian* etc.); *Cirisium* spp.; *Crotalaria* spp.; *Quercus nutgall;Matricaria recutita; Saussurea medusa; Lantana montevidensis* Briq.	4′,5,7-Trihydroxyflavone	+	−	−	24–26
Arctigenin* and demehyltraxillagenin		*Arctium lappa; Centaurea* spp.; *Torreya nucifera*		+	−	−	27
Auraptene	Citrus fruits (hassaku, grapefruit, natudaidai)	*Citrus* spp.	7-Geranyl-oxycoumarin	−	−	−	28
Baicalein* and its derivatives[a]	Skullcap	*Scutellaria* spp. (included in Chinese herbal mixture, PCSPES, *Huang-Qi; Qingkailing; Shuanghuanglian*, etc.)	5,6,7-Trihydroxyflavone	+	+	+	29–31
Blueberry and berry mix	Blueberry and berry mix	*Rubus* spp.; *Vaccinium* spp.; *Fragaria ananassa*		+	−	−	32–34
Cannabinol*	Hemp seed oil, marijuana	*Cannabis* spp.	6,6,9-Trimethyl-3-pentyl-6H-dibenzo[b,d]pyran-1-ol;	+	−	+	35–37

Catalposide*	Catalpa spp.; Veronica spp.	Oxireno[4,5]cyclopenta[1,2-c]pyran-β-D-glucopyranoside		+	—	—	38
Catechins* (and theaflavins*)[b]	Green tea (including fermented, i.e., black teas), spotted knapweed, shea kernels, cocoa	Camellia sinensis; Centaurea maculosa; Vitellaria paradoxa; Theobroma cacao	3',4',5,7-Tetrahydroxy-2,3-trans-flavan-3-ol	+	+	+	39-41
Cirsimaritin*	Basil, sage, rosemary	Cirsium maritimum; Ocimum sanctum; Salvia officinalis; Rosmarinus officinalis	5,4'-Dihydroxy-6,7-dimethoxyflavone	—	—	—	42
Curcumin*	Turmeric (haldi)	Curcuma longa	Diferuloylmethane	+	+	+	43
Ellagic acid*	Strawberries, raspberries, blackberries, bayberries, fuejioa, pomegranates, pineapple, walnuts	Fragaria ananassa; Rubus idaeus; Punica granatum; Juglans regia	4,4',5,5',6,6'-Hexahydroxy-diphenic acid dilactone	—	+	—	34,44
Emodin*	Aloe vera	Polygonum spp.; Cassia spp.; Glossostemon bruguieri (moghat); Rheum spp. (rhubarb); Hovenia acerba	3-Methyl-1,6,8-Trihydroxy-anthraquinone	+	+	—	45
Flavopiridol[c]			5,7-Dihydroxy-8-(4-N-methyl-2-hydroxypyridyl)-6'-chloroflavone hydrochloride	+	—	—	46
Genistein*	Soybeans, chickpea, kudzu root	Glycine max; Cicer arietinum; Pueraria lobata radix; Desmodium spp.	3-(4-Hydroxyphenyl)-5,7-dihydroxy-chromen-4-one	+	+	+	47,48
Glossogyne tenuifolia[d]	Herb	Glossogyne tenuifolia		+	—	—	49

TABLE 21.1 (continued)
Natural Products from Plants that Exhibit Chemopreventive and Therapeutic Activities against Cancer

Compound	Source	Botanical Name	Structure	↓NF-κB	P	T	Ref.
Hematein*	Natural dye from logwood			+	—	—	50,51
Hesperidine	Oranges	*Citrus* spp.	Hesperitin-7-rutinoside	—	—	—	52
HMP*	Black fruit or galangal	*Alpinia* spp.	7-(4'-Hydroxy-3'-methoxyphenyl)-phenylheptenone	+	—	—	53
Hypericin*	St. John's wort	*Hypericum* spp.		+	+	+	54,55
Isothymusin	Basil	*Ocinum* spp.; *Limnophila geoffrayi*; *Becium grandiflorum*	6,7-Dimethoxy-5,8,4'-trihydroxyflavone	—	—	—	56
Isomallotochromanol* and isomallotochromene		*Mallotus japonicus*		+	—	—	57
Kaempferol	Fruits and vegetables e.g., tomato, onions	*Lycopersicon esculentum, Ginkgo biloba*	3,5,7,4'-Tetrahydroxyflavone	—	—	—	58
Luteolin*	Fruits and vegetables, tea	*Camellia sinensis*; *Scutellaria* spp.	2-(3,4-Dihydroxyphenyl)-5,7-dihydroxy-chromen-4-one	+	—	—	25,59
Morin	Guava, almond	*Psidium guajava*; *Prunus dulcis*	2',3,4',5,7-Pentahydroxy-flavone	—	+	+	60
Myricetin	Fruits and vegetables		2-(3,4-5-Trihydroxyphenyl)-3,5,7-Trihydroxy-chromen-4-one	—	—	—	61

Compound	Common source	Scientific name	Chemical name			Ref.
Nasunin	Eggplant	Solanum melongena	Delphinidin-3-(p-coumaroylrutinoside)-5-glucoside	—	—	62
Nobiletin	Citrus	Citrus spp.	5,6,7,8,3',4'-Hexamethoxy flavone	—	—	28
Nordihydroguaiaretic acid*		Guaiacum officinale		+	—	63
Ochna macrocalyx ext.		Ochna macrocalyx		+	—	64
Oenothein B		Oenothera spp.; Eugenia uniflora	Hydrolyzable ellagitannin	—	—	65
Panduratin A		Kaempferia pandurata		+	—	66
Procyanidins	Tea, cranberries, apple, grape seeds, pear	Camellia sinensis, Vaccinum spp.; Prunus spp.	Condensed tannins	—	—	67
Purpurogallin		Quercus sp. Nutgall	2,3,4,6-Tetrahydroxy-5H-bezocyclohepten-5-one	—	—	68,69
Pycnogenol	Maritime pine bark extract	Pinus maritima	Bioflavanoid extract	+	—	70,71
Quercetin	Fruits and vegetables	Malus spp.; Lycopersicon esculentum	2-(3,4-Dihydroxyphenyl)-3,5,7-trihydroxy-chromen-4-one	+	—	72
Rhein		Daylilies (Hemerocallis spp.); Rheum officinale (dahuang)	1,8-Dihydroxy-3-carboxyanthraquinone	+	—	73
Sanggenon C*	Mulberry	Morus spp.		+	—	74
Silymarin^e	Milk thistle, artichokes	Silybum marianum; Cynara scolymus	Flavonolignan extract	+	+	75,76,77
Saucerneols*, sauchinone, and manassantins*		Saururus spp.		+	—	78,79
Tangeretin	citrus fruits	Citrus spp.	5,6,7,8,4'-Pentamethoxyflavone	—	—	80

TABLE 21.1 (continued)
Natural Products from Plants that Exhibit Chemopreventive and Therapeutic Activities against Cancer

Compound	Source	Botanical Name	Structure	↓NF-κB	P	T	Ref.
Wedelolactone*	Wedelia spp.; Eclipta alba	1,8,9-Trihydroxy-3-methoxy-6H-benzofuro[3,2][1]-benzopyran-6-one,	+	—	—	81
Yakuchinones* A and B	Alpinia oxyphylla	1-(4'-Hydroxy-3'-methoxyphenyl)-7-phenyl-3-heptanone,1-(4'-hydroxy-3'-methoxyphenyl)-7-phenylhept-1-en-3-one	+	—	—	82
Terpenes							
Andalusol*	Sideritis foetens		+	—	—	83
Anethol*f and analogs	Brassica oleracea; Illicum verum; Ocimum spp.; Syzygium aromaticum; Anacardium occidentale; Hibiscus sabdariffa	4-Methoxypropenylbenzene	+	+	+	84,85
Artemisinin (qinghaosu)	Artemisia annua L. spp.		+	—	—	86
Avicins*g	Acasia victoriae		+	—	—	87
Azadirachtinh	Azadirachta indica, A. Jussieu		—	—	—	88
β-carotene	Daucus carota sativus; Citrus unshiu mar; Curcurbita moschata		—	—	—	89,90
β-cryptoxanthin	Carica papaya L. Physalis	β,β-Caroten-3-ol	—	—	—	89,91
Bakuchiol (drupanol)*	Psoralea corylifolia (bemchi); Otholobium pubescens	4-(3-ethenyl-3,7-dimethyl-1,6-octadienyl)-phenol	+	—	—	92

Compound	Source	Botanical source	Chemical name				Ref.
Betulinic acid*	Birch tree, almond hulls	Betula spp.; Quisqualis Fructus; Coussarea paniculata; Alangium lamarckii	3-β-Hydroxy-1up-20(29)en-28-acid	+	—	—	93
Carnosol*	Rosemary, sage	Rosmarinus officinalis; Salvia officinalis	2H-9,4a-(epoxymethano)-phenanthren-12-one	+	—	—	94
Celastrol*		Celastrus orbiculatus		+	—	—	95
Costunolide*		Magnolia grandiflora; Tsoongiodendron odorum; Saussurea lappa		+	—	—	96
Cucurbitacins[i]	Cucurbitaceae	Cucurbita andreana; Trichosanthes kirilowii; Elaeocarpus mastersii	3-β-(β-D-Glucosyloxy)-16-,23-α-epoxycucurbita-5,24-diene-11-one	—	—	—	97,98
Ergolide*		Inula spp.	Dihydrobigelovin	+	—	—	99
Excisanin A*		Isodon (Rabdosia) spp.		+	—	—	100
Foliol*		Sideritis spp.		+	—	—	101
Germacranolides * and Eudesmanolides		Carpesium divaricatum; Montanoa hibiscifolia		+	—	—	102,103
Ginkgo biloba ext.		Gingko biloba		+	—	—	104
Ginsenoside Rg3*		Panax spp.		+	—	—	105
Glycyrrhizin*	Licorice root	Glycyrrhiza glabra; Glycyrrhiza uralens		+	—	—	106
Guaianolides*		Viguiera gardneri		+	—	—	107
Helenalin*		Arnicae spp.; Helenium aromaticum		+	—	—	108
Hypoestoxide		Hypoestes rosea		+	—	—	109,110
Kamebacetal A*		Isodon (Rabdosia) spp.		+	—	—	100
Kamebakaurin		Isodon (Rabdosia) spp.		+	—	—	100
Kaurenic acid*		Sideritis spp.	ent-kaur-16-ene-19-oic acid	+	—	+	101
Limonene	Citrus fruits	Citrus spp.	4-Isopropenyl-1-methyl-1-cyclohexane	—	+	+	111

TABLE 21.1 (continued)
Natural Products from Plants that Exhibit Chemopreventive and Therapeutic Activities against Cancer

Compound	Source	Botanical Name	Structure	↓NF-κB	P	T	Ref.
Linearol*		Sideritis spp.		+	—	—	101
Lutein	Tomato	Lycopersicon esculentum		—	—	—	112,113
Lycopene	Tomato	Lycopersicon esculentum	ψ,ψ-Carotene	—	—	—	112–114
Oleandrin*		Nerium oleander; Plumeria obsta		+	—	—	115
Oxoacanthospermoldes^k		Milleria quinqueflora		+	—	—	116
Parthenolide*	Feverfew	Tanacetum parthenium and T. larvatum; Michelia champaca; Talauma ovat; Magnolia grandiflora; Artemisia myriantha		+	—	—	117,118
Pristimerin*		Hippocratea spp.; Maytenus spp.;Celastrus orbiculatus; Reissantia buchananii; Salacia beddomei; Heisteria pallida		+	—	—	119
Triptolide** (PG 490)		Tripterygium wilfordii		+	—	—	120,121
Ursolic acid*	Basil, rosemary, berries	Rosmarinus officinalis; Ocimum sanctum; Aronia melanocarpa; Oxycoccus quadripetalus; Origanum majorana; Diospyros melanoxylon; Salvia przewalskii Maxim		+	—	—	122
Withanolides	Solanaceae	Withania spp.; Physalis angulata; Sorpichroa origanifolia		—	—	—	123,124

Alkaloids

Compound	Common source	Scientific name	Active principle				References
Capsaicinoids*l	Pepper, red chili, paprika fruits	*Capsicum* spp.	8-Methyl-N-vanillyl-trans-6-noneamide	+	+	+	125–127
Cepharanthine*		*Stephania cepharantha*		+	−	−	128
Conophylline*		*Tabernaemontana spp;; Ervatamia microphylla*		+	−	−	129
Higenamine	Ranunculaceae	*Aconitum japonicum; Argemone mexicana; Gnetum parvifolium*		−	−	−	130,131
Mahanimbine	Rutaceae	*Murraya koenigii; Clausena dunniana; Murraya siamensis*		−	+	−	132,133
Mahanine	Rutaceae	*Murraya koenigi; Micromelum minutum*		−	−	−	132
Morphine^m and its analogs	Opium poppy	*Papaver* spp.		+	−	+	134–136
Murrayanol	Rutaceae	*Murraya koenigii*		−	−	−	133
Piperine	Black pepper	*Garcinia xanthochymus; Piper* spp.		−	−	−	115,137
Rocaglamides*		*Aglaia* spp.		+	−	−	138
Tetrandine* (sinomenine A)		*Stephania tetrandra*		+	−	−	139
Thionia diversifolia ext.		*Thionia diversifolia*		−	−	−	140,141

Allylthiosulfinate

Compound	Common source	Scientific name	Active principle				References
Allicin	Garlic	*Allium sativum*	2-Propene-1-sulfinothioic acid-S-2-propenyl ester	−	−	−	142,143

Benz[a]phenazine

Compound	Common source	Scientific name	Active principle				References
Lapachone	Ginseng, lapacho tree, trunkwood	*Tabebuia* spp.		+	+	+	21

TABLE 21.1 (continued)
Natural Products from Plants that Exhibit Chemopreventive and Therapeutic Activities against Cancer

Compound	Source	Botanical Name	Structure	↓NF-κB	P	T	Ref.
Benzopyrene							
Rotenone		Derris spp.		+	—	—	144
Caffeic acid phenylethyl ester							
CAPE	Honey bee propolis	Apis mellifera capensis		+	—	—	21
Chlorophyll Catabolite							
Pheophorbide A		Solanum diflorum		+	—	—	145
Glucosinolate							
Sulphoraphane	Brassicaceae, e.g., broccoli, cauliflower	Brassica oleracea	4-Methylsulphinyl butyl-isothiocyanate	—	—	—	146
Indoles							
Indole-3-carbinol	Brassicaceae, e.g., onions, cabbage	Allium cepa; Brassica spp.	3-Indolemethanol	—	—	—	147
Iridoid glycoside							
Aucubin*	Algae	Eucommia spp.; Veronica spp.; Vitex spp.; Globularia spp.		+	—	—	148
Naptoquinone							
Plumbagin		Plumbago zeylanica	5-Hydroxy-2-methyl-1,4-naphtoquinone	—	—	—	149

Phenyl Propanoid

1'-Acetoxychavicol acetate	Zingiberaceae	Zingiber officinale; Languas galanga		−	−	−	28,150
Phenolics							
Ethyl gallate	Grapes, tea, red maple	Paeonia spp.; Sophora japonica; Vitis vinifera; Vitellaria paradoxa; Camellia sinensis		+	−	−	151
Gallic acid*	Fruits, e.g., guava	Psidium guajava; Erodium glaucophyllum; Melaleuca quinquenervia	3,4,5-Trihydroxybenzoic acid	−	+	−	152
Gingerol	Ginger	Zingiber officinale	(1-[4'-Hydroxy-3'methylphenyl-]-5-hydroxy-3-decane	−	−	−	153
Morellin	Indica fruit	Garcinia spp.		−	−	−	154
Sphondin*		Heracleum laciniatum;Ruta graveolens		+	−	−	155
Phenolic Acid							
Rosemarinic acid	Rosemary, sage	Rosmarinus officinalis; Saliva officinalis		−	−	−	156
Synapic acid		Sida acuta	4-Hydroxy-3,5-dimethoxycinnamic acid	−	−	−	157
Syringic acid		Radix isatidis	4-Hydroxy-3,5-dimethoxybenzoic acid	−	−	−	158,159
Polysaccharide							
Ganoderma lucidum ext.	Reishi	Ganoderma lucidum		+	+	+	160

TABLE 21.1 (continued)
Natural Products from Plants that Exhibit Chemopreventive and Therapeutic Activities against Cancer

Compound	Source	Botanical Name	Structure	↓NF-κB	P	T	Ref.
Polyisoprenylated Benzophenone Derivatives							
Garcinol[a] and its analogs	Garcinia indica fruit	Garcinia spp.		−	−	−	161
Phytoalexin							
Allixin	Garlic	Allium sativum	3-Hydroxy-5-methoxy-6-methyl-2-penthyl-4h-pyran-4-one	−	−	−	162
Saponin							
Calagualine		Polypodium spp.		+	−	−	163
Stilbene							
Resveratrol*[o] and analogs	Japanese knotweed; berry fruits, grapes, cranberries, etc.	Polygonum cuspidatum; Veratrum spp.; Vitis spp.; Vaccinum spp.	trans-3,4',5'-Trihydroxystilbene	+	+	+	164–166
Others							
Aged garlic ext.	Garlic	Allium sativum		+	−	−	167,168
α-lipoic acid*[p]	Asparagus, wheat, potatoes		1,2-Dithiolanepentanoic acid	+	−	−	169–171
Apple ext. (juice)	Apple juice	Malus spp.		+	−	−	172
Astaxanthin*	Microalga, algae	Haematococcus pluvialis	3,3'-Dihydroxy-β,β,carotene-4,4'-dione	+	−	−	173

Name	Source	Species	Compound				Reference
β-Glucan	Barley, soy bean, mushroom	Avena sativa; Hordeum vulgare; Agaricus blazei		—	—	—	174
β-sitosterol	Plants, nuts, lapacha tree, cactus	Glycine max; Arachis spp.; Miconia rubiginosa; Opuntia ficus-indica	α–Dihydrofucosterol	—	—	—	175
Cat's claw	Uncaria tomentosa			+	—	—	176
Cirsilineol	Basil, thyme	Ocimum sanctum; Lantana montevidensis Briq.; Thymus vulgaris		—	—	—	56
Diallylsulfide	Garlic, Chinese leek	Allium sativum	Diallylsulfide	—	—	—	177
Flavokawains (Kava lactones)	Kava kava	Piper methysticum		—	—	—	178
Germinated barley				+	—	—	179
Persenone A	Avocado	Persea americana	1-(Acetyloxy)-2-hydroxy-5,12,15-heneicosatrien-4-one	—	—	—	180
Pomegranate wine		Punica granatum		—	—	—	181
S-allylcysteine*		Allium sativum		+	—	—	182
Stinging nettle ext.		Urtica dioica		—	—	—	183
Trans-Asarone	Carrot	Daucus carota L.	trans-1-Prophenyl-2,4,5-trimethoxybenzene	—	—	—	184
Vitamin C*	Fruits and vegetables		Ascorbic acid	+	—	—	185,186
Vitamin E	Plant seeds and vegetables		α-Tocopherol	+	—	—	187,188

Note: T and P refer to therapy and prevention, respectively; the asterisk indicates that the chemical structure is shown in Figure 21.2 (A to D); + indicates inhibition of NF-κB; — indicates no activity reported.

[a] Baicalein and its derivatives include baicalin, wogonin, and wogonin, 6-methoxy-baicalein (oroxylin A).

[b] Catechins include catechin, epicatechin, epicatechingallate, epigallocatechin, and epigallocatechingallate; theaflavins are polyphenols found in fermented green tea, i.e., black tea.

[c] Synthetic compound closely related to a polyphenol isolated from the Indian plant, Dysoxylum binectariferum.

TABLE 21.1 (continued)
Natural Products from Plants that Exhibit Chemopreventive and Therapeutic Activities against Cancer

d   *Glossogyne tenuifolia* is Chinese medicine Hiang-ju.

e   Silymarin includes silybin, silibinin, silidian, and silychrist.

f   Anethol and analogs include eugenol, bis-eugenol, isoeugenol, and anetholdithiolthione.

g   Avicins include avicin D and avicin G.

h   Azadirachtin analogs include axadirachtin A,B, D, H, I, etc.

i   Curcubutacins analogs include curcubitacin B, D, E, etc.

j   Germacranolides include 2b,5-epoxy-5,10-dihydroxy-6a-angeloyloxy-9b-isobutyloxy-germacran-8a,12-olide.

k   Oxoacanthospermoldes include methoxymiller-9Z-enolide.

l   Capsaicinoids include capsaicin* and analogs, e.g., resiniferatoxin* (daphnetoxin).

m   Morphine and its analogs include KT 90 and sanguinarine.

n   Garcinol and its analogs include isogarcinol.

o   Resveratrol and analogs include piceatannol.

p   α-Lipoic acid includes dihydrolipoic acid.

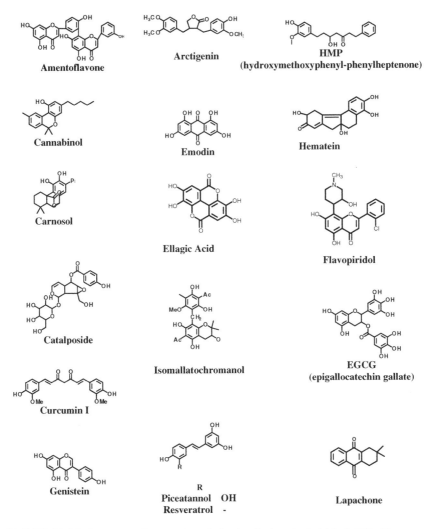

FIGURE 21.2A Structure of some plant-derived polyphenols that block NF-κB.

that the synergies and additive effects of such treatments, coupled with their mild nature, allow for their safe use. The increase in the popularity and awareness of complementary medicine has resulted in its use for the therapy of a variety of human diseases and by a large percentage of cancer patients. For many such patients, the use of naturally derived plant compounds and plant extracts is an essential part of their treatment. In addition, people are now more aware of the health benefits associated with the use of natural products and plant-derived compounds and have turned to these because of the negative perceptions associated with synthetic compounds. Most of the compounds outlined in Table 21.1 are commonly used in the form of concentrated plant extracts, and combinations

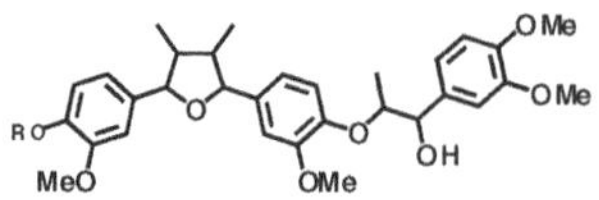

Mana-ssantin A

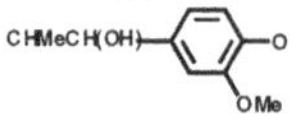

Mana-ssantin B

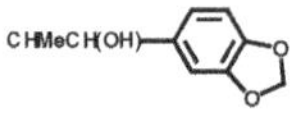

Saucerneol

Silybin

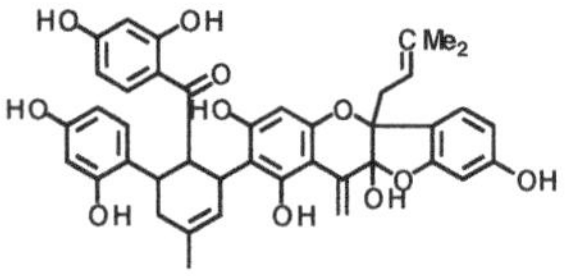

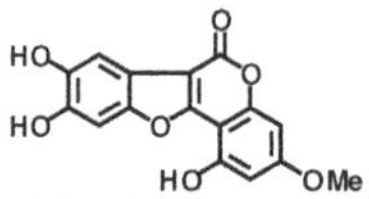

Wedelolactone

Apigenin

Baicalein

Luteolin

Oroxylin-A

Wogonin

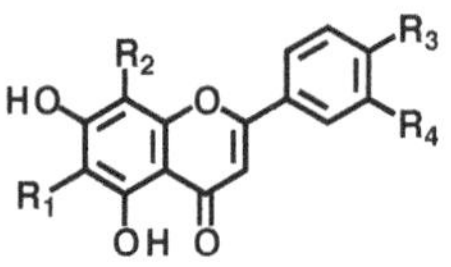

	R_1	R_2	R_3	R_4
Apigenin	-	-	OH	-
Baicalein	OH	-	-	-
Luteolin	-	-	OH	OH
Oroxylin-A	OCH₃	-	-	-
Wogonin	-	OCH₃	-	-

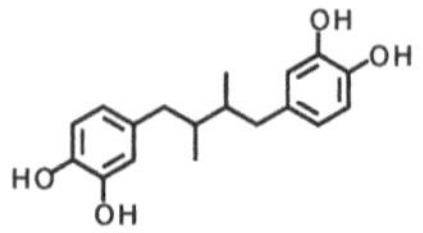

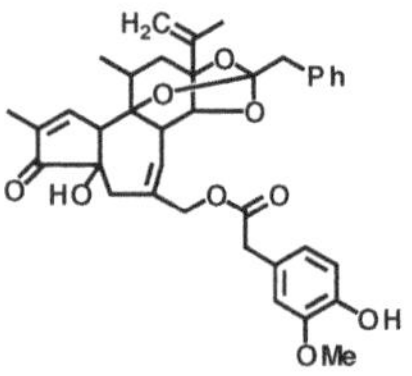

Nordihydroguaiaritic acid

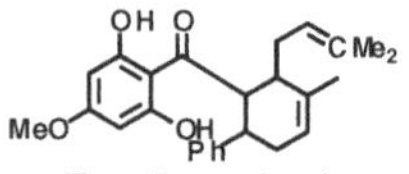

Panduratin A

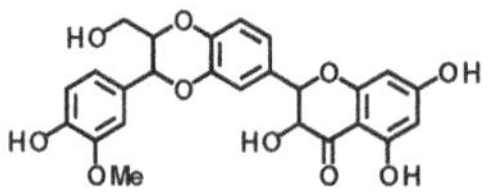

Sanggenon C

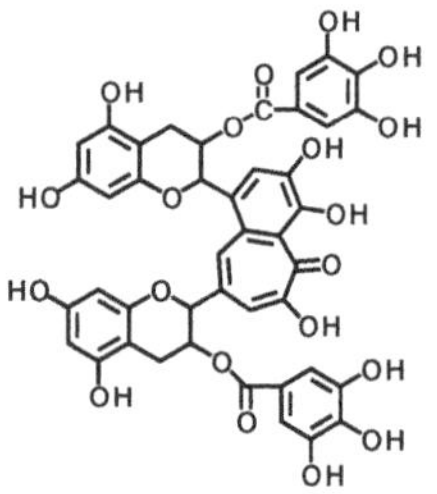

Theaflavin-3,3-digallate

Resinferatoxin

FIGURE 21.2A (continued) Structure of some plant-derived polyphenols that block NF-κB.

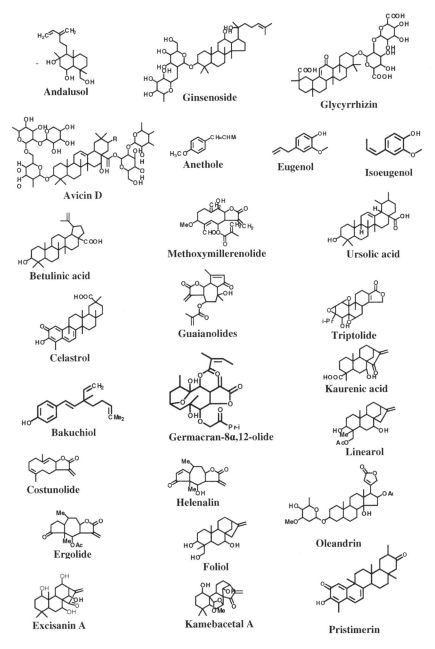

FIGURE 21.2B (continued) Structure of some plant-derived triterpenoids that block NF-κB.

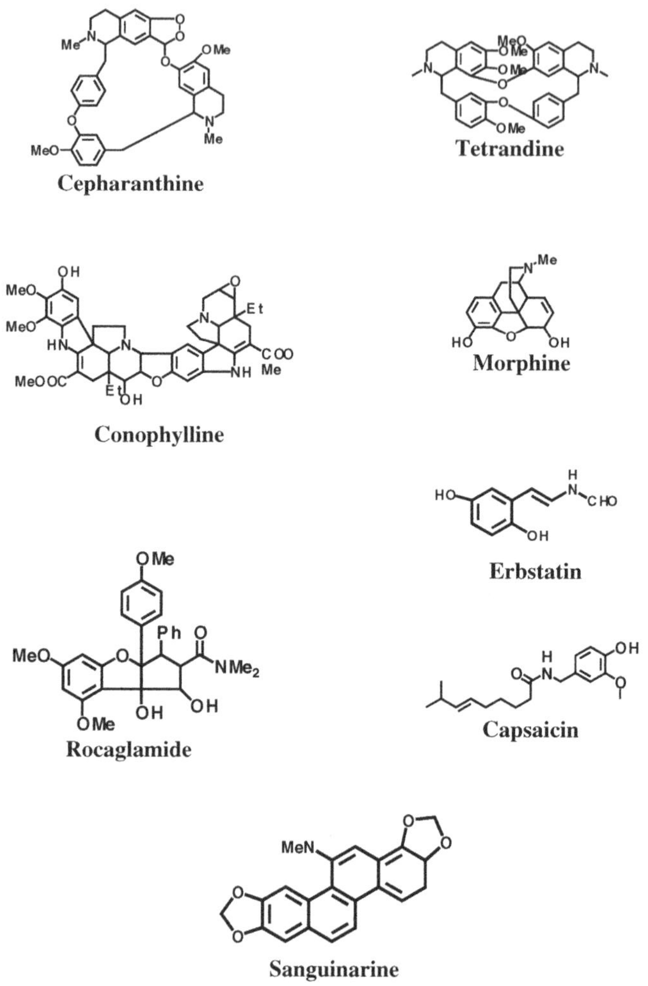

FIGURE 21.2C (continued) Structure of some plant-derived alkaloids that block NF-κB.

of these extracts can be very complex. However, although humans have been using complex mixtures for much longer than they have been using single isolated compounds or drugs, there is a need for careful standardization of dietary supplements and effective regulatory control to ensure human safety.

ACKNOWLEDGMENTS

We would like to thank Walter Pagel for a careful review of the manuscript. The contributions of Aggarwal, a Ransom Horne, Jr., Distinguished Professor of Cancer Research, have been supported by the Clayton Foundation for Research, a Department of Defense U.S. Army Breast Cancer Research Program grant

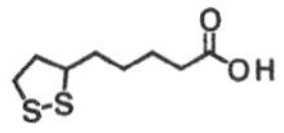

α-lipoic acid

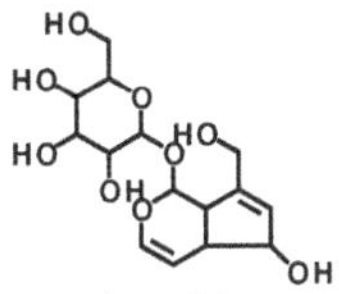

Aucubin

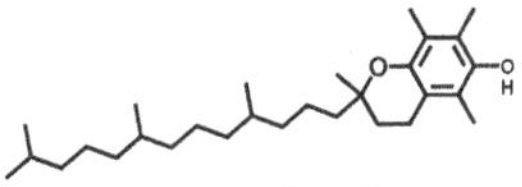

α-tocopherol

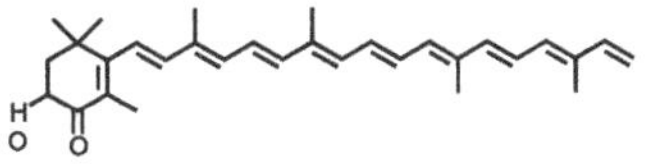

Astaxanthin

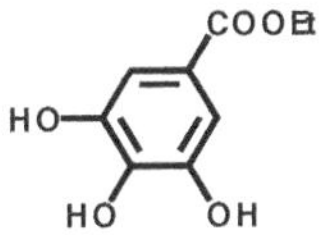

Ethylgallate

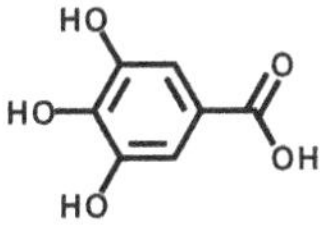

Gallic acid

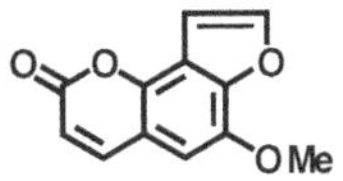

Sphondin

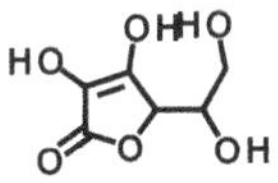

Vitamin C

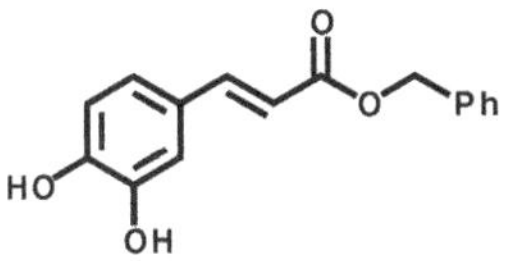

CAPE
(caffeic acid phenylethyl ester)

FIGURE 21.2D (continued) Structure of some plant-derived miscellaneous compounds that block NF-κB.

(BC010610), a PO1 grant (CA91844) from the National Institutes of Health on lung chemoprevention, and a P50 Head and Neck SPORE grant from the National Institutes of Health.

ABBREVIATIONS

IκBα: inhibitory subunit of NF-κB
IKK: IκB kinase

REFERENCES

1. Garg A, Aggarwal BB. Nuclear transcription factor-κB as a target for cancer drug development. *Leukemia* 2002; 16: 1053–1068.
2. Kumar A, Takada Y, Boriek AM, Aggarwal BB. Nuclear factor-κB: Its role in health and diseases. *J Mol Med* 2004; 82: 434–448.
3. Shishodia S, Aggarwal BB. Nuclear factor (NF)-κB regulates the expression of genes involved in transformation, proliferation, invasion, angiogenesis and metastasis of cancer. In *Molecular Targeting and Signal Transduction*. Ed., Rakesh Kumar PD; 2003; Kluwer Publishers: Belgium.
4. Lin A, Karin M. NF-κB in cancer: a marked target. *Semin Cancer Biol* 2003; 13: 107–114.
5. Orlowski RZ, Baldwin AS, Jr. NF-κB as a therapeutic target in cancer. *Trends Mol Med* 2002; 8: 385–389.
6. Valen G, Yan ZQ, Hansson GK. Nuclear factor κ-B and the heart. *J Am Coll Cardiol* 2001; 38: 307–314.
7. Jones WK, Brown M, Ren X, He S, McGuinness M. NF-κB as an integrator of diverse signaling pathways: the heart of myocardial signaling? *Cardiovasc Toxicol* 2003; 3: 229–254.
8. Shoelson SE, Lee J, Yuan M. Inflammation and the IKK beta/I kappa B/NF-κB axis in obesity- and diet-induced insulin resistance. *Int J Obes Relat Metab Disord* 2003; 27 Suppl. 3: S49–52.
9. Yang L, Cohn L, Zhang DH, Homer R, Ray A, Ray P. Essential role of nuclear factor κB in the induction of eosinophilia in allergic airway inflammation. *J Exp Med* 1998; 188: 1739–1750.
10. Das J, Chen CH, Yang L, Cohn L, Ray P, Ray A. A critical role for NF-κB in GATA3 expression and TH2 differentiation in allergic airway inflammation. *Nat Immunol* 2001; 2: 45–50.
11. Gagliardo R, Chanez P, Mathieu M, et al. Persistent activation of nuclear factor-κB signaling pathway in severe uncontrolled asthma. *Am J Respir Crit Care Med* 2003; 168: 1190-1198.
12. Roshak AK, Callahan JF, Blake SM. Small-molecule inhibitors of NF-κB for the treatment of inflammatory joint disease. *Curr Opin Pharmacol* 2002; 2: 316–321.
13. van Heel DA, Udalova IA, De Silva AP, et al. Inflammatory bowel disease is associated with a TNF polymorphism that affects an interaction between the OCT1 and NF(-κ)B transcription factors. *Hum Mol Genet* 2002; 11: 1281-1289.
14. Huang CJ, Nazarian R, Lee J, Zhao PM, Espinosa-Jeffrey A, de Vellis J. Tumor necrosis factor modulates transcription of myelin basic protein gene through nuclear factor-κB in a human oligodendroglioma cell line. *Int J Dev Neurosci* 2002; 20: 289–296.
15. Mattson MP, Camandola S. NF-kappaB in neuronal plasticity and neurodegenerative disorders. *J Clin Invest* 2001; 107: 247–254.
16. Kaltschmidt B, Uherek M, Volk B, Baeuerle PA, Kaltschmidt C. Transcription factor NF-κB is activated in primary neurons by amyloid beta peptides and in neurons surrounding early plaques from patients with Alzheimer disease. *Proc Natl Acad Sci USA* 1997; 94: 2642-2647.
17. Burke JR. Targeting I kappa B kinase for the treatment of inflammatory and other disorders. *Curr Opin Drug Discov Dev* 2003; 6: 720–728.

18. Yamamoto Y, Gaynor RB. Role of the NF-κB pathway in the pathogenesis of human disease states. *Curr Mol Med* 2001; 1: 287–296.

19. Yamamoto Y, Gaynor RB. Therapeutic potential of inhibition of the NF-κB pathway in the treatment of inflammation and cancer. *J Clin Invest* 2001; 107: 135–42.

20. Shobana S, Naidu KA. Antioxidant activity of selected Indian spices. *Prostaglandins Leukot Essent Fatty Acids* 2000; 62: 107–110.

21. Natarajan K, Singh S, Burke TR, Jr., Grunberger D, Aggarwal BB. Caffeic acid phenethyl ester is a potent and specific inhibitor of activation of nuclear transcription factor NF-κB. *Proc Natl Acad Sci USA* 1996; 93: 9090–9095.

22. Banerjee T, Valacchi G, Ziboh VA, van der Vliet A. Inhibition of TNFalpha-induced cyclooxygenase-2 expression by amentoflavone through suppression of NF-κB activation in A549 cells. *Mol Cell Biochem* 2002; 238: 105-110.

23. Gupta S, Afaq F, Mukhtar H. Involvement of nuclear factor-κB, Bax and Bcl-2 in induction of cell cycle arrest and apoptosis by apigenin in human prostate carcinoma cells. *Oncogene* 2002; 21: 3727–3738.

24. Shukla S, Gupta S. Molecular mechanisms for apigenin-induced cell-cycle arrest and apoptosis of hormone refractory human prostate carcinoma DU145 cells. *Mol Carcinog* 2004; 39: 114–126.

25. Choi JS, Choi YJ, Park SH, Kang JS, Kang YH. Flavones mitigate tumor necrosis factor-alpha-induced adhesion molecule upregulation in cultured human endothelial cells: role of nuclear factor-kappaB. *J Nutr* 2004; 134: 1013–1019.

26. Kang BY, Chung SW, Kim SH, Cho D, Kim TS. Involvement of nuclear factor-kappaB in the inhibition of interleukin-12 production from mouse macrophages by baicalein, a flavonoid in *Scutellaria baicalensis*. *Planta Med* 2003; 69: 687–691.

27. Cho MK, Park JW, Jang YP, Kim YC, Kim SG. Potent inhibition of lipopolysaccharide-inducible nitric oxide synthase expression by dibenzylbutyrolactone lignans through inhibition of I-κB α-phosphorylation and of p65 nuclear translocation in macrophages. *Int Immunopharmacol* 2002; 2: 105–116.

28. Murakami A, Matsumoto K, Koshimizu K, Ohigashi H. Effects of selected food factors with chemopreventive properties on combined lipopolysaccharide- and interferon-gamma-induced IκB degradation in RAW264.7 macrophages. *Cancer Lett* 2003; 195: 17–25.

29. Krakauer T, Li BQ, Young HA. The flavonoid baicalin inhibits superantigen-induced inflammatory cytokines and chemokines. *FEBS Lett* 2001; 500: 52–55.

30. Kim H, Kim YS, Kim SY, Suk K. The plant flavonoid wogonin suppresses death of activated C6 rat glial cells by inhibiting nitric oxide production. *Neurosci Lett* 2001; 309: 67–71.

31. Chen Y-C, Shen S-C, Hsu F-L. Biological activities of flavonoids. In: *Oriental Foods and Herbs,* ACS Symposium Series, 2003; Vol. 859.

32. Huang C, Huang Y, Li J, et al. Inhibition of benzo(a)pyrene diol-epoxide-induced transactivation of activated protein 1 and nuclear factor κB by black raspberry extracts. *Cancer Res* 2002; 62: 6857–6863.

33. Atalay M, Gordillo G, Roy S, et al. Anti-angiogenic property of edible berry in a model of hemangioma. *FEBS Lett* 2003; 544: 252–7.

34. Kresty LA, Morse MA, Morgan C, et al. Chemoprevention of esophageal tumorigenesis by dietary administration of lyophilized black raspberries. *Cancer Res* 2001; 61: 6112–6119.

35. Herring AC, Koh WS, Kaminski NE. Inhibition of the cyclic AMP signaling cascade and nuclear factor binding to CRE and kappaB elements by cannabinol, a minimally CNS-active cannabinoid. *Biochem Pharmacol* 1998; 55: 1013–1023.

36. Herring AC, Kaminski NE. Cannabinol-mediated inhibition of nuclear factor-κB, cAMP response element-binding protein, and interleukin-2 secretion by activated thymocytes. *J Pharmacol Exp Ther* 1999; 291: 1156–1163.

37. Hollister LE. Interactions of marijuana and THC with other drugs: what we don't, but should know. *Marihuana and Medicine*, New York, March 20–21, 1998.

38. An SJ, Pae HO, Oh GS, et al. Inhibition of TNF-alpha, IL-1beta, and IL-6 productions and NF-κB activation in lipopolysaccharide-activated RAW 264.7 macrophages by catalposide, an iridoid glycoside isolated from Catalpa ovata G. Don (Bignoniaceae). *Int Immunopharmacol* 2002; 2: 1173–1181.

39. Mackenzie GG, Carrasquedo F, Delfino JM, Keen CL, Fraga CG, Oteiza PI. Epicatechin, catechin, and dimeric procyanidins inhibit PMA-induced NF-κB activation at multiple steps in Jurkat T cells. *FASEB J* 2004; 18: 167–169.

40. Pianetti S, Guo S, Kavanagh KT, Sonenshein GE. Green tea polyphenol epigallocatechin-3 gallate inhibits Her-2/neu signaling, proliferation, and transformed phenotype of breast cancer cells. *Cancer Res* 2002; 62: 652–655.

41. Pan MH, Lin-Shiau SY, Ho CT, Lin JH, Lin JK. Suppression of lipopolysaccharide-induced nuclear factor-kappaB activity by theaflavin-3,3'-digallate from black tea and other polyphenols through down-regulation of IκB kinase activity in macrophages. *Biochem Pharmacol* 2000; 59: 357–367.

42. Ibanez E, Kubatova A, Senorans FJ, Cavero S, Reglero G, Hawthorne SB. Subcritical water extraction of antioxidant compounds from rosemary plants. *J Agric Food Chem* 2003; 51: 375–382.

43. van't Land B, Blijlevens NM, Marteijn J, et al. Role of curcumin and the inhibition of NF-κB in the onset of chemotherapy-induced mucosal barrier injury. *Leukemia* 2004; 18: 276–284.

44. Rao CV, Tokumo K, Rigotty J, Zang E, Kelloff G, Reddy BS. Chemoprevention of colon carcinogenesis by dietary administration of piroxicam, alpha-difluoromethylornithine, 16 alpha-fluoro-5-androsten-17-one, and ellagic acid individually and in combination. *Cancer Res* 1991; 51: 4528–4534.

45. Kumar A, Dhawan S, Aggarwal BB. Emodin (3-methyl-1,6,8-trihydroxyanthraquinone) inhibits TNF-induced NF-kappaB activation, IκB degradation, and expression of cell surface adhesion proteins in human vascular endothelial cells. *Oncogene* 1998; 17: 913–918.

46. Takada Y, Aggarwal BB. Flavopiridol inhibits NF-kappaB activation induced by various carcinogens and inflammatory agents through inhibition of IκBα kinase and p65 phosphorylation: abrogation of cyclin D1, cyclooxygenase-2, and matrix metalloprotease-9. *J Biol Chem* 2004; 279: 4750–4759.

47. Muraoka K, Shimizu K, Sun X, et al. Flavonoids exert diverse inhibitory effects on the activation of NF-kappaB. *Transplant Proc* 2002; 34: 1335–1340.

48. Gong L, Li Y, Nedeljkovic-Kurepa A, Sarkar FH. Inactivation of NF-κB by genistein is mediated via Akt signaling pathway in breast cancer cells. *Oncogene* 2003; 22: 4702–4709.

49. Wu MJ, Wang L, Ding HY, Weng CY, Yen JH. Glossogyne tenuifolia acts to inhibit inflammatory mediator production in a macrophage cell line by downregulating LPS-induced NF-κB. *J Biomed Sci* 2004; 11: 186–199.

50. Choi JH, Jeong TS, Kim DY, et al. Hematein inhibits atherosclerosis by inhibition of reactive oxygen generation and NF-κB-dependent inflammatory mediators in hyperlipidemic mice. *J Cardiovasc Pharmacol* 2003; 42: 287–295.

51. Hong JJ, Jeong TS, Choi JH, et al. Hematein inhibits tumor necrotic factor-alpha-induced vascular cell adhesion molecule-1 and NF-kappaB-dependent gene expression in human vascular endothelial cells. *Biochem Biophys Res Commun* 2001; 281: 1127–1133.

52. Behar A, Pujade-Lauraine E, Maurel A, et al. The pathophysiological mechanism of fluid retention in advanced cancer patients treated with docetaxel, but not receiving corticosteroid comedication. *Br J Clin Pharmacol* 1997; 43: 653–658.

53. Yadav PN, Liu Z, Rafi MM. A diarylheptanoid from lesser galangal (*Alpinia officinarum*) inhibits proinflammatory mediators via inhibition of mitogen-activated protein kinase, p44/42, and transcription factor nuclear factor-κB. *J Pharmacol Exp Ther* 2003; 305: 925–931.

54. Bork PM, Bacher S, Schmitz ML, Kaspers U, Heinrich M. Hypericin as a non-antioxidant inhibitor of NF-κB. *Planta Med* 1999; 65: 297–300.

55. Agostinis P, Vantieghem A, Merlevede W, de Witte PA. Hypericin in cancer treatment: more light on the way. *Int J Biochem Cell Biol* 2002; 34: 221–241.

56. Kelm MA, Nair MG, Strasburg GM, DeWitt DL. Antioxidant and cyclooxygenase inhibitory phenolic compounds from Ocimum sanctum Linn. *Phytomedicine* 2000; 7: 7–13.

57. Ishii R, Horie M, Saito K, Arisawa M, Kitanaka S. Inhibition of lipopolysaccharide-induced pro-inflammatory cytokine expression via suppression of nuclear factor-κB activation by *Mallotus japonicus* phloroglucinol derivatives. *Biochim Biophys Acta* 2003; 1620: 108–118.

58. Lombardi-Boccia G, Lucarini M, Lanzi S, Aguzzi A, Cappelloni M. Nutrients and antioxidant molecules in yellow plums (*Prunus domestica* L.) from conventional and organic productions: a comparative study. *J Agric Food Chem* 2004; 52: 90–94.

59. Kim SH, Shin KJ, Kim D, et al. Luteolin inhibits the nuclear factor-kappa B transcriptional activity in Rat-1 fibroblasts. *Biochem Pharmacol* 2003; 66: 955–963.

60. Ono K, Yoshiike Y, Takashima A, Hasegawa K, Naiki H, Yamada M. Potent anti-amyloidogenic and fibril-destabilizing effects of polyphenols in vitro: implications for the prevention and therapeutics of Alzheimer's disease. *J Neurochem* 2003; 87: 172–181.

61. Hollman PC, Katan MB. Health effects and bioavailability of dietary flavonols. *Free Radic Res* 1999; 31 Suppl.: S75–80.

62. Noda Y, Kneyuki T, Igarashi K, Mori A, Packer L. Antioxidant activity of nasunin, an anthocyanin in eggplant peels. *Toxicology* 2000; 148: 119–23.

63. Brennan P, O'Neill LA. Inhibition of nuclear factor kappaB by direct modification in whole cells — mechanism of action of nordihydroguaiaritic acid, curcumin and thiol modifiers. *Biochem Pharmacol* 1998; 55: 965–973.

64. Tang S, Bremner P, Kortenkamp A, et al. Biflavonoids with cytotoxic and anti-bacterial activity from *Ochna macrocalyx*. *Planta Med* 2003; 69: 247–253.

65. Aoki K, Maruta H, Uchiumi F, Hatano T, Yoshida T, Tanuma S. A macrocircular ellagitannin, oenothein B, suppresses mouse mammary tumor gene expression via inhibition of poly(ADP-ribose) glycohydrolase. *Biochem Biophys Res Commun* 1995; 210: 329–337.

66. Yun JM, Kwon H, Hwang JK. *In vitro* anti-inflammatory activity of panduratin A isolated from *Kaempferia pandurata* in RAW264.7 cells. *Planta Med* 2003; 69: 1102–1108.

67. Llopiz N, Puiggros F, Cespedes E, et al. Antigenotoxic effect of grape seed procyanidin extract in fao cells submitted to oxidative stress. *J Agric Food Chem* 2004; 52: 1083–1087.

68. Wu TW, Zeng LH, Wu J, et al. Molecular structure and antioxidant specificity of purpurogallin in three types of human cardiovascular cells. *Biochem Pharmacol* 1996; 52: 1073-1080.

69. O'Coinceanainn M, Astill C, Baderschneider B. Coordination of aluminium with purpurogallin and theaflavin digallate. *J Inorg Biochem* 2003; 96: 463–468.

70. Peng Q, Wei Z, Lau BH. Pycnogenol inhibits tumor necrosis factor-alpha-induced nuclear factor kappa B activation and adhesion molecule expression in human vascular endothelial cells. *Cell Mol Life Sci* 2000; 57: 834–841.

71. Packer L. Antioxidant and biological activity of a French maritime pine-bark extract. *219th ACS National Meeting, 2000*.

72. Cho SY, Park SJ, Kwon MJ, et al. Quercetin suppresses proinflammatory cytokines production through MAP kinases and NF-κB pathway in lipopolysaccharide-stimulated macrophage. *Mol Cell Biochem* 2003; 243: 153–160.

73. Martin G, Bogdanowicz P, Domagala F, Ficheux H, Pujol JP. Rhein inhibits interleukin-1β-induced activation of MEK/ERK pathway and DNA binding of NF-κB and AP-1 in chondrocytes cultured in hypoxia: a potential mechanism for its disease-modifying effect in osteoarthritis. *Inflammation* 2003; 27: 233–246.

74. Li LC, Shen F, Hou Q, Cheng GF. Inhibitory effect and mechanism of action of sanggenon C on human polymorphonuclear leukocyte adhesion to human synovial cells. *Acta Pharmacol Sin* 2002; 23: 138–142.

75. Dhanalakshmi S, Singh RP, Agarwal C, Agarwal R. Silibinin inhibits constitutive and TNFα-induced activation of NF-κB and sensitizes human prostate carcinoma DU145 cells to TNFα-induced apoptosis. *Oncogene* 2002; 21: 1759–1767.

76. Manna SK, Mukhopadhyay A, Van NT, Aggarwal BB. Silymarin suppresses TNF-induced activation of NF-κB, c-Jun N-terminal kinase, and apoptosis. *J Immunol* 1999; 163: 6800–6809.

77. Gallo D, Giacomelli S, Ferlini C, et al. Antitumour activity of the silybin-phosphatidylcholine complex, IdB 1016, against human ovarian cancer. *Eur J Cancer* 2003; 39: 2403–2410.

78. Hwang BY, Lee JH, Nam JB, Hong YS, Lee JJ. Lignans from *Saururus chinensis* inhibiting the transcription factor NF-kappaB. *Phytochemistry* 2003; 64: 765–771.

79. Hwang BY, Lee JH, Jung HS, et al. Sauchinone, a lignan from Saururus chinensis, suppresses iNOS expression through the inhibition of transactivation activity of RelA of NF-κB. *Planta Med* 2003; 69: 1096–1101.

80. Bracke ME, Depypere HT, Boterberg T, et al. Influence of tangeretin on tamoxifen's therapeutic benefit in mammary cancer. *J Natl Cancer Inst* 1999; 91: 354–359.

81. Kobori M, Yang Z, Gong D, et al. Wedelolactone suppresses LPS-induced caspase-11 expression by directly inhibiting the IKK complex. *Cell Death Differ* 2004; 11: 123–130.

82. Chun KS, Kang JY, Kim OH, Kang H, Surh YJ. Effects of yakuchinone A and yakuchinone B on the phorbol ester-induced expression of COX-2 and iNOS and activation of NF-kappaB in mouse skin. *J Environ Pathol Toxicol Oncol* 2002; 21: 131–139.

83. de las Heras B, Navarro A, Diaz-Guerra MJ, et al. Inhibition of NOS-2 expression in macrophages through the inactivation of NF-κB by andalusol. *Br J Pharmacol* 1999; 128: 605–612.

84. Chainy GB, Manna SK, Chaturvedi MM, Aggarwal BB. Anethole blocks both early and late cellular responses transduced by tumor necrosis factor: effect on NF-κB, AP-1, JNK, MAPKK and apoptosis. *Oncogene* 2000; 19: 2943–2950.

85. Murakami Y, Shoji M, Hanazawa S, Tanaka S, Fujisawa S. Preventive effect of bis-eugenol, a eugenol ortho dimer, on lipopolysaccharide-stimulated nuclear factor κB activation and inflammatory cytokine expression in macrophages. *Biochem Pharmacol* 2003; 66: 1061–1066.

86. Aldieri E, Atragene D, Bergandi L, et al. Artemisinin inhibits inducible nitric oxide synthase and nuclear factor NF-κB activation. *FEBS Lett* 2003; 552: 141–144.

87. Haridas V, Arntzen CJ, Gutterman JU. Avicins, a family of triterpenoid saponins from *Acacia victoriae* (Bentham), inhibit activation of nuclear factor-kappaB by inhibiting both its nuclear localization and ability to bind DNA. *Proc Natl Acad Sci USA* 2001; 98: 11557–11562.

88. Akudugu J, Gade G, Bohm L. Cytotoxicity of azadirachtin A in human glioblastoma cell lines. *Life Sci* 2001; 68: 1153–1160.

89. Adom KK, Sorrells ME, Liu RH. Phytochemical profiles and antioxidant activity of wheat varieties. *J Agric Food Chem* 2003; 51: 7825–7834.

90. Prakash P, Liu C, Hu KQ, Krinsky NI, Russell RM, Wang XD. Beta-carotene and beta-apo-14′-carotenoic acid prevent the reduction of retinoic acid receptor beta in benzo[a]pyrene-treated normal human bronchial epithelial cells. *J Nutr* 2004; 134: 667–673.

91. Uchiyama S, Yamaguchi M. Inhibitory effect of beta-cryptoxanthin on osteoclast-like cell formation in mouse marrow cultures. *Biochem Pharmacol* 2004; 67: 1297–1305.

92. Pae HO, Cho H, Oh GS, et al. Bakuchiol from *Psoralea corylifolia* inhibits the expression of inducible nitric oxide synthase gene via the inactivation of nuclear transcription factor-kappaB in RAW 264.7 macrophages. *Int Immunopharmacol* 2001; 1: 1849–1855.

93. Takada Y, Aggarwal BB. Betulinic acid suppresses carcinogen-induced NF-kappa B activation through inhibition of I kappa B alpha kinase and p65 phosphorylation: abrogation of cyclooxygenase-2 and matrix metalloprotease-9. *J Immunol* 2003; 171: 3278–3286.

94. Lo AH, Liang YC, Lin-Shiau SY, Ho CT, Lin JK. Carnosol, an antioxidant in rosemary, suppresses inducible nitric oxide synthase through down-regulating nuclear factor-κB in mouse macrophages. *Carcinogenesis* 2002; 23: 983–991.

95. Jin HZ, Hwang BY, Kim HS, Lee JH, Kim YH, Lee JJ. Antiinflammatory constituents of Celastrus orbiculatus inhibit the NF-κB activation and NO production. *J Nat Prod* 2002; 65: 89–91.

96. Koo TH, Lee JH, Park YJ, et al. A sesquiterpene lactone, costunolide, from Magnolia grandiflora inhibits NF-κB by targeting I κB phosphorylation. *Planta Med* 2001; 67: 103–107.

97. Sun IC, Kashiwada Y, Morris-Natschke SL, Lee KH. Plant-derived terpenoids and analogues as anti-HIV agents. *Curr Top Med Chem* 2003; 3: 155–169.

98. Blaskovich MA, Sun J, Cantor A, Turkson J, Jove R, Sebti SM. Discovery of JSI-124 (cucurbitacin I), a selective Janus kinase/signal transducer and activator of transcription 3 signaling pathway inhibitor with potent antitumor activity against human and murine cancer cells in mice. *Cancer Res* 2003; 63: 1270–1279.

99. Whan Han J, Gon Lee B, Kee Kim Y, et al. Ergolide, sesquiterpene lactone from *Inula britannica*, inhibits inducible nitric oxide synthase and cyclo-oxygenase-2 expression in RAW 264.7 macrophages through the inactivation of NF-kappaB. *Br J Pharmacol* 2001; 133: 503–512.

100. Hwang BY, Lee JH, Koo TH, et al. Kaurane diterpenes from *Isodon japonicus* inhibit nitric oxide and prostaglandin E2 production and NF-κB activation in LPS-stimulated macrophage RAW264.7 cells. *Planta Med* 2001; 67: 406–410.

101. Castrillo A, de Las Heras B, Hortelano S, Rodriguez B, Villar A, Bosca L. Inhibition of the nuclear factor kappa B (NF-κB) pathway by tetracyclic kaurene diterpenes in macrophages. Specific effects on NF-κB-inducing kinase activity and on the coordinate activation of ERK and p38 MAPK. *J Biol Chem* 2001; 276: 15854–15860.

102. Kim EJ, Jin HK, Kim YK, et al. Suppression by a sesquiterpene lactone from *Carpesium divaricatum* of inducible nitric oxide synthase by inhibiting nuclear factor-kappaB activation. *Biochem Pharmacol* 2001; 61: 903–910.

103. Muller S, Murillo R, Castro V, Brecht V, Merfort I. Sesquiterpene lactones from *Montanoa hibiscifolia* that inhibit the transcription factor NF-κB. *J Nat Prod* 2004; 67: 622–630.

104. Wei Z, Peng Q, Lau BH, Shah V. Ginkgo biloba inhibits hydrogen peroxide-induced activation of nuclear factor κB in vascular endothelial cells. *Gen Pharmacol* 1999; 33: 369–375.

105. Keum YS, Han SS, Chun KS, et al. Inhibitory effects of the ginsenoside Rg3 on phorbol ester-induced cyclooxygenase-2 expression, NF-κB activation and tumor promotion. *Mutat Res* 2003; 523–524: 75–85.

106. Wang JY, Guo JS, Li H, Liu SL, Zern MA. Inhibitory effect of glycyrrhizin on NF-κB binding activity in CCl4- plus ethanol-induced liver cirrhosis in rats. *Liver* 1998; 18: 180–185.

107. Schorr K, Garcia-Pineres AJ, Siedle B, Merfort I, Da Costa FB. Guaianolides from *Viguiera gardneri* inhibit the transcription factor NF-κB. *Phytochemistry* 2002; 60: 733–740.

108. Lyss G, Schmidt TJ, Merfort I, Pahl HL. Helenalin, an anti-inflammatory sesquiterpene lactone from Arnica, selectively inhibits transcription factor NF-kappaB. *Biol Chem* 1997; 378: 951–961.

109. Ojo-Amaize EA, Kapahi P, Kakkanaiah VN, et al. Hypoestoxide, a novel anti-inflammatory natural diterpene, inhibits the activity of IκB kinase. *Cell Immunol* 2001; 209: 149–157.

110. Ojo-Amaize EA, Nchekwube EJ, Cottam HB, et al. Hypoestoxide, a natural nonmutagenic diterpenoid with antiangiogenic and antitumor activity: possible mechanisms of action. *Cancer Res* 2002; 62: 4007–4014.

111. Gould MN, Moore CJ, Zhang R, Wang B, Kennan WS, Haag JD. Limonene chemoprevention of mammary carcinoma induction following direct in situ transfer of v-Ha-ras. *Cancer Res* 1994; 54: 3540–3543.

112. Chew BP, Park JS. Carotenoid action on the immune response. *J Nutr* 2004; 134: 257S–261S.

113. Khachik F, Beecher GR, Smith JC, Jr. Lutein, lycopene, and their oxidative metabolites in chemoprevention of cancer. *J Cell Biochem Suppl* 1995; 22: 236–246.

114. Liu C, Lian F, Smith DE, Russell RM, Wang XD. Lycopene supplementation inhibits lung squamous metaplasia and induces apoptosis via up-regulating insulin-like growth factor-binding protein 3 in cigarette smoke-exposed ferrets. *Cancer Res* 2003; 63: 3138–3144.

115. Selvendiran K, Senthilnathan P, Magesh V, Sakthisekaran D. Modulatory effect of Piperine on mitochondrial antioxidant system in Benzo(a)pyrene-induced experimental lung carcinogenesis. *Phytomedicine* 2004; 11: 85–89.

116. Castro V, Rungeler P, Murillo R, et al. Study of sesquiterpene lactones from *Milleria quinqueflora* on their anti-inflammatory activity using the transcription factor NF-κB as molecular target. *Phytochemistry* 2000; 53: 257–263.

117. Hehner SP, Heinrich M, Bork PM, et al. Sesquiterpene lactones specifically inhibit activation of NF-kappa B by preventing the degradation of I kappa B-alpha and I kappa B-beta. *J Biol Chem* 1998; 273: 1288–1297.

118. Kwok BH, Koh B, Ndubuisi MI, Elofsson M, Crews CM. The anti-inflammatory natural product parthenolide from the medicinal herb Feverfew directly binds to and inhibits IκB kinase. *Chem Biol* 2001; 8: 759–766.

119. Dirsch VM, Kiemer AK, Wagner H, Vollmar AM. The triterpenoid quinonemethide pristimerin inhibits induction of inducible nitric oxide synthase in murine macrophages. *Eur J Pharmacol* 1997; 336: 211–217.

120. Sylvester J, Liacini A, Li WQ, Dehnade F, Zafarullah M. *Tripterygium wilfordii* Hook F extract suppresses proinflammatory cytokine-induced expression of matrix metalloproteinase genes in articular chondrocytes by inhibiting activating protein-1 and nuclear factor-κB activities. *Mol Pharmacol* 2001; 59: 1196–1205.

121. Qiu D, Kao PN. Immunosuppressove and anti-inflammatory mechanisms of triptolide, the principal active diterpenoid from the Chinese medicinal herb tripterygium wilfordii Hook. f. *Drugs in R&D* 2003; 4: 1–18.

122. Shishodia S, Majumdar S, Banerjee S, Aggarwal BB. Ursolic acid inhibits nuclear factor-κB activation induced by carcinogenic agents through suppression of IκBα kinase and p65 phosphorylation: correlation with down-regulation of cyclooxygenase 2, matrix metalloproteinase 9, and cyclin D1. *Cancer Res* 2003; 63: 4375–4383.

123. Park EJ, Pezzuto JM. Botanicals in cancer chemoprevention. *Cancer Metastasis Rev* 2002; 21: 231–255.

124. Misico RI, Song LL, Veleiro AS, et al. Induction of quinone reductase by withanolides. *J Nat Prod* 2002; 65: 677–680.

125. Sancho R, Lucena C, Macho A, et al. Immunosuppressive activity of capsaicinoids: capsiate derived from sweet peppers inhibits NF-kappaB activation and is a potent antiinflammatory compound in vivo. *Eur J Immunol* 2002; 32: 1753–1763.

126. Chen CW, Lee ST, Wu WT, Fu WM, Ho FM, Lin WW. Signal transduction for inhibition of inducible nitric oxide synthase and cyclooxygenase-2 induction by capsaicin and related analogs in macrophages. *Br J Pharmacol* 2003; 140: 1077–1087.

127. Singh S, Natarajan K, Aggarwal BB. Capsaicin (8-methyl-N-vanillyl-6-nonenamide) is a potent inhibitor of nuclear transcription factor-κB activation by diverse agents. *J Immunol* 1996; 157: 4412–4420.

128. Okamoto M, Ono M, Baba M. Suppression of cytokine production and neural cell death by the anti-inflammatory alkaloid cepharanthine: a potential agent against HIV-1 encephalopathy. *Biochem Pharmacol* 2001; 62: 747–753.

129. Gohda J, Inoue J, Umezawa K. Down-regulation of TNF-α receptors by conophylline in human T-cell leukemia cells. *Int J Oncol* 2003; 23: 1373–1379.

130. Chang YC, Chang FR, Khalil AT, Hsieh PW, Wu YC. Cytotoxic benzophenanthridine and benzylisoquinoline alkaloids from *Argemone mexicana*. *Z Naturforsch* [C] 2003; 58: 521–526.

131. Kang YJ, Lee YS, Lee GW, et al. Inhibition of activation of nuclear factor κB is responsible for inhibition of inducible nitric oxide synthase expression by higenamine, an active component of aconite root. *J Pharmacol Exp Ther* 1999; 291: 314–320.

132. Tachibana Y, Kikuzaki H, Lajis NH, Nakatani N. Antioxidative activity of carbazoles from *Murraya koenigii* leaves. *J Agric Food Chem* 2001; 49: 5589–5594.

133. Ramsewak RS, Nair MG, Strasburg GM, DeWitt DL, Nitiss JL. Biologically active carbazole alkaloids from *Murraya koenigii*. *J Agric Food Chem* 1999; 47: 444–447.

134. Sueoka E, Sueoka N, Kai Y, et al. Anticancer activity of morphine and its synthetic derivative, KT-90, mediated through apoptosis and inhibition of NF-κB activation. *Biochem Biophys Res Commun* 1998; 252: 566–570.

135. Rajagopal A, Vassilopoulou-Sellin R, Palmer JL, Kaur G, Bruera E. Hypogonadism and sexual dysfunction in male cancer survivors receiving chronic opioid therapy. *J Pain Symptom Manage* 2003; 26: 1055–1061.

136. Chaturvedi MM, Kumar A, Darnay BG, Chainy GB, Agarwal S, Aggarwal BB. Sanguinarine (pseudochelerythrine) is a potent inhibitor of NF-kappaB activation, IκBα phosphorylation, and degradation. *J Biol Chem* 1997; 272: 30129–30134.

137. Rauscher FM, Sanders RA, Watkins JB, 3rd. Effects of piperine on antioxidant pathways in tissues from normal and streptozotocin-induced diabetic rats. *J Biochem Mol Toxicol* 2000; 14: 329–334.

138. Baumann B, Bohnenstengel F, Siegmund D, et al. Rocaglamide derivatives are potent inhibitors of NF-κB activation in T-cells. *J Biol Chem* 2002; 277: 44791–44800.

139. Ye J, Ding M, Zhang X, Rojanasakul Y, Shi X. On the role of hydroxyl radical and the effect of tetrandrine on nuclear factor — kappaB activation by phorbol 12-myristate 13-acetate. *Ann Clin Lab Sci* 2000; 30: 65–71.

140. Bork PM, Schmitz ML, Kuhnt M, Escher C, Heinrich M. Sesquiterpene lactone containing Mexican Indian medicinal plants and pure sesquiterpene lactones as potent inhibitors of transcription factor NF-κB. *FEBS Lett* 1997; 402: 85–90.

141. Bork PM, Schmitz ML, Weimann C, Kist M, Heinrich M. Nahua Indian medicinal plants (Mexico). Inhibitory activity on NF-κB antibacterial effects. *Phytochemistry* 1996; 3: 263–269.

142. Hirsch K, Danilenko M, Giat J, et al. Effect of purified allicin, the major ingredient of freshly crushed garlic, on cancer cell proliferation. *Nutr Cancer* 2000; 38: 245–254.

143. Oommen S, Anto RJ, Srinivas G, Karunagaran D. Allicin (from garlic) induces caspase-mediated apoptosis in cancer cells. *Eur J Pharmacol* 2004; 485: 97–103.

144. Suzuki YJ, Mizuno M, Packer L. Signal transduction for nuclear factor-kappa B activation. Proposed location of antioxidant-inhibitable step. *J Immunol* 1994; 153: 5008–5015.

145. Heinrich M, Bork PM, Schmitz ML, Rimpler H, Frei B, Sticher O. Pheophorbide A from *Solanum diflorum* interferes with NF-κB activation. *Planta Med* 2001; 67: 156–157.

146. van Lieshout EM, Posner GH, Woodard BT, Peters WH. Effects of the sulforaphane analog compound 30, indole-3-carbinol, D-limonene or relafen on glutathione S-transferases and glutathione peroxidase of the rat digestive tract. *Biochim Biophys Acta* 1998; 1379: 325–336.

147. Chinni SR, Li Y, Upadhyay S, Koppolu PK, Sarkar FH. Indole-3-carbinol (I3C) induced cell growth inhibition, G1 cell cycle arrest and apoptosis in prostate cancer cells. *Oncogene* 2001; 20: 2927–2936.

148. Jeong HJ, Koo HN, Na HJ, et al. Inhibition of TNF-α and IL-6 production by Aucubin through blockade of NF-κB activation RBL-2H3 mast cells. *Cytokine* 2002; 18: 252–259.

149. Srinivas G, Annab LA, Gopinath G, Banerji A, Srinivas P. Antisense blocking of BRCA1 enhances sensitivity to plumbagin but not tamoxifen in BG-1 ovarian cancer cells. *Mol Carcinog* 2004; 39: 15–25.

150. Nakamura Y, Murakami A, Ohto Y, Torikai K, Tanaka T, Ohigashi H. Suppression of tumor promoter-induced oxidative stress and inflammatory responses in mouse skin by a superoxide generation inhibitor 1′-acetoxychavicol acetate. *Cancer Res* 1998; 58: 4832–4839.

151. Murase T, Kume N, Hase T, et al. Gallates inhibit cytokine-induced nuclear translocation of NF-κB and expression of leukocyte adhesion molecules in vascular endothelial cells. *Arterioscler Thromb Vasc Biol* 1999; 19: 1412–1420.

152. Milner JA. Mechanisms by which garlic and allyl sulfur compounds suppress carcinogen bioactivation. Garlic and carcinogenesis. *Adv Exp Med Biol* 2001; 492: 69–81.

153. Bode AM, Ma WY, Surh YJ, Dong Z. Inhibition of epidermal growth factor-induced cell transformation and activator protein 1 activation by [6]-gingerol. *Cancer Res* 2001; 61: 850–853.

154. Sani BP, Rao PL. Antibiotic principles of *Garcinia morella*. VII. Antiprotozoal activity of morellin, neomorellin & other insoluble neutral phenols of the seed coat of *Garcinia morella*. *Indian J Exp Biol* 1966; 4: 27–28.

155. Yang LL, Liang YC, Chang CW, et al. Effects of sphondin, isolated from *Heracleum laciniatum*, on IL-1β-induced cyclooxygenase-2 expression in human pulmonary epithelial cells. *Life Sci* 2002; 72: 199–213.

156. Kohda H, Takeda O, Tanaka S, et al. Isolation of inhibitors of adenylate cyclase from dan-shen, the root of *Salvia miltiorrhiza*. *Chem Pharm Bull (Tokyo)* 1989; 37: 1287–90.

157. Bylka W, Matlawska I. Flavonoids and free phenolic acids from *Phytolacca americana* L. leaves. *Acta Pol Pharm* 2001; 58: 69–72.

158. Liu Y, Fang J, Lei T, Wang W, Lin A. Anti-endotoxic effects of syringic acid of *Radix isatidis*. *J Huazhong Univ Sci Technol Med Sci* 2003; 23: 206–208.

159. Kampa M, Alexaki VI, Notas G, et al. Antiproliferative and apoptotic effects of selective phenolic acids on T47D human breast cancer cells: potential mechanisms of action. *Breast Cancer Res* 2004; 6: R63–R74.

160. Sliva D, Labarrere C, Slivova V, Sedlak M, Lloyd FP, Jr., Ho NW. *Ganoderma lucidum* suppresses motility of highly invasive breast and prostate cancer cells. *Biochem Biophys Res Commun* 2002; 298: 603–612.

161. Tanaka T, Kohno H, Shimada R, et al. Prevention of colonic aberrant crypt foci by dietary feeding of garcinol in male F344 rats. *Carcinogenesis* 2000; 21: 1183–1189.

162. Borek C. Antioxidant health effects of aged garlic extract. *J Nutr* 2001; 131: 1010S–1015S.

163. Manna SK, Bueso-Ramos C, Alvarado F, Aggarwal BB. Calagualine inhibits nuclear transcription factors-κB activated by various inflammatory and tumor promoting agents. *Cancer Lett* 2003; 190: 171–182.

164. Manna SK, Mukhopadhyay A, Aggarwal BB. Resveratrol suppresses TNF-induced activation of nuclear transcription factors NF-κB, activator protein-1, and apoptosis: potential role of reactive oxygen intermediates and lipid peroxidation. *J Immunol* 2000; 164: 6509–6519.

165. Holmes-McNary M, Baldwin Jr. AS. Chemopreventive properties of trans-resveratrol are associated with inhibition of activation of the IκB kinase. *Cancer Res* 2000; 60: 3477–3483.

166. Ashikawa K, Majumdar S, Banerjee S, Bharti AC, Shishodia S, Aggarwal BB. Piceatannol inhibits TNF-induced NF-κB activation and NF-κB-mediated gene expression through suppression of IκBα kinase and p65 phosphorylation. *J Immunol* 2002; 169: 6490–6497.

167. Ide N, Lau BH. Garlic compounds minimize intracellular oxidative stress and inhibit nuclear factor-κB activation. *J Nutr* 2001; 131: 1020S–1026S.

168. Keiss HP, Dirsch VM, Hartung T, et al. Garlic (Allium sativum L.) modulates cytokine expression in lipopolysaccharide-activated human blood thereby inhibiting NF-κB activity. *J Nutr* 2003; 133: 2171–2175.

169. Horakova O, Kalamar J, Sopinska M, Horak F. Presence of α-lipoic acid in natural raw materials. *Cesko-Slovenska Farmacie* 1964; 13: 107–110.

170. Sen CK, Tirosh O, Roy S, Kobayashi MS, Packer L. A positively charged alpha-lipoic acid analogue with increased cellular uptake and more potent immunomodulatory activity. *Biochem Biophys Res Commun* 1998; 247: 223–228.

171. Suzuki YJ, Aggarwal BB, Packer L. Alpha-lipoic acid is a potent inhibitor of NF-kappa B activation in human T cells. *Biochem Biophys Res Commun* 1992; 189: 1709–1715.

172. Shi D, Jiang BH. Antioxidant properties of apple juice and its protection against Cr(VI)-induced cellular injury. *J Environ Pathol Toxicol Oncol* 2002; 21: 233–242.

173. Lee SJ, Bai SK, Lee KS, et al. Astaxanthin inhibits nitric oxide production and inflammatory gene expression by suppressing I(kappa)B kinase-dependent NF-kappaB activation. *Mol Cells* 2003; 16: 97–105.

174. Bobek P, Galbavy S. Effect of pleuran (beta-glucan from *Pleurotus ostreatus*) on the antioxidant status of the organism and on dimethylhydrazine-induced precancerous lesions in rat colon. *Br J Biomed Sci* 2001; 58: 164–168.

175. Nigro ND, Bull AW, Wilson PS, Soullier BK, Alousi MA. Combined inhibitors of carcinogenesis: effect on azoxymethane-induced intestinal cancer in rats. *J Natl Cancer Inst* 1982; 69: 103–107.

176. Aguilar JL, Rojas P, Marcelo A, et al. Anti-inflammatory activity of two different extracts of *Uncaria tomentosa* (Rubiaceae). *J Ethnopharmacol* 2002; 81: 271–276.

177. Wargovich MJ, Woods C, Eng VW, Stephens LC, Gray K. Chemoprevention of N-nitrosomethylbenzylamine-induced esophageal cancer in rats by the naturally occurring thioether, diallyl sulfide. *Cancer Res* 1988; 48: 6872–6875.

178. Hashimoto T, Suganuma M, Fujiki H, Yamada M, Kohno T, Asakawa Y. Isolation and synthesis of TNF-alpha release inhibitors from Fijian kawa (*Piper methysticum*). *Phytomedicine* 2003; 10: 309–317.

179. Kanauchi O, Serizawa I, Araki Y, et al. Germinated barley foodstuff, a prebiotic product, ameliorates inflammation of colitis through modulation of the enteric environment. *J Gastroenterol* 2003; 38: 134–141.

180. Kim OK, Murakami A, Takahashi D, et al. An avocado constituent, persenone A, suppresses expression of inducible forms of nitric oxide synthase and cyclooxygenase in macrophages, and hydrogen peroxide generation in mouse skin. *Biosci Biotechnol Biochem* 2000; 64: 2504–2507.

181. Schubert SY, Neeman I, Resnick N. A novel mechanism for the inhibition of NF-kappaB activation in vascular endothelial cells by natural antioxidants. *Faseb J* 2002; 16: 1931–1933.

182. Geng Z, Rong Y, Lau BH. S-allyl cysteine inhibits activation of nuclear factor κB in human T cells. *Free Radic Biol Med* 1997; 23: 345–350.

183. Riehemann K, Behnke B, Schulze-Osthoff K. Plant extracts from stinging nettle (Urtica dioica), an antirheumatic remedy, inhibit the proinflammatory transcription factor NF-κB. *FEBS Lett* 1999; 442: 89–94.

184. Momin RA, De Witt DL, Nair MG. Inhibition of cyclooxygenase (COX) enzymes by compounds from *Daucus carota* L. Seeds. *Phytother Res* 2003; 17: 976–979.

185. Carcamo JM, Pedraza A, Borquez-Ojeda O, Golde DW. Vitamin C suppresses TNF alpha-induced NF kappa B activation by inhibiting I kappa B alpha phosphorylation. *Biochemistry* 2002; 41: 12995–13002.

186. Bowie AG, O'Neill LA. Vitamin C inhibits NF-kappa B activation by TNF via the activation of p38 mitogen-activated protein kinase. *J Immunol* 2000; 165: 7180–7188.

187. Siler U, Barella L, Spitzer V, et al. Lycopene and Vitamin E interfere with autocrine/paracrine loops in the Dunning prostate cancer model. *Faseb J* 2004.

188. Egger T, Schuligoi R, Wintersperger A, Amann R, Malle E, Sattler W. Vitamin E (alpha-tocopherol) attenuates cyclo-oxygenase 2 transcription and synthesis in immortalized murine BV-2 microglia. *Biochem J* 2003; 370: 459–467.

22 Ascorbic Acid and Oxidative Damage to DNA: Interrelationships and Possible Relevance to Lifestyle-Related Diseases

I.F.F. Benzie, C.K. Chan, S.W. Choi, and Y.T. Szeto
The Hong Kong Polytechnic University, Hong Kong, China

CONTENTS

22.1 INTRODUCTION

Ascorbic acid (vitamin C) is the major water-soluble antioxidant in the body and, for humans, must be supplied in the diet. The U.S.-recommended daily intake of this dietary derived antioxidant is 75 mg/d for women and 90 mg/d for men, with higher intakes recommended for smokers and pregnant women. Ascorbic acid is present in fasting blood plasma of healthy adults at around 25 to 80 µmol/l, but its concentration within nucleated cells and intraocular fluids is 50 to 100 times higher than the plasma level. Deficiency of vitamin C causes a syndrome known as scurvy, in which collagen and blood vessels are weakened. This potentially fatal condition is rare in developed societies nowadays, as very few individuals have an intake of less than 10 mg/d of vitamin C, a level that has been shown to meet minimal requirements to avoid deficiency. However, avoiding deficiency is not synonymous with adequacy. A large body of epidemiological evidence indicates that risk of various lifestyle-related diseases is lower in those whose intake of vitamin C is high. Therefore, the question of how much is needed to achieve optimal ascorbic acid status has been the focus of much research and discussion in recent years. To date, the observational evidence of benefit in relation to lowered risk of lifestyle-related disease, such as coronary heart disease and cancer, accompanying high intakes of vitamin C is strong and credible. However, results of most supplementation trials published to date have not shown the expected benefits. Moreover, there have been well-publicized reports of potentially damaging, pro-oxidant activity of vitamin C. In this chapter, observational evidence of the benefit of high vitamin C intake and its apparently paradoxical *in vitro* pro-oxidant activity will be reviewed briefly; new data in relation to both this and the acute *in vivo* effects of high-dose vitamin C ingestion will be discussed, and preliminary data will be presented in regard to interrelationships between plasma ascorbic acid and DNA damage in fasting blood of healthy subjects.

22.2 ASCORBIC ACID, HEALTH, AND MORTALITY: THE BIG PICTURE

There are several large observational studies that show a strong protective association between plasma ascorbic acid and/or vitamin C intake and mortality (1–6). The EPIC-Norfolk Prospective Study in the U.K. followed 19,496 men and women aged 45 to 79 yr for 4 yr and showed that men and women in the highest quintile of plasma ascorbic acid concentration [mean (SD): 72.6 (11.5) μmol/l for men and 85.1 (13.7) μmol/l for women] within 1 yr of entry had significantly ($p < .0001$) lower all-cause mortality than those in the lowest quintile [20.8 (7.1) and 30.3 (10.1) μmol/l, respectively, for men and women] (1). In men and women in the highest quintile, all-cause mortality was 50% lower compared with those in the corresponding lowest quintile, and it was estimated that mortality risk decreased by 20% for every 20 μmol/l increase in plasma ascorbic acid. A 50% decrease in mortality was seen also in "The Medical Research Council Trial of Assessment and Management of Older People in the Community," a prospective trial in the U.K. of 1214 elderly subjects followed for a median of 4.4 yr (2). In the U.S. NHANES II prospective study, a 43% decrease in mortality was seen in association with higher plasma ascorbic acid levels in more than 3000 men followed for up to 16 yr (3), and a recent follow-up report of the U.S. Nurses' Health Study showed a 28% reduction in coronary heart disease (CHD) risk in women who took vitamin C supplements compared with those who did not (4). The Kuopio IHD Risk Factor Study followed 1605 men for 5 yr and reported a risk reduction of almost 90% for acute myocardial infarction in men with high plasma ascorbic acid concentrations (5). In a detailed analysis of cohort and case-control studies, high plasma ascorbic acid was reported overall to be a powerful predictor of freedom from CHD in the follow-up period (6).

There is evidence also of decreased risk of cancer of the mouth, pharynx, stomach, pancreas, lung, cervix, and breast in association with increased vitamin C intake, though not all studies find this (7). The EPIC study (1) showed an average 15% reduction in mortality risk for a 20 μmol/l increase in plasma ascorbic acid. In men, there was an average reduction of >50% in cancer mortality risk in the highest quintile of plasma ascorbic acid concentrations compared to the lowest quintile. However, in women, who averaged around 10 μmol/l higher plasma ascorbic acid concentration across quintiles, the decrease in risk was not statistically significant. The NHANES II study (3) reported a reduction of >60% in cancer mortality risk in men in the highest quartile of ascorbic acid during 12 yr of follow-up, compared with the lowest quartile. However, this relationship was not seen in women, who again had higher ascorbic acid levels than men.

Overall, the observational evidence of lower mortality in association with high plasma ascorbic acid is strong and consistent (1–8). However, the findings in relation to the source (food or supplements) of the vitamin are conflicting. This may reflect the difficulty in obtaining accurate dietary information and also suggests that different individuals need different intakes to achieve certain plasma levels of ascorbic acid and to reach complete saturation of tissues with the vitamin (8–10). In addition, it is likely that there are other dietary components that work in cooperation with vitamin C (9). However, the findings of the strong effect of supplements suggests that vitamin C itself has a beneficial effect, rather than vitamin C being simply a coincidental cotraveler with other, as yet unknown, key components (1,7,8).

22.3 PRO-OXIDANT EFFECTS OF ASCORBIC ACID: PARADOX, ARTIFACT, OR PHYSIOLOGICAL CONCERN?

Ascorbic acid is known to be an important antioxidant *in vivo* (10–12). However, ascorbic acid can, apparently paradoxically, exert pro-oxidant effects (12–14). In a mixture of copper or iron/ascorbic acid/hydrogen peroxide, the oxidized form of the transition metal ion (Cu^{2+} or Fe^{3+}) is reduced by ascorbic acid. The Cu^+ or Fe^{2+} formed is reoxidized upon interaction with hydrogen peroxide (H_2O_2), which forms hydroxyl ion and the highly reactive hydroxyl radical ($^\bullet OH$). This is known as the Haber–Weiss reaction (referred to as the Fenton reaction when the transition metal is iron) (14). The hydroxyl radical can cause oxidative damage to DNA, lipid, and protein (14,15). Without reduction by ascorbic acid or another antioxidant, Fe^{3+} ions react with H_2O_2 slowly and form small amounts of $^\bullet OH$. Therefore, the action of ascorbic acid, by reducing Fe^{3+} to Fe^{2+}, sets in motion a vicious cycle of redox reactions that greatly increases formation of $^\bullet OH$. Thus, ascorbic acid and other similarly acting electron-donating antioxidants can stimulate free radical damage *in vitro* by interaction with free transition metal ions.

This action of ascorbic acid has been employed in many *in vitro* experimental studies to induce and study the effects of oxidative damage and possible antioxidant protection of key biological sites. However, the physiological relevance of this finding is debatable (16–20). Simply put, is there credible evidence that increased ascorbic acid within body fluids damages DNA, leading to downstream deleterious effects on health? To address this, we performed four sets of experiments, each addressing a specific question, as follows:

1. Does ascorbic acid damage DNA *in vitro*? To address this question, lymphocytes were harvested from the venous blood of healthy, consenting adults, pooled, and incubated *ex vivo* with various amounts of ascorbic acid and, separately, various amounts of another antioxidant, epigallocatechin gallate (EGCG, a flavonoid found in high concentration in teas); DNA damage was measured using the comet assay and following established protocols (21,22).
2. Does ascorbic acid *in vitro* produce the oxidant hydrogen peroxide? To address this question, aqueous solutions of ascorbic acid were measured at timed intervals to see if they contained (generated) hydrogen peroxide using a modification of the xylenol orange method (23).
3. Does DNA damage in circulating lymphocytes increase in response to an acute increase in plasma ascorbic acid caused by ingestion of vitamin C? To address this question, a double-blinded, placebo-controlled intervention trial of cross-over design was performed on healthy, fasting, consenting subjects who ingested a single dose of 500 mg vitamin C. Venous blood was collected at timed intervals afterwards to investigate the acute response in terms of plasma ascorbic acid and DNA damage using a modification of the ferric reducing/antioxidant power (FRAP) assay, referred to as FRASC, and the comet assay (22,23,25).
4. Is there a direct correlation between plasma ascorbic acid and DNA damage in circulating lymphocytes? To address this question, fasting blood was collected from healthy, consenting adults, and plasma ascorbic acid and the DNA damage in lymphocytes was measured using FRASC and the comet assay, respectively (22,23).

The results of these four sets of experiments are presented in Figure 22.1 to Figure 22.5.

The trigger for a pro-oxidant effect of ascorbic acid is the presence of redox-active transition metal ions and hydrogen peroxide. Iron and copper ions *in vivo* are largely bound to transferrin, ferritin, and ceruloplasmin, in which form they are unable to catalyze free radical reactions (13,14,19,20). If the iron or copper ions leave the binding protein, they can catalyze free radical damage. However, the binding is very efficient and effective, and the amount of free metal ions is normally vanishingly small (13,14,26). Furthermore, peroxide levels are controlled by the action of catalase and glutathione peroxidase (14,26). However, antioxidants such as catechins and ascorbic acid can autoxidize *in vitro* with production of hydrogen peroxide, and this can damage DNA (16,23,26,27). This is illustrated in Figure 22.1 and Figure 22.2. This apparent pro-oxidant damaging

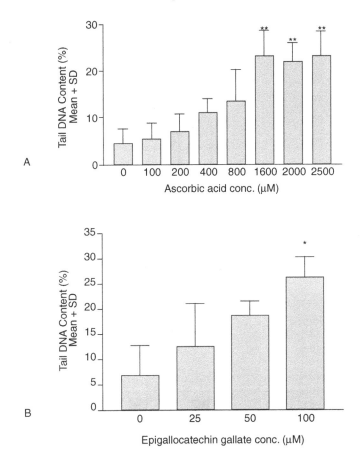

FIGURE 22.1 *In vitro* effects of increasing doses of (A) ascorbic acid, and (B) epigallo-catechin gallate (EGCG) on DNA damage in human lymphocyte. **p < .01; *p < .05 compared to control cells.

effect of ascorbic acid *in vitro* is well known and is attenuated by catalase (which destroys hydrogen peroxide) (Figure 22.3) and by desferrioxamine (which binds and inactivates iron). Can this damaging effect happen *in vivo*? Under certain circumstances, for example in iron overload or in an area of decreased tissue pH with metal ion displacement, the Haber–Weiss reaction is possible, but under normal circumstances *in vivo*, with a coordinated antioxidant system, it is not likely to happen. Therefore, the antioxidant/pro-oxidant effect of ascorbic acid and other antioxidants is paradoxical but real. However, it is situation dependent, and the situation or conditions that give rise to a damaging pro-oxidant effect of ascorbic acid is highly unlikely to occur *in vivo*.

This view is supported by the wealth of observational data described above and also by the finding that, following ingestion of a high dose (500 mg) of

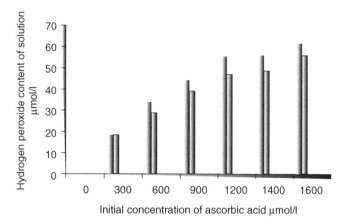

FIGURE 22.2 Ascorbic acid in aqueous solution generates hydrogen peroxide *in vitro*: hydrogen peroxide content at 0.5 h (horizontal lines) and 2 h (diagonal lines) after preparation of ascorbic acid solutions.

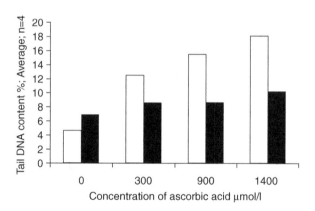

FIGURE 22.3 Ascorbic-acid-induced *in vitro* effects on DNA (open bars) are attenuated by removal of hydrogen peroxide by catalase (shaded bars).

vitamin C, there is a marked increase in plasma ascorbic acid concentration but there is no evidence of increased DNA damage in peripheral lymphocytes in the same blood sample (Figure 22.4) (25). Furthermore, in a small pilot study we found that those subjects with higher fasting plasma ascorbic acid levels have lower, not higher, level of DNA damage in peripheral lymphocytes collected at the same time (Figure 22.5).

22.4 SUMMARY AND CONCLUSION

There is a range of varied and convincing data that supports the observation that diets rich in ascorbic acid have strongly protective effects against lifestyle-related

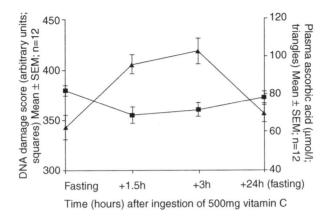

FIGURE 22.4 DNA damage in peripheral lymphocytes of healthy subjects (n = 12) following ingestion of 500 mg vitamin C; plasma ascorbic increased as expected, but no evidence of increase in DNA damage was seen.

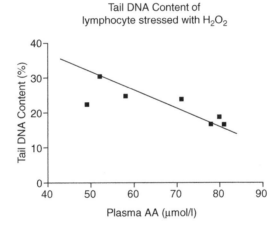

FIGURE 22.5 Significant inverse association between plasma ascorbic acid (AA) and lymphocyte DNA damage in healthy, fasting subjects ($r = .796$; $p < .05$; n = 7).

diseases such as cancer and coronary heart disease. The scenario of damaging Fenton chemistry occurring *in vivo* is very unlikely. The absence of an increase in DNA damage in circulating lymphocytes after an acute increase in plasma ascorbic acid resulting from ingestion of a large dose of ascorbic acid is reassuring. The finding in a small, preliminary study of an inverse association seen between fasting plasma ascorbic acid and lymphocytic DNA damage is new and interesting. Together, available evidence provides a substantial framework that supports strongly the view that a damaging pro-oxidant effect of ascorbic acid within healthy body tissues is not an issue of physiological concern. Furthermore, data are supportive, although not conclusively so, of vitamin C being a key protective agent. The optimal intake of vitamin C is still uncertain and under debate. However, it is certain that lifestyles that encompass intake of high quantities of vitamin C are associated with decreased disease and mortality. Therefore, in conclusion, current evidence in regard to the interrelationship between ascorbic acid and oxidative damage to DNA is not supportive of a pro-oxidant effect that damages human health and indicates that the molecular effects of ascorbic acid have an overall protective effect against lifestyle-related diseases.

REFERENCES

1. Khaw KT, Bingham S, Welch A, Luben R, Wareham N, Oakes S, Day N. Relation between plasma ascorbic acid and mortality in men and women in EPIC-Norfolk prospective study: a prospective population study. *Lancet* 2001; 357: 657–663.
2. Fletcher AE, Breeze E, Shetty PS. Antioxidant vitamins and mortality in older persons: findings from the nutrition add-on study to the Medical Research Council Trial of Assessment and Management of Older People in the Community. *Am J Clin Nutr* 2003; 78: 999–1010.
3. Loria CM, Klag MJ, Caulfield LE, Whelton PK. Vitamin C status and mortality in US adults. *Am J Clin Nutr* 2000; 72: 139–145.
4. Osganian SK, Stampfer MJ, Rimm E, Spiegelman D, Hu FB, Manson JE, Willett WC. Vitamin C and risk of coronary heart disease in women. *J Am Coll Cardiol* 2003; 42: 246–252.
5. Rissanen TH, Voutilainen S, Virtanen JK, Venho B, Vanharanta M, Mursu J, Salonen JT. Low intakes of fruits, berries and vegetables is associated with excess mortality in men: the Kuopio Ischaemic Heart Disease Risk Factor (KIHD) study. *J Nutr* 2003; 133: 199–204.
6. Asplund K. Antioxidant vitamins in the prevention of cardiovascular disease: a systematic review. *J Intern Med* 2002; 251: 372–392.
7. Benzie IFF. Observational Epidemiology. In *The Encyclopedia of Human Nutrition*, Sadler M, Strain JJ, Cabellero B, Eds., Academic Press: London, 1999, pp. 106–115.
8. Frei B. To C or not to C, that is the question! *J Am Coll Cardiol* 2003; 42(2): 253–255.
9. Gey KF. Vitamins E plus C and interacting co-nutrients required for optimal health. *Biofactors* 1998; 7: 113–275.
10. Padayatty SJ, Katz A, Wang Y, Eck P, Kwon O, Lee JH, Chen S, Corpe C, Dutta A, Dutta S, Levine M. Vitamin C as an antioxidant: evaluation of its role in disease prevention. *J Am Coll Nut* 2003; 22; 18–35.
11. Carr AC, Frei B. Toward a new recommended dietary allowance for vitamin C based on antioxidant and health effects in humans. *Am J Clin Nutr* 1999; 69: 1086–1107.
12. Tosa CS. An overview of Ascorbic Acid Chemistry and Biochemistry. In *Vitamin C in Health and Disease*, Packer L, Fuchs J, Eds., Marcel Dekker: New York, 1997, pp. 25–58.
13. Benzie IFF, Chung WY, Strain JJ. 'Antioxidant' (reducing) efficiency of ascorbate in plasma is not affected by concentration. *J Nutr Biochem* 1999; 10: 146–150.

14. Halliwell B, Gutteridge JMC. *Free Radicals in Biology and Medicine*, 3rd Ed., Clarendon Press: Oxford, 1999.

15. Collins AR. Oxidative DNA damage, antioxidants, and cancer. *BioEssays* 1999; 21: 238–246.

16. Podmore ID, Griffiths HR, Herbert KE, Mistry N, Mistry P, Lunec J. Vitamin C exhibits pro-oxidant properties. *Nature* 1998; 392: 559–561.

17. Lee SH, Oe T, Blair IA. Vitamin C-induced decomposition of lipid hydroperoxides to endogenous genotoxins. *Science* 2001; 292: 2083–2086.

18. Levine M, Daruwala RC, Park JB, Rumsey SC, Wang Y. Does vitamin C have a pro-oxidant effect? *Nature* 1998; 395: 231.

19. Carr A, Frei B. Does vitamin C act as a pro-oxidant under physiological conditions? *FASEB J* 1999; 13: 1007–1024.

20. Poulsen HE, Weimann A, Salonen JT, Nyyssonen K, Loft S, Cadet J, Douki T, Ravanat JL. Does vitamin C have a pro-oxidant effect? *Nature* 1998; 395: 231–232.

21. Collins AR, Dusinska M, Franklin, Somorovska M, Petrovska H, Duthia S, Fillio L, Panayiotidis M, Raslova K, Vaughan N. Comet assay in human biomonitoring studies: reliability, validation, and application. *Environ Mol Mutagen* 1997; 30: 139–146.

22. Szeto YT, Benzie IFF. Effect of dietary antioxidants on human DNA *ex vivo*. *Free Radic Res* 2002; 36: 113–118.

23. Yuen JWM, Benzie IFF, Hydrogen peroxide in urine as potential biomarker of oxidative stress. *Free Radic Res* 2003; 37: 1209–1213.

24. Benzie IFF, Strain JJ. Ferric reducing (antioxidant) power as a measure of antioxidant capacity: the FRAP assay. In *Oxidants and Antioxidants,* Vol. 299, Methods in Enzymology, Packet L, Ed., Academic Press: Orlando, 1999, pp. 15–27.

25. Choi SW, Benzie IFF, Collins AR, Hannigan BM, Strain JJ. Vitamins C and E: main and interactive effects on biomarkers of antioxidant defence and oxidative stress. *Mutat Res* 2004. [In press].

26. Benzie IFF. Evolution of antioxidant defence mechanisms, *Eur J Nutr* 2000; 39: 53–61.

27. Halliwell B, Clement MV, Ramalingam J, Long LH. Hydrogen peroxide. Ubiquitous in cell culture and in vivo? *Life* 2000; 251–257.

23 The Role of Lipid Peroxidation in Chromosomal DNA Fragmentation Associated with Cell Death Induced by Glutathione Depletion

Yoshihiro Higuchi, Hideji Tanii, Yuji Mizukami, and Tanihiro Yoshimoto
Kanazawa University, Kanazawa, Japan

CONTENTS

23.1 INTRODUCTION

Cellular genomes are continually subjected to endogenous and environmentally induced structural alterations, damage that can manifest itself as gross chromosomal abnormalities resulting in cell death. Although it has proven difficult to classify cell death based on the morphology of dying cells and on the DNA fragmentation or damages, two distinct patterns of cell death have been identified. These have been termed *necrosis* and *apoptosis* [1]. Mammalian cell death can be induced through chromosomal DNA damage by oxidative stresses such as ionizing and ultra violet (UV) radiation, anticancer drugs, and various DNA-adduct-inducing substances. Under oxidative stress, reactive oxygen species (ROS) such as hydroxyl radicals ($\cdot$OH), superoxide anions ($O_2^{\cdot}$), and hydrogen peroxide (H_2O_2) have been shown to damage chromosomal DNA and other cellular components, resulting in DNA degradation, protein denaturation, and lipid peroxidation. However, the mechanisms behind these cellular effects are rather complex and are not yet fully understood. DNA damage induced by oxygen radicals occurs by oxidative nucleic acid base modification and scission of DNA strands, the latter resulting in single- and double-strand breaks [2]. Some agents producing ROS induce cell death, including apoptosis, and cause lipid peroxidation and DNA damage [3]. However, the implications of lipid peroxidation in ROS-induced DNA damage remain to be elucidated. Furthermore, it is not clear whether ROS are a part of the signal transduction cascade triggered by various inducers of apoptosis or whether they are generated in a parallel pathway that can independently trigger apoptosis. It is important, therefore, to distinguish between ROS molecules involved in such signaling pathways and those that mediate general cellular damage, including giant DNA and high-molecular-weight DNA fragmentation. These DNA fragmentations in chromatin, estimated by size and emerging time of the fragments, might reveal either apoptosis or necrosis.

Apoptosis and necrosis are two distinct forms of cell death that have profoundly different implications for the surrounding tissues. Apoptosis is characterized by cell shrinkage, chromatin condensation, some caspase activations, and fragmentation of DNA at internucleosomal linker sites [4]. Necrosis is an uncontrolled event resulting from loss of homeostasis and dispersal of cell contents, which may then have adverse effects on neighboring tissue [5].

Our aim is to review chromosomal DNA fragmentations such as giant DNA, high-molecular-weight (HMW) DNA, and internucleosomal DNA fragmentations and to reflect upon their significance in apoptosis or necrosis induced by oxidative stress under GSH depletion.

23.2 CHROMATIN STRUCTURE OF MAMMALIAN CELLS

A mammalian cell nucleus contains almost 50 cm of DNA, requiring more than a 50,000-fold reduction in length to fit in the nucleus and nuclear matrix. The nuclear matrix is an important structural component in a variety of nuclear functions in nuclear morphology, DNA organization, DNA replication, RNA synthesis, and nuclear regulation. DNA loop domains of chromatin are attached at their base to the nuclear matrix, an organization that is maintained throughout both interphase and metaphase. These loops are 50 to 150 kbp long, equivalent in size to a replicon [6]. The DNA loop domain was first proposed by Cook et al., who suggested that loop structures are involved in the superhelical organization of eukaryotic DNA [7]. The actual structure of these loop domains has only recently begun to be understood (Figure 23.1). The haploid human genome contains 3000 megabase pairs (Mbp) of DNA with a mean chromosomal size of 130 Mbp.

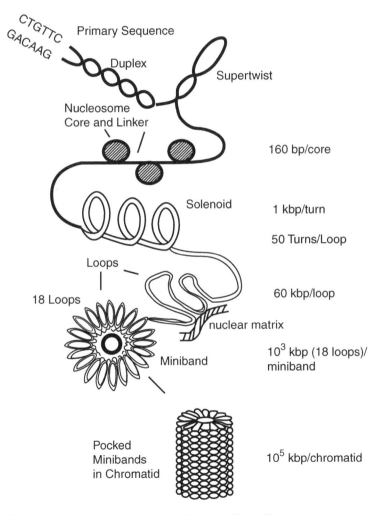

FIGURE 23.1 Chromatin structure model of mammalian cells.

23.3 PATTERNS OF CHROMOSOMAL DNA FRAGMENTATION

Chromosomal DNA fragmentation is caused by two types of DNA breaks, classified into single-strand and double-strand DNA breaks. Some single-strand DNA breaks may be repaired by repair enzyme systems [8]. Double-strand DNA breaks are generally thought to have a greater biological consequence than single-strand DNA breaks because they can lead directly to chromosomal aberrations and, more frequently, to the loss of genetic information [9]. Electrophoretic techniques, particularly pulsed-field gel electrophoresis (PFGE), have been extensively applied; these have broad applications in biochemistry and genetics, owing to their sensitivity, for the analysis of DNA molecules with lengths up to 10 Mbp [10,11]. PFGE has been used to analyze the precise molecular nature of DNA fragments produced by oxidative stress in cultured animal cells [3,12–14].

23.3.1 GIANT DNA FRAGMENT CLEAVAGE

A chromosomal DNA fragment more than 1 Mbp in size is classified as a giant DNA fragment [15]. DNA degradation accompanied by DNA fragmentation producing 1 to 2 Mbp and 200 to 800 kbp DNA fragments was observed during cell death in T-24 human bladder cells treated with agents that produce ROS [3,15], as well as under GSH depletion in C6 rat glioma cells [12]. Chromosomal DNA fragments of 200 to 800 kbp and 50 to 300 kbp sizes are called *high molecular weight* DNA fragments. The 1 to 2 Mbp giant DNA and 200 to 800 kbp HMW fragmentations, which may represent features of high-order chromatin structures such as minibands and loops of DNA [16], lead to apoptosis, as ascertained by internucleosomal DNA fragmentation in C6 cells [17]. However, little is known about the mechanism of giant DNA and HMW DNA fragmentation during apoptosis induced by ROS or other triggers.

23.3.2 LARGE DNA AND INTERNUCLEOSOMAL DNA FRAGMENT CLEAVAGE

Wyllie [18] has popularized the notion that DNA fragmentation is a component of apoptosis. Studies using glucocorticoids and rat thymocytes as a model system showed that DNA was fragmented into 180 to 200 kbp prior to cell death. This form of DNA degradation has been very widely observed in apoptosis, although exceptions do exist. One of the hallmarks of apoptosis is the digestion of genomic DNA by an endonuclease, generating a ladder of small fragments of double-stranded DNA. Single-strand nicks were frequently found in internucleosomal regions but also occurred in core-particle-associated DNA. DNA fragmentation induced during apoptosis is not due to a double-strand cutting enzyme as previously postulated, but rather is the result of single-strand breaks. This ensures dissociation of the DNA molecule at sites in which cuts are found within close proximity [19]. In apoptosis, there is a two-step process of DNA fragmentation: DNA is first cleaved into large fragments of 50 to 300 kbp, and these are subsequently cleaved into smaller oligonucleosomes in some, but not all, cells. Significantly, only the first stage is considered essential for cell death because some cells, such as human MCF7 breast carcinoma and human NT2 neuronal cells, do not show this behavior but still display normal nuclear morphological apoptotic changes [20]. Three types of DNA fragmentation occurring during apoptosis can thus be distinguished. Recently, other types of DNA fragmentation have been reported during apoptosis, in the presence or absence of the characteristic internucleosomal DNA cleavage (ladder-like) pattern [21]. Internucleosomal DNA is cleaved with fragmentation into large (50 to 300 kbp) lengths and single-strand cleavage events. These enzymatic events encompass a vast array of chromosomal degradation states in cells with the universal consequence of cell death [22].

23.4 CHROMOSOMAL DNA FRAGMENTATION IN APOPTOSIS OR NECROSIS

Apoptosis is characterized by chromatin condensation, activation of some caspases, and fragmentation of DNA at internucleosomal linker sites, giving rise to discrete bands of multiples of 180 to 200 base pairs [4]. In contrast, necrosis is a passive process, typified by cell and organelle swelling with spillage of the intracellular contents into the extracellular milieu. Necrosis is an uncontrolled event resulting from the loss of homeostasis and dispersal of cellular contents, which may then have adverse effects on neighboring tissues [23].

Some inducers of apoptosis, such as etoposide and glucocorticoids, have provided formations of 50 to 300 kbp HMW DNA fragments prior to internucleosomal DNA fragmentation induced by etoposide in apoptotic MCF-7 cells [20] and by tumor necrosis factor (TNF-α) in mouse L-929 cells [24]. These DNA fragment formations induced by serum deprivation have been observed in several human epithelial cells [25], as well as in HeLa nuclei treated with apoptosis-inducing factor (AIF) [21]. Apoptosis has also been widely observed in some cells treated with anticancer drugs [26], as well as in other cell death processes induced by some biological events, such as depletion of nutrients [27]. However, little has been reported about the involvement of not only 1 to 2 Mbp giant DNA fragmentation but also HMW DNA fragmentation of 200 to 800 kbp and that of 50 to 300 kbp fragments, and their significance or role in apoptosis. In some cases of apoptosis, ROS may be involved not only as inducers of DNA damage but also as specific second messengers in the signal transduction pathway; however, in others, they may manifest as side effects of either an experimental system or changes in the cellular redox status as a result of ROS-independent apoptosis signaling pathways [28]. Therefore, it is still unclear whether endogenous ROS are really involved in DNA degradation leading to apoptosis.

23.5 GSH DEPLETION INDUCES GIANT DNA FRAGMENTATION

When C6 rat glioma cells were treated with glutamate, intracellular glutathione (GSH) was reduced to approximately one seventh of the initial level, and the reduction of GSH-induced cytolysis was accompanied by apoptosis [12]. Although the source was obscure, extracellular hydrogen peroxide was released without decaying and accumulated in C6 cells under glutamate-induced GSH depletion [12]. In nerve cells such as glial and neuronal cells, hydrogen peroxide production from the metabolism of catecholamines and indoleamines by monoamine oxidase (MAO) has been considered [29,30]. Under GSH depletion, hydrogen peroxide endogenously produced by MAO enhances arachidonic acid release through cellular phospholipase A_2 activation in membranes [31] and, furthermore, might be converted to OH radicals by metals such as iron and copper. The accumulated OH radicals might attack chromosomal DNA, leading to 8-hydroxy-2'-deoxyguanosine (8-OH-dG) formation. Besides OH radicals, lipid hydroperoxides or lipid alkoxyl radicals contributed 8-OH-dG formation under GSH depletion through oxidative modification of nuclear membrane integrity by lipid peroxidation [32]. Chromosomal giant DNA of 1 to 2 Mbp and HMW DNA (200 to 800 kbp) fragments were observed during apoptotic cell death [12]. In glial cells, glutamate induces GSH depletion, and consequently apoptosis, through endogenously produced active oxygen species. Apoptosis is accompanied by 1 to 2 Mbp giant DNA fragmentation prior to internucleosomal DNA fragmentation. Apoptosis is also observed under GSH depletion induced by L-buthionine-(S,R)-sulfoximine (BSO), an inhibitor of γ-glutamyl cysteine synthetase. BSO-induced C6 cell death has been associated with caspase-3 activation [49]. Caspase-3 activation is an early biochemical marker of apoptosis in some types of cells, induced by various triggers [33]. Furthermore, giant DNA fragments of approximately 2 Mbp were observed in BSO-treated cells. Giant DNA fragmentation was followed by approximately 30 to 700 kbp and then less than 100 kbp, including internucleosomal, DNA fragmentations.

23.6 EFFECT OF POLYUNSATURATED FATTY ACIDS OR LIPID HYDROPEROXIDES ON THE PROMOTION OF GIANT DNA FRAGMENTATION

Polyunsaturated fatty acids (PUFAs) such as arachidonic acid, γ-linolenic acid, and linoleic acid markedly enhanced lipid hydroperoxide formation in C6 cells treated with glutamate or BSO, but did not affect glutamate- or BSO-induced GSH depletion. Oleic acid did not affect GSH depletion but conversely suppressed both cytolysis and lipid peroxidation caused by glutamate or BSO [13]. The 1 to 2 Mbp giant DNA fragmentation induced by BSO was enhanced by exogenous addition of PUFAs, depending on the species of PUFAs. The 200 to 800 kbp HMW DNA fragments produced in both glutamate- and BSO-treated C6 cells were also increased by the addition of PUFAs (Figure 23.2). However, some antioxidants, ROS scavengers, and metal chelaters suppressed not only the 1 to 2 Mbp giant DNA fragmentation but also HMW DNA fragmentation, even in the presence of PUFA. These findings imply that 1 to 2 Mbp giant DNA fragmentation is modulated by ROS or ROS-mediated lipid peroxidation.

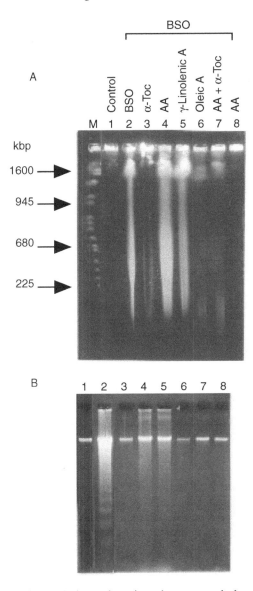

FIGURE 23.2 Pulsed-field gel electrophoresis and agarose gel electrophoresis analysis of chromosomal DNA of C6 cells treated with BSO. In panel A, each lane indicates the control (lane 1), samples from the cells treated with 10 mM BSO (lane 2) or 50 μM AA alone (lane 8) for 24 h and with 10 mM BSO for 18 h in the presence of 100 μM α-Toc (lane 3), 50 μM AA (lane 4), 50 μM γ-linolenic acid (lane 5), 50 μM oleic acid (lane 6), or 50 μM AA plus 100 μM α-Toc (lane 7). In panel B, the amount of 1 to 2 Mbp giant DNA fragments obtained according to panel A was determined and is expressed as a percentage of the control, which indicates the DNA amount in the sample treated with BSO for 18 h. The values are the mean ± SD of three independent experiments. Panel C shows

(continued on next page)

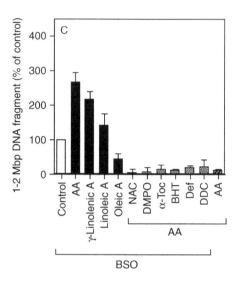

FIGURE 23.2 (continued)   AGE analysis of C6 cells treated with 10 m*M* BSO. The lane number corresponds to the same lane number in panel A. AA, arachidonic acid; BHT, butylated hydroxytoluene; DDC, sodium N-N-diethyldithio-carbamate; Def, deferoxamine; DMPO, 5,5′-dimethyl-1-pyrroline-N-oxide; NAC, *N*-acetylcysteine; α-Toc, α-tocopherol. (From Higuchi Y. Glutathione depletion induces giant DNA fragmentation associated with apoptosis or necrosis. In *Micronutrients and Health: Molecular Biological Mechanisms*, Nesaretnam K, Packer L, Eds., Champaign: AOCS Press, 2001, pp. 202–216. With permission.)

Nuclear DNA from C6 cells treated with both glutamate and BSO had DNA fragments with 3′-OH termini associated with the giant DNA fragments. The chromosomal DNA fragments with free 3′-OH termini caused by the GSH depletion were enhanced by PUFAs in C6 cells [34]. Thus, PUFAs could promote 1 to 2 Mbp giant DNA fragmentation with 3′-OH termini, causing rapid necrotic cell death through lipid peroxidation. This suggestion is supported by a report in which *N*-methyl-D-aspartate and kainite induced terminal deoxynucleotidyl-transferase-mediated dUTP-biotin nick end-labeling (TUNEL) positive internucleosomal DNA cleavage associated with both apoptotic and necrotic cell death in neonatal rat brain [35].

PUFAs are easily oxidized by •OH radicals and by the enzymatic metabolisms of lipoxygenase and thereby induce lipid peroxidation of cells without diminishing the lipid hydroperoxides generated under GSH depletion and abundant oxygen [36]. PUFAs are particularly vulnerable to free radical attack, because the carbon double bond in PUFAs within membranes allows for the easy removal of hydrogen atoms by ROS such as •OH radicals [36]. Under aerobic conditions, lipid peroxidation continues as carbon double bonds with doubly allylic hydrogen combine with O2 to form additional organic LOO• (Figure 23.3). The oxidizability of PUFAs may be linearly dependent on the number of doubly allylic positions at 37°C [37,38]. Oleic acid can act as an antioxidant in GSH-depletion-induced cell

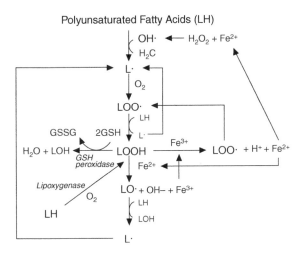

FIGURE 23.3 Lipid peroxidation of polyunsaturated fatty acids (PUFA) with lipid chain reaction. (From Higuchi Y. Glutathione depletion induces giant DNA fragmentation associated with apoptosis or necrosis. In *Micronutrients and Health: Molecular Biological Mechanisms*, Nesaretnam K, Packer L, Eds., Champaign: AOCS Press, 2001, pp. 202–216. With permission.)

death [39]. Lipid free radicals such as L•, LO•, and LOO• are most likely produced during lipid peroxidation through a chain reaction and are amplified in the presence of iron or copper not only intracellularly but also in the cell-cultured medium. LOO• can combine with an abstracted hydrogen atom from adjacent fatty acid acyl chains to form lipid hydroperoxides (LOOH) that are decomposed to LO• by Fenton's reaction in the presence of Fe^{2+} or Cu^+ under conditions of GSH depletion. LOOH produced in this way is ordinarily decayed to lipid hydroxides (LOH) and H_2O by GSH peroxidase, using endogenous GSH as a substrate. LOO• can be converted to cycloperoxide, hydrolyzing to lipid aldehydes, which are detectable in the 2-thiobarbituric acid assay [40]. Thus, the action of •OH molecules initiates a chain reaction that generates numerous toxic reactants that then disrupt membrane integrity and damage membrane proteins, consequently leading to cell death [41].

Lipid peroxidation producing lipid free radicals might proceed not only in plasma membranes but also in the nuclear membranes close to chromosomes, to the loss of membrane integrity in cell membranes consisting of phospholipids. This may thereby lead to circumstances suitable for oxygen radicals such as •OH radicals to attack the chromatin DNA. In addition to •OH radicals, LO• or LOO• radicals may also cleave DNA strands, leading to giant DNA fragmentation in chromatin. LO• radicals are produced from LOOH by Fenton's reaction in the presence of iron or copper. Thus, intracellular metal ions such as iron and copper may play an important role in the generation of hydroxyl radicals from H_2O_2 or O_2^- in or around chromatin [42,43]. 13-Hydroperoxy-octadecadienoic acid (LOOH), a metabolite of linoleic acid, could cleave double-strand DNA at the position of guanosine nucleotides in pBR322, but neither linoleic acid (LH) nor

13-hydroxyoctadecadienoic acid (LOH) was effective in the cleavage [44]. In the circumstance of membrane integrity loss induced by lipid peroxidation, ROS such as •OH radicals might attack and directly cleave chromosomal DNA into single-strand forms, consequently leading to double-strand DNA breaks. Hydroxyl radicals caused single-strand breaks with nicked forms, leading to double-strand breaks predominantly in the 5′-G-G-3′ or 5′-G-G-G-3′ sequences of pZ189 [45]. The occurrence of these relatively simple sequences in mammalian chromosomal DNA appears to be too high to account for the low frequency of nicking induced by treatments with these drugs. Therefore, the location of the specific DNA target sites attacked by ROS in this cell system is still unclear. The target sites in the chromosomal DNA may occur at a relatively regular distance of 1 to 2 Mbp. ROS-mediated action may be more probable than, and indeed preferable to, lipid-radical-mediated action for providing 3′-OH-termini in single- or double-strand breakage.

23.7   LIPID PEROXIDATION CONVERTS APOPTOSIS TO NECROSIS UNDER GSH DEPLETION

Lipoxygenase activity of lymphocytes and endogenous 15-hydroxyeicosa tetraenoic acid (15-HETE), an arachidonate metabolite, is increased by x-ray irradiation in rats, stimulating internucleosomal DNA fragmentation [46]. Li et al. [47] have reported that 12-lipoxygenase is activated and plays an important role in nerve cell death in HT-22-hippocampus-derived cells under BSO-induced GSH depletion. The expression and distribution of 12-lipoxygenase have been reported to be specific to cell types or tissues in animals [48]. The specific activity and the amount of 12-lipoxygenase in C6 cells treated with BSO were increased by the addition of arachidonic acid [49]. The involvement of 12-lipoxygenase in BSO-caused cell death is likely, because 12-hydroperoxy eicosatetraenoic acid (12-HPETE), a product of 12-lipoxygenase, may act as an initiator and amplify lipid peroxidation through a chemical chain reaction under GSH depletion *in vitro*. Arachidonic acid induced the disappearance of internucleosomal DNA fragmentation under BSO-induced GSH depletion in rat glioma cells [34]. A decrease of GSH triggered an activation of neuronal 12-lipoxygenase leading to the production of peroxides, the influx of Ca^{2+}, and ultimately to cell death of H-22 hippocampas-derived cells [47]. Thus, exogenous arachidonic acid may potentiate cell death under conditions of GSH depletion. Lipid metabolites, such as arachidonic-acid-derived eicosanoids, may play a role in regulating cell survival [49]. A possible chain reaction for lipid peroxidation is initiated by lipid hydroperoxides produced through enzyme reactions of lipoxygenases under GSH depletion.

BSO-induced C6 cell death is associated with caspase-3 activation. Ac-DEVD-CHO, a cell permeable specific inhibitor of caspase-3, inhibited BSO-induced apoptosis, but not in the presence of PUFA, in which the caspase-3 activity was decreased [50]. The caspase-3 activation may be involved in the mild cell death induced by BSO but not in the intense cell death in the presence of AA during GSH depletion. Caspase-activated DNase (CAD) has been shown to be involved in the internucleosomal DNA fragmentation during apoptosis induced by various apoptosis triggers [51]. However, it is not yet obvious which ROS or redox catastrophe induces caspase-3 activation and DNA ladder fragmentation at low levels or the depletion of GSH. Armstrong et al. [52] have reported activation of caspase-3 in cerebellar granule cells undergoing apoptosis, but not necrosis.

PUFAs such as arachidonic acid cause necrosis at high concentrations under GSH depletion, promoting 1 to 2 Mbp giant DNA fragmentation of chromatin and suppressing internucleosomal DNA fragmentation through lipid peroxidation, inducing membrane integrity loss [34]. Therefore, it is likely that BSO-induced GSH depletion initiates apoptosis through the caspase activation signaling system in C6 cells, whereas high concentrations of PUFA turns apoptosis into necrosis, estimated by alterations of not only caspase-3 activation but also internucleosomal DNA fragmentations (Figure 23.4) [4,51]. Activation of some phospholipases during necrosis, particularly cytosolic Ca^{2+}-dependent phospholipase A_2 (PLA_2), has been demonstrated [53]. PLA_2 is specific to substrate with AA at the *sn*-2 position and its translocation to cell membranes. Lysophosphatidic acid, which is produced from phospholipids by removing arachidonic acid of PLA_2, also

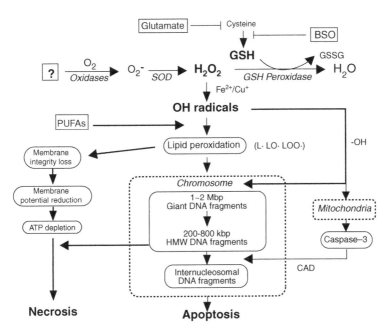

FIGURE 23.4 Possible mechanism underlying the glutathione-depletion-induced apoptosis or necrosis in glioma cells. (From Higuchi Y. Chromosomal DNA fragmentation in apoptosis and necrosis induced by oxidative stress. *Biochem Pharmacol* 2003; 66: 1527–1535. With permission.)

induced both apoptosis and necrosis in hippocampal neurons by unknown mechanisms [54]. Thus, it seems clear that not only lysophosphatidic acid but also other PUFAs cannot be ignored in the cell death associated with lipid peroxidation. Lysophosphatidic acid induces both apoptosis and necrosis in hippocampal neurons [54]. Both mild and intensive insults with NMDA or nitric oxide/superoxide induced two distinct events of apoptosis and necrosis in cortical cell cultures [55]. The GSH-induced oxidative cell death showed characteristics of both apoptosis and necrosis in HT-22-hippocampus-derived cells [56]. Excess accumulation of lipid hydroperoxides produced by lipid peroxidation under intracellular GSH depletion may promote cell death through necrosis. The change from apoptosis to necrosis accompanied by a reduction of internucleosomal DNA fragmentation might be dependent on the intensity of oxidative stress and lipid peroxidation.

23.8 ROLE OF ATP IN APOPTOSIS AND NECROSIS

It has been reported that chromatin condensation in apoptosis depends on intracellular ATP levels or ATP synthesis [57–59]. Furthermore, the intracellular ATP level could be responsible for converting necrosis to apoptosis in oxidant-induced endothelial cell death [8]. Cell damage accompanies mitochondria membrane dysfunction not only in apoptosis but also in necrosis. A decrease in the mitochondrial membrane potential in the absence of a functional electron transport system results mainly from the leakage of protons through the membrane and precedes apoptotic cell death [60].

ACKNOWLEDGMENTS

This work was supported in part by a grant-in-aid for scientific research (C15590268) from the Ministry of Education, Culture, Sport, Science, and Technology of Japan.

REFERENCES

1. Wyllie AH, Kerr JFR, Currie AR. Cell death: the significance of apoptosis. *Int Rev Cytol* 1990; 68: 251–306.
2. Halliwell B, Aruoma, OI. DNA damage by oxygen-derived species-Its mechanism and measurement in mammalian systems. *FEBS Lett* 1991; 281: 9–19.
3. Higuchi Y, Matsukawa S. Appearance of 1-2 Mb giant DNA fragments as an early common response leading to cell death induced by various substances which cause oxidative stress. *Free Radic Biol Med* 1997; 23: 0–99.
4. Carson DA, Ribeiro JM. Apoptosis and disease. *Lancet* 1993; 341: 1251–1254.
5. Schwartz LM, Smith SW, Jone MEE, Osbone BA. Do all programmed cell deaths occur via apoptosis? *Proc Natl Acad Sci USA* 1993; 90: 980–984.
6. Pardoll DM, Vogelstein B, Coffey DS. A fixed site of DNA replication in eucaryotic cells. *Cell* 1980; 19: 527–536.

7. Cook PR, Brazell IA, Jost E. Characterization of nuclear structures containing superhelical DNA. *J Cell Sci* 1976; 22: 303–324.

8. Bohr VA, Wassermann K. DNA-repair at the level of the gene. *Trends Biochem Sci* 1988; 13: 429–433.

9. Bryant PE. Enzymatic restriction of mammalian cell DNA using Pvu II and Bam HI; evidence for the double strand break origin of chromosomal aberrations. *Int J Radiat Biol* 1984; 46: 57–65.

10. Schwartz DC, Cantor CR. Separation of yeast chromosome-size DNAs by pulsed field gradient gel electrophoresis. *Cell* 1984; 37: 67–75.

11. Higuchi Y. Measurement of DNA double-strand breaks with giant DNA and high molecular weight DNA fragments by pulsed-field gel electrophoresis. *Method Mol Biol* 2002; 186: 161–170.

12. Higuchi Y, Matsukawa S. Active oxygen-mediated chromosomal 1-2 Mbp giant DNA fragmentation into internucleosomal DNA fragmentation in apoptosis of glioma cells induced by glutamate. *Free Radic Biol Med* 1998; 24: 418–426.

13. Higuchi Y. Glutathione depletion induces giant DNA fragmentation associated with apoptosis or necrosis. In *Micronutrients and Health: Molecular Biological Mechanisms*, Nesaretnam K, Packer L, Eds., Champaign: AOCS Press, 2001, pp. 202–216.

14. Higuchi Y, Yoshimoto T. Chromosomal DNA fragmentation in oxidative stress-induced mammalian cell death. In *Recent Res Develop Biophys Biochem*, Vol. I, Pandalai SE, Ed., Research Signpost: Trivandrum, India, 2001, pp. 161–171.

15. Matsukawa S, Higuchi Y. The nature of giant DNA molecules produced from nuclear chromosome DNA by active oxygen producing agents. In *Oxidative Damage and Repair*, Devies KJA, Ed., Pergamon: New York, 1991, pp. 197-201.

16. Cohen GM, Sun X, Fearnhead H, MacFarlane M, Brown DG, Snowden RT, Dinsdale D. Formation of large molecular weight fragments of DNA is a key committed step of apoptosis in thymocytes. *J Immunol* 1994; 153: 507–516.

17. Higuchi Y, Matsukawa S. Glutathione depletion induces giant DNA and high molecular weight DNA fragmentation associated with apoptosis through lipid peroxidation and protein kinase C activation in C6 glioma cells. *Arch Biochem Biophys* 1999; 363: 33–42.

18. Wyllie AH. Glucocorticoid-induced apoptosis is associated with endogeneous endonuclease activation. *Nature* 1980; 284: 555–556.

19. Peitsch MC, Muller C, Tschopp J. DNA fragmentation during apoptosis is caused by frequent single-strandcuts. *Nucleic Acids Res* 1993; 21: 4206–4209.

20. Oberhammer F, Wilson JW, Dive C, Morris ID, Hickman JA, Wakeling E, Walker PR, Sikorska M. Apoptotic death in epithelial cells: cleavage of DNA to 300 and/or 50 kb fragments prior to or in the absence of internucleosomal fragmentation. *EMBO J* 1993; 12: 3679–3684.

21. Susin SA, Lorenzo HK, Zamzami N, Marzo I, Snow BE, Brothers GM, Mangion J, Jacotot E, Costantini P, Loeffler M, Larochette N, Goodlett DR, Aebersold R, Siderovski DP, Penhinger JM, Kroemer G. Molecular characterization of mitochondrial apoptosis-inducing factor. *Nature* 1999; 397: 441–446.

22. Bortner CD, Oldenburg NE, Cidlowski JA. The role of DNA fragmentation in apoptosis. *Trends Cell Biol* 1995; 5: 21-26.

23. Schwartz LM, Smith SW, Jones ME, Osborne BA. Do All Programmed Cell Deaths Occur Via Apoptosis? *Proc Natl Acad Sci USA* 1993; 90: 980–984.

24. Lagarkova MA, Iarovaia OV, Razin SV. Large-scale fragmentation of mammalian DNA in the course of apoptosis proceeds via excision of chromosomal DNA loops and their oligomers. *J Biol Chem* 1995; 270: 20239–20241.

25. Zhivotovsky B, Orrenius S. Involvement of Ca2+ in the formation of high molecular weight DNA fragments in thymocyte apoptosis. *Biochem Biophys Res Commun* 1994; 202: 120–127.

26. Kaufmann SH. Induction of endonuclolytic DNA cleavage in human acute myelogeneous leukemia cells by etoposide, camptothecin and other cytotoxic anticancer drugs. *Cancer Res* 1989; 49: 5870–5875.

27. Batistatou A, Greene LA. Aurintricarboxylic acid rescues PC 12 cells and sympathetic neurons from cell death caused by nerve growth factor deprivation: Correlation with suppression of endonuclease activity. *J Cell Biol* 1991; 115: 461–471.

28. Jacobson MD. Reactive oxygen species and programmed cell death. *Trends Biochem Sci* 1996; 21: 83–86.

29. Maher P, Davis JM. The role of monoamine metabolism in oxidative glutamate toxicity. *J Neurosci* 1996; 16: 6394–6401.

30. Cohen G, Farooqui R, Kesler N. Parkinson disease: A new link between monoamine oxidase and mitochondrial electron flow. *Proc Natl Acad Sci USA* 1997; 94: 4890–4894.

31. Samanta S, Perkinton MS, Morgan M, Williams RJ. Hydrogen peroxide enhances signal-responsive arachidonic acid release from neurons: Role of mitogen-activated protein kinase. *J Neurochem* 1998; 70: 2082–2090.

32. Higuchi Y. Polyunsaturated fatty acids promote 8-hydroxy-2-deoxyguanosine formation through lipid peroxidation under the glutamate-induced GSH depletion in rat glioma cells. *Arch Biochem Biophys* 2001; 292: 65–70.

33. Ochu EE, Rothwell NJ, Waters CM. Caspases mediate 6-hydroxydopamine-induced apoptosis but not necrosis in PC12 cells. *J Neurochem* 1998; 70: 2637–2640.

34. Higuchi Y, Yoshimoto T. Promoting effect of polyunsaturated fatty acids on chromosomal giant DNA fragmentation associated with cell death induced by glutathion depletion. *Free Radic Res* 2004; 38: 649–658.

35. Campagne ML, Lucassen PJ, Vermeulen JP, Balazs R. NMDA and kainate induce internucleosomal DNA cleavage associated with both apoptotic and necrotic cell death in the neonatal rat brain. *Eur J Neurosci* 1995; 7: 1627–1640.

36. Halliwell B, Gutteridge JMC. Lipid peroxidation: a radical chain reaction. In *Free Radicals in Biology and Medicine*, Halliwell B and Gutteridge JMC, Eds., Clarendon Press: Oxford, 1989, pp. 188–276.

37. Cosgrove JP, Church DF, Pryor WA. The kinetics of the autooxidation of polyunsaturated fatty acids. *Lipids* 1987; 22: 299–304.

38. Hawkins RH, Sangster K, Arends AM. Apoptotic death of pancreatic cancer cells induced by polyunsaturated fatty acids varies with double bond number and involves an oxidative mechanism. *J Pathol* 1998; 185: 61–70.

39. Sola R, La Ville AE, Richard JL, Motta C, Bargallo MT, Girona J, Masana L, Jacotot B. Oleic acid rich diet protects against the oxidative modification of high density lipoprotein. *Free Radic Biol Med* 1997; 22: 1037–1045.

40. Buege JA, Aust SD. Microsomal lipid peroxidation. *Methods Enzymol* 1978; 52: 302–310.

41. Halliwell B, Gutteridge MC. Oxygen toxicity, oxygen radicals, transition metals and disease. *Biochem J* 1984; 219: 1–14.

42. Mello Filho AC, Meneghini R. *In vivo* formation of single-strand breaks in DNS by hydrogen peroxide is mediated by the Haber-Weiss reaction. *Biochim Biophys Acta* 1984; 781: 56–63.

43. Mello Filho AC, Hoffman ME, Meneghini R. Cell killing and DNA damage by hydrogen peroxide are mediated by intracellular iron. *Biochem J* 1984; 218: 273–275.

44. Inoue S. Site-specific cleavage of double-strand DNA by hydroperoxide of linoleic acid. *FEBS Lett* 1984; 172: 231–234.

45. Sagripanti J, Kraemer KH. Site-specific oxidative DNA damage at polyguanosines produced by copper plus hydrogen peroxide. *J Biol Chem* 1989; 264: 1729–1734.

46. Matyshevskaia OP, Pastukh VN, Solodushko VA. Inhibition of lipoxygenase activity reduces radiation-induced DNA fragmentation in lymphocytes. *Radiat Biol Radioecol* 1999; 39: 282–286.

47. Li Y, Maher P, Schubert D. A role for 12-lipoxygenase in nerve cell death caused by glutathione depletion. *Neuron* 1997; 19: 453–463.

48. Yamamoto S. Mammalian lipoxygenases; molecular structures and functions. *Biochim Biophys Acta* 1992; 1128: 117–131.

49. Tang D, Chen Y Q, Honn KV. Arachidonate lipoygenases as essential regulations of cell survival and apoptosis. *Proc Natl Acad Sci USA* 1996; 93: 5241.

50. Higuchi Y, Yoshimoto T. Archidonic acid converts the glutathione depletion-induced apoptosis to necrosis by promoting lipid peroxidation in glioma cells. *Arch Biochem Biophys* 2002; 400: 133–140.

51. Enari M, Sakahira H, Yokoyama H, Okawa K, Iwamatsu A, Nagata S. A caspase-activated DNase that degrades DNA during apoptosis and its inhibitor ICAD. *Nature* 1998; 391: 43–50.

52. Armstrong RC, Aja TJ, Hoang KD, Gaur S, Bai X, Alnemri ES, Litwack G, Karanewsky DS, Fritz C, Tomaselli KJ. Activation of the CED3/ICE-related protease CPP32 in cerebellar granule neurons undergoing apoptosis but not necrosis. *J Neurosci* 1997; 17: 553–562.

53. Cummings BS, Mchowat J, Schnellmann RG. Phospholipase A2 in cell injury and death. *J Pharmacol Exp Ther* 2000; 294: 793–799.

54. Holtsberg FW, Steiner MR, Keller JN, Mark RJ, Mattson MP, Steiner SM. Lyso-phosphatidic acid induces necrosis and apoptosis in hippocampal neurons. *J Neurochem* 1998; 70: 66–76.

55. Bonfoco E, Krainc D, Ankarcrona M, Nicotera P, Lipton SA. Apoptosis and necrosis; two distinct events induced, respectively, by mild and intense insults with DMDA or nitric oxide/superoxide in cortical cultures. *Proc Natl Acad Sci USA* 1995; 92: 7162–7166.

56. Tan S, Wood M, Maher P. Oxidative stress induces a form of programmed cell death with characteristics of both apoptosis and necrosis in neuronal cells. *J Neurochem* 1998; 71: 95–105.

57. Lelli JL, Becks LL, Dabrowska MI, Hinshaw DB. ATP converts necrosis to apoptosis in oxidant-injured endotherial cells. *Free Radic Biol Med* 1998; 25: 694–702.

58. Kass GEN, Eriksson JE, Weis M, Orrenius S, Chow SC. Chromatin condensation during apoptosis requires ATP. *Biochem J* 1996; 318: 749–752.

59. Eguchi Y, Srinivasan A, Tomaselli KJ, Shimizu S, Tsujimoto Y. ATP-dependent steps in apoptotic signal transduction. *Cancer Res* 1999; 59: 2174–2181.
60. Ankarcrona M, Dypbukt JM, Bonfoco E, Zhivotovsky B, Orrenius S, Lipton SA, Nicotera P. Glutamate-induced neuronal death: A succession of necrosis or apoptosis depending on mitochondrial function. *Neuron* 1995; 15: 961–973.
61. Higuchi Y. Chromosomal DNA fragmentation in apoptosis and necrosis induced by oxidative stress. *Biochem Pharmacol* 2003; 66: 1527–1535.

24 Antioxidant Properties of Select Indian Medicinal Plants in Relation to Their Therapeutic Effects

J.C. Tilak and S. Adhikari
Bhabha Atomic Research Centre, Mumbai, India

B. Lakshmi and K.K. Janardhanan
Amala Cancer Research Centre, Kerala, India

T.P.A. Devasagayam
Bhabha Atomic Research Centre, Mumbai, India

CONTENTS

24.1 OVERVIEW

Indian systems of medicine such as ayurveda, siddha, and unani have been practiced for several centuries for the prevention and therapy of chronic diseases. A number of medicinal plants have been identified for this purpose with different therapeutic properties. We have chosen three medicinal plants known for their various therapeutic effects and studied their antioxidant abilities at different levels as a possible mechanism to explain the reported therapeutic properties. The plants used in our studies and their respective effects are: *Terminalia arjuna* (arjun) for cardioprotection, *Plumbago zeylanica* (chitrak) for neuroprotection, and *Pleurotus roseus* (a medicinal mushroom) for anticancer properties. The antioxidant effects of the preparations used for therapeutic applications as well as the active ingredients (wherever possible) were examined at different levels using studies on radical formation, radical scavenging, and inhibition of membrane damage. Mechanisms of protection were also studied by pulse radiolysis of active components. The results obtained showed that aqueous and ethanolic extracts of *P. zeylanica* have high activities in FRAP (ferric reducing antioxidant power), DPPH (1,1'-diphenyl-2-picrylhydrazyl), and ABTS (2,2'-azinobis-3-ethylbenz-thiazoline-6-sulfonic acid) assays. The ethanolic extract of *P. roseus* has high activities in DPPH and ABTS assays. The various extracts of *T. arjuna* showed moderate antioxidant activities in these assays. The studies on the inhibitory effect on lipid peroxidation induced by radiation, peroxynitrite, ascorbate-Fe^{2+}, and AAPH [2,2'-azobis (amidinopropane) dihydrochloride] in rat liver mitochondria, brain mitochondria, or cardiac homogenate also revealed significant protective abilities. Extracts of *T. arjuna* were effective against peroxidation induced by AAPH and radiation, and the extracts of *P. zeylanica* were protective against damage induced by ascorbate-Fe^{2+}. When we examined the antioxidant effects of baicalein from *T. arjuna* and plumbagin from *P. zeylanica* in different model systems, they proved to be very potent antioxidants. Their reaction with biologically relevant radicals was also examined by the special technique of pulse radiolysis. Plumbagin was more reactive towards thiyl radicals, whereas baicalein was more effective against hydroxyl radical. They were also effective against peroxyl radicals derived from different sources. The results obtained showed that the various extracts and active ingredients from these medicinal plants have significant antioxidant effects up to varying degrees in the relevant model systems at different levels of protection. These antioxidant abilities may at least in part explain the reported therapeutic effects of these medicinal plants.

24.2 INTRODUCTION

In recent years, the possible role of nutrition in prevention of human ailments has been shown to be important. In this context, antioxidants, especially those derived from natural sources, require special attention. As Indian medicinal plants are a rich source of these beneficial compounds, they require a thorough investigation, especially in terms of their antioxidant activities. Antioxidants neutralize the toxic and volatile *free radicals* that are defined as atoms or groups of atoms having an unpaired electron. These also include related reactive species such as "excited states" that lead to free radical generation or that result from free radical reactions. In general, free radicals are very short lived, with half-lives in milli, micro, or nanoseconds. *Antioxidants* are substances capable of neutralizing free radicals and their actions. In a normal, healthy organism or human body, the generation of pro-oxidants in the form of, reactive oxygen species (ROS) and RNS are effectively kept in check by the various levels of antioxidant defense.[1,2] Cellular damage induced by oxidative stress has been implicated in the etiology of a large number (>100) of human diseases as well as in the process of aging. The degenerative diseases in which free radicals have been implicated include cardiovascular ailments, neurodegenerative diseases such as Parkinson's disease, Alzheimer's disease, multistage process of carcinogenesis, etc. Other pathological conditions implicating ROS/RNS are diabetes, rheumatoid arthritis, hemorrhagic shock, gastrointestinal ulcerogenesis, AIDS, lung diseases such as adult respiratory distress syndrome, asthma, etc., radiation injury, toxicity due to physicochemical agents or pollutants or toxicants, inflammation resulting from infection, and ischemia–reperfusion type injuries associated with heart, brain, kidney, liver, and gastrointestinal tract.

Traditional Indian systems of medicine such as ayurveda, siddha, and unani have identified many plants and their respective phytochemicals that play a significant role in disease prevention and therapy. A significant part of Ayurvedic therapeutics aims to promote positive health.[3,4] Indian medicinal plants are a rich source of substances that have several therapeutic properties with cardioprotective, neuroprotective, chemopreventive, and other effects. We have chosen three medicinal plants known for their various therapeutic effects and studied their antioxidant abilities at different levels as a possible mechanism to explain the reported therapeutic properties. The plants used in our studies with their respective effects are *Terminalia arjuna* (arjun) for cardioprotection, *Plumbago zeylanica* (chitrak) for neuroprotection, and *Pleurotus roseus* (a medicinal mushroom) for anticancer properties.

P. zeylanica has been credited with therapeutic properties to treat bronchitis, anemia, liver diseases, diseases of the spleen, and fevers, besides bacterial and microbial infections.[5,6] It brings about regression of atheroma and hinders plaque formation.[5,7] The major portion of the plant used is the root that contains bioactive constituents such as naphthoquinones, binaphthoquinones, coumarin and anthroquinones.[6] Plumbagin, a naphthoquinone and a major component, is considered as the active ingredient responsible for therapeutic effects.[8] It possesses cardiotonic, hypolipidemic, antiatherosclerotic, anticoagulant, anticarcinogenic, antitumor, antimutagenic, wound healing, antifungal, antibacterial, and antifertility properties.[9,10] Though there are many reports, as detailed above, on the beneficial properties of *P. zeylanica* and plumbagin, their antioxidant effects have not been studied. The extracts used were aqueous, boiled aqueous, ethanolic, and boiled ethanolic ones. The aqueous extract corresponds to the paste prepared for rheumatoid arthritis, paralysis, leprosy, and other skin diseases.[11] Boiled aqueous extract is similar to preparations used for fertility control.[6] Ethanolic extract simulates the paste prepared in milk and vinegar or oil and applied for leprosy, other skin diseases, rheumatism, paralysis, glandular tumors, and abscesses.[5] The boiled ethanolic extract is similar to oil-based preparation for topical applications in muscular pains and arthritis.[6]

Terminalia arjuna is one of these plants credited for its cardiotonic and cardioprotective properties. The bark of the tree has been used in the ayurvedic system of medicines for over three centuries, primarily as a cardiac tonic as well as a cure for hemorrhages, fractures, diarrhea, ulcers, and acne.[5] It has also been known to possess antimutagenic,[12] antiischemic,[13] hypocholesterolemic,[14] cardio-protective,[15] and antiradical and antilipoperoxidative[16] abilities. Hence, the radical scavenging activities and membrane protective abilities of *T. arjuna* extracts, as well as baicalein, an active ingredient, were examined. The different extracts used for the assays were benzene, chloroform, acetone, methanolic, methanolic-HCl, and aqueous fractions of Sohxlet extraction. Baicalein (5,6,7-trihydroxy-2-phe-nyl-4H-1-benzopyran-4-one) is a naturally occurring flavonoid found in the bark of *T. arjuna*.

Mushrooms are traditionally used in Indian and Chinese medicine and are commonly utilized for pharmaceutical purposes and in health foods. Increasing experimental evidence indicates that mushrooms contain a large number of bio-logically active components that offer health benefits and protection against degenerative diseases. A number of medicinal mushrooms have recently been reported to possess significant antioxidant activity.[17–20] *Phellinus rimosus* have been used by some local tribes for the treatment of ailments like mumps.[18] Earlier investigations showed that ethyl acetate extracts of *P. rimosus* possessed antioxidant, antitumor, antiinflammatory, hepatoprotective, and nephroprotective, activities.[21] Oyster mushrooms (*Pleurotus* spp.) are widespread throughout the hardwood forests of the world; they are edible, nutritious, and rank second among the cultivated mushrooms in the world. They are known to be efficient blood-pressure-lowering agents, diuretics, cholesterol reducers, adjuvants, and aphrodisiacs. They have also been found to modulate the immune system.

Antioxidants exhibit their effects at different levels. These include ability to bind iron that can prevent radical formation, the scavenging of primary and secondary radicals, and ability to inhibit free-radical-induced membrane damage. To study radical scavenging abilities, the model radicals DPPH and ABTS can be used.[22,23] Among the subcellular organelles, mitochondria are crucial targets for oxidative damage.[24] Hence, in this study, we have examined the protection afforded by various extracts of PZ in FRAP, DPPH, and ABTS assays besides the ability to inhibit lipid peroxidation in mitochondria from rat liver induced by pathophysiologically relevant agents.

Radical scavenging and lipid peroxidation inhibitory effects of pure com-pounds such as plumbagin and baicalein were also examined. To study their possible mechanisms of antioxidant action, we studied its reaction with various biologically relevant radicals by pulse radiolysis. The radicals examined were hydroxyl ($^\bullet$OH), trichloromethylperoxyl ($CCl_3OO^\bullet$), linoleate peroxyl (LOO$^\bullet$), and glutathiyl (GS•). The •OH is extremely reactive and found to degrade DNA, proteins, and lipids, and forms DNA-protein, DNA-lipid, and DNA-DNA cross-links.[2] The lipid peroxyl radicals (LOO•) are intermediates in lipid peroxidation and play an important role during chain propagation.[2,25] Thiyl radicals (RS•) are formed by the reaction of different reactive species with sulfur-containing compounds that play a central role in the structure and activity of compounds of biological interest.[26]

24.3 MATERIALS AND METHODS

24.3.1 MATERIALS

Ascorbic acid, ABTS diammonium salt, butylated hydroxytoluene (BHT), DPPH, ethylene diamine tetra acetic acid (EDTA), ferric chloride, hydrogen peroxide, methanol (HPLC grade), myoglobin, potassium ferricyanide, sodium acetate, Trolox (6-hydroxy-2,5,7,8-tetra methyl chroman-2-carboxylic acid), 2,4,6-tripyridyl-s-triazine (TPTZ), 2-thiobarbituric acid, triphenyl phospene (TPP), trichloro acetic acid, and xylenol orange were purchased from Sigma Chemical Co. AAPH was from Aldrich Chemical Co. Other chemicals used in the studies were of analytical grade from reputed manufacturers.

The roots of *P. zeylanica* were obtained from Gnanadhass Devasahayam, professor, Government Siddha Medical College, Tirunelveli, Tamil Nadu, India. *T. arjuna* bark was obtained from H.S. Palep, Palep's Research Foundation, Mumbai, India. Fruiting bodies of *Pleurotus florida* and *Pleurotus sajor-caju* were obtained from the small-scale cultivation unit of Integrated Rural Technology Centre, Palakkad, Kerala, India. *Ganoderma lucidum* and *P. rimosus* were collected locally from the outskirts of Thrissur, Kerala, India.

24.3.2 METHODS

24.3.2.1 Preparation of the Extracts of Medicinal Plants

24.3.2.1.1 Preparation of the Extracts from the Roots of P. zeylanica

The roots of PZ were sun dried for several days and powdered. Extracts were prepared so as to simulate conditions of their extraction pertaining to their use in medicinal preparations. The concentrations of the extracts prepared were 1%. The aqueous extract (PZ1) was prepared by adding the root powder to distilled water (D/W) and stirring it for 60 min. For boiled water extract (PZ2), the powder was boiled in D/W for 30 min. The ethanolic extract (PZ3) was prepared by stirring the root powder in ethanol for 60 min and the boiled ethanolic extract (PZ4) was prepared by boiling it in ethanol for 30 min.

24.3.2.1.2 Preparation of the Extracts from the Bark of T. arjuna

The Soxhlet extracts of *T. arjuna* were prepared by taking the bark, free of dust and other impurities. It was sun dried and finely powdered. Depending on the increasing order of polarity of the solvents, sequential extraction was done using benzene, chloroform, acetone, methanol, and methanol-HCl using Soxhlet apparatus. The bark powder was extracted for 8 to 10 h with each solvent to remove the soluble matter. The aqueous extract was prepared by stirring the bark powder in distilled water for 4 h and then filtering it.

24.3.2.1.3 Preparation of Mushroom Extracts

Fruiting bodies of *P. rimosus* were dried at 45 to 50°C for 48 h and powdered. The powdered material was defatted by extracting with petroleum ether in Soxhlet apparatus for 8 to 10 h. The defatted material was then extracted with ethyl acetate. The extracts were concentrated at low temperature under vacuum and solvents evaporated completely. The residues thus obtained were used for experiments. For *in vitro* antioxidant activity studies, the ethyl acetate extract of *P. rimosus* was dissolved in methanol. Methanolic extracts of *P. rimosus* were prepared by dissolving 0.01 g ethyl acetate extract in 10 ml of methanol and stirring for 30 min. All the extracts were centrifuged for 15 min and supernatants were stored at 20°C.

24.3.2.2 Determination of Antioxidant Activity

Antioxidant activity of the extracts were assayed by ferric-reducing antioxidant power (FRAP), DPPH, and ABTS methods. For the antioxidant activity assays, 0.1%, 0.5%, and 1% concentrations of the extracts were used. The results were expressed as Trolox equivalent antioxidant capacity (TEAC) and ascorbic acid equivalent antioxidant capacity (AEAC). TEAC is the concentration of Trolox (%) required to give the same antioxidant capacity as test substance and AEAC is that of ascorbic acid.[27]

24.3.2.2.1 FRAP Assay

The ferric-reducing ability was measured at low pH.[28,29] The stock solution of 10 mM TPTZ in 40 mM HCl, 20 mM FeCl$_3$.6H$_2$O, and 0.3 M acetate buffer (pH3.6) were prepared. The FRAP reagent contained 2.5 ml TPTZ solution, 2.5 ml ferric chloride solution, and 25 ml acetate buffer. It was prepared fresh and warmed to 37°C. Then, 900 µl of D/W and 30 µl test sample/methanol/D/W standard solutions were added. The reaction mixture was then incubated at 37°C for 30 min and absorbance was recorded at 595 nm. An intense blue color complex was formed when ferric tripyridyl triazine (Fe^{3+}TPTZ) complex was reduced to the ferrous (Fe^{2+}) form and the absorption at 595 nm was recorded. The calibration curve was plotted with absorbance at 595 nm vs. concentration of FeSO$_4$ in the range of 0 to 1 mM (both aqueous and methanolic solutions). The concentrations of FeSO$_4$ were in turn plotted against concentrations of standard antioxidants (L-ascorbic acid or Trolox).

24.3.2.2.2 DPPH Radical Scavenging Assay

In this method, a commercially available and stable free radical DPPH$^+$, soluble in methanol, was used.[22] DPPH in its radical form has an absorption peak at

515 nm, which disappeared on reduction by an antioxidant compound. An aliquot (37.5 µl) of the extract was added to 1.5 ml of freshly prepared DPPH solution (0.25 g/l in methanol). Absorbance was measured at 515 nm, 20 min after the reaction was started. The DPPH concentration in the reaction medium was calculated from the calibration curve of percentage of DPPH scavenged vs. concentration of the standard antioxidant (L-ascorbic acid or Trolox).

24.3.2.2.3 Ferrylmyoglobin/ABTS Assay

In this assay, the radical scavenging activities of extracts were determined by using ferryl myoglobin/ABTS protocol.[23] The stock solutions of 500 µM ABTS diammonium salt, 400 µM myoglobin, 740 µM potassium fericyanide, and 450 µM H_2O_2 were prepared in phosphate-buffered saline (PBS) (pH7.4). Metmyoglobin (MbIII) was prepared by mixing equal amounts of myoglobin and potassium ferricyanide solutions. The reaction mixture (total volume 2 ml) contained the following substances (final concentrations in the reaction mixture): ABTS (150 µM), MbIII (2.5 µM), 16.8 µl of the sample, and 978 µl PBS. The reaction was initiated by adding 75 µM H_2O_2 (330 µl) and the lag time (in seconds) before absorbance of ABTS$^+$ at 734 nm began to increase was recorded. The calibration curve was plotted with lag time in seconds vs. concentration of the standard antioxidants (L-ascorbic acid or Trolox).

24.3.2.3 Isolation of Mitochondrial Fraction from Rat Liver

Three-month-old female Wistar rats (weighing about 250 ± 20 g) were used for the preparation of mitochondria. The mitochondrial pellet was washed three times with 0.15 M Tris-HCl buffer, pH 7.4, to remove sucrose.[24] Protein was estimated and pellets were suspended in the above buffer at the concentration of 10 mg protein/ml.

24.3.2.4 Exposure of Rat Liver Mitochondria to Radiation and Agents for Inducing Oxidative Stress

Oxidative damage was induced by exposure to γ-rays from a ^{60}Co source (dose rate 65 Gy/min, Bhabha Atomic Research Centre, Mumbai).[30] The mitochondria (final conc. 2 mg/ml) were suspended in 5 mM phosphate buffer (pH 7.4) and exposed to radiation with and without extracts or baicalein. The dose selected was 450 Gy at which optimum damage is obtained. The unexposed samples served as controls.

Peroxyl radical–induced lipid peroxidation was observed using azobis-amidino-propane hydrochloride.[31] In the system for treating mitochondria to AAPH, mitochondria (final conc. 0.2 mg/ml) were incubated with AAPH (final conc. 10 mM) at 37°C for 30 min in a shaker-water bath with continuous bubbling of oxygen.

24.3.2.5 Pulse Radiolysis Studies

The pulse radiolysis system using 7 MeV electrons has been described earlier.[32] The dosimetry was carried out using an air-saturated aqueous solution containing

5×10^{-2} mol dm^{-3} KSCN (Gε = 23,889 dm^3 mol^{-1} cm^{-1} per 100 eV at 500 nm). The kinetic spectrophotometric detection system covered the wavelength range from 250 to 800 nm. The optical path length of the cell was 1.0 cm. The width of the electron pulse was 50 ns and the dose was 16 Gy per pulse. Alkaline pH was obtained by adding NaOH only. High purity (>99.9%) N$_2$O, from BOC India was used. The following reactions occur after irradiation in the reaction medium used.

$$H_2O \rightarrow e_{aq}^-, H^\bullet, {}^\bullet OH, H_2, H_2O_2, \text{etc.}$$

$$e_{aq}^- + N_2O + H_2O \rightarrow {}^\bullet OH + OH^- + N_2$$

$$LH + {}^\bullet OH \rightarrow L^\bullet + H_2O$$

$$L^\bullet + O_2 \rightarrow LOO^\bullet$$

$$e_{aq}^- + CCl_4 \rightarrow CCl_3^\bullet$$

$$CCl_3^\bullet + O_2 \rightarrow CCl_3OO^\bullet$$

The bimolecular rate constants were calculated by plotting pseudo first-order rate of formation of the transient against the concerned solute concentration. The uncertainty in the measurement in bimolecular rate constant is ±10%. The transients obtained in the pulse radiolysis study were used to characterize the product radical. The rate constants determined and presented in the text are not mere radiation chemical parameters but reflect the efficiency of scavenging free radicals and the ease with which competing reactions occur.

24.4 RESULTS AND DISCUSSION

Recent clinical evidence suggests that reactive oxygen and nitrogen species are involved in pathophysiology of many harmful human diseases, and it is also known that antioxidants help in prevention of these diseases. Indian systems of medicine have identified many plants and herbs that have beneficial therapeutic effects. The mechanism behind these properties could be their antioxidant abilities, which help them quench harmful free radicals or inhibit damage to cellular moieties. Although organisms are bestowed with antioxidant and repair systems that have evolved to protect them against oxidative damage, these systems are insufficient to prevent the damage totally.[2] Hence, antioxidants in the diet are of importance as possible protective agents to help the human body reduce oxidative damage. There are some antioxidant enzymes such as superoxide dismutase, catalase, and glutathione peroxidase, which are present endogenously as the body's first line of defense. Apart from this, nonenzymatic antioxidants such as vitamin C, vitamin E, carotenoids, and other natural flavonoids inactivate pro-oxidants by a redox reaction in which the antioxidant acts as a reducing agent.

Hence, the antioxidant power can be referred to as *reducing ability*. In FRAP assay, an easily reducible oxidant Fe III is used in excess. Thus, reduction of Fe^{III}-TPTZ complex by antioxidant leads to blue-colored Fe^{II}-TPTZ complex, which can be measured spectrophotometrically at 595 nm.[28] The first line of defense is the preventive antioxidants, which suppress the formation of free radicals. In the ferrylmyoglobin/ABTS assay, on addition of antioxidant, reaction of ferrylmyoglobin and ABTS is delayed with prolongation of ABTS radical formation, which can be measured as the lag time in seconds.[23] The second line of defense is scavenging free radicals to suppress chain initiation and/or break the chain propogation reactions. DPPH is a purple-colored, stable primary radical, which is commercially available. The ability of an antioxidant to scavenge DPPH is tested by its depolarization spectrophotometrically at 515 nm.[22]

The concentration of Trolox required for giving the same radical scavenging capacity as 1 mmol/l extracts from root of *P. zeylanica* is shown in Table 24.1. The concentration of the extract used for the assays was 0.05, 0.1, and 1% (w/v), and concentration-dependent effect of the extracts in terms of their antioxidant potential was observed. In the case of the FRAP assay, boiled ethanolic extract (PZ4) is the most significant reducing agent with TEAC = 0.44. This was followed by ethanolic (PZ3) and boiled aqueous (PZ2) with 0.39 and 0.37 TEAC values, respectively. In the case of the DPPH radical scavenging assay, PZ4 is the most potent radical scavenger (TEAC = 0.57) followed by PZ3 (TEAC = 0.21). However, in the ferrylmyoglobin/ABTS assay, boiled aqueous extract (PZ2) is the most significant in inhibition of ABTS radical formation (TEAC = 1.02), whereas PZ4 is the next potent extract (TEAC = 0.57).

Lipid peroxidation is a major chain reaction, which damages the biological membranes. In the present study, we have examined the effect of the various extracts of medicinal plants and pure compounds on lipid peroxidation induced

TABLE 24.1
Antioxidant Activity of *Plumbago zeylanica* Root Extracts Evaluated by Three Methods

1% (w/v) P. zeylanica Extracts	FRAP Assay	DPPH Assay	Ferrylmyoglobin/ABTS Assay
		Antioxidant Activity (TEAC) mmol/l	
PZ1 (Aqueous)	0.27 ± 0.02	0.11 ± 0.01	0.55 ± 0.01
PZ2 (Boiled aqueous)	0.37 ± 0.02	0.18 ± 0.01	1.02 ± 0.05
PZ3 (Ethanolic)	0.39 ± 0.01	0.21 ± 0.06	0.50 ± 0.02
PZ4 (Boiled ethanoilc)	0.44 ± 0.01	0.57 ± 0.05	0.57 ± 0.02

Note: PZ1: Aqueous extract of root of *Plumbago zeylanica*; PZ2: Boiled aqueous extract of root of *Plumbago zeylanica*; PZ3: Ethanolic extract of root of *Plumbago zeylanica*; PZ4: Boiled ethanolic extract of root of *Plumbago zeylanica*.

Data presented is mean ± S.E. from four different experiments.

by different free-radical generators. These include γ-radiation, endogenous free-radical generator ascorbate-Fe^{2+}, peroxynitrite, and azo-initiator of peroxyl radicals AAPH. Lipid damage was assessed by the thiobarbituric acid reactive substances (TBARS) assay. In the case of ascorbate-Fe^{2+}-induced TBARS formation in rat liver mitochondria, as shown in Figure 24.1a, PZ3 and PZ4 extracts give maximum protection followed by the aqueous ones. In Figure 24.1b, plumbagin also inhibits ascorbate-Fe^{2+}-induced lipid peroxidation in rat liver mitochondria completely.

Our results, in general, indicate that ethanolic preparations (pZ3 and PZ4) are more effective in giving antioxidant protection at various levels, reduction of iron complexes, radical scavenging, and in membrane protection (as assayed by lipid peroxidation). The mechanisms involved in many human diseases such as hepatotoxicities, carcinogenesis, diabetes, malaria, acute myocardial infarction,

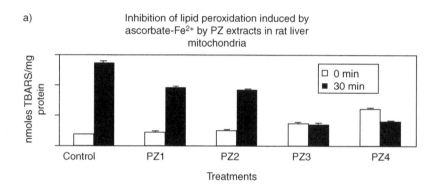

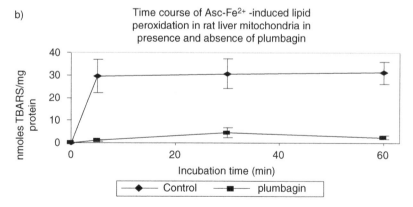

FIGURE 24.1 (a) Effect of different *P. zeylanica* (PZ) root extracts on ascorbate-Fe^{2+}-induced lipid peroxidation in rat liver mitochondria. (b) Data on time course of ascorbate-Fe^{2+}-induced lipid peroxidation in rat liver mitochondria with and without plumbagin incorporated during isolation. The concentration of plumbagin in rat liver mitochondria was 156 μM. The formation of TBARS was estimated. The values are ± SEM from four experiments.

TABLE 24.2
Antioxidant Activity of Sohxlet Fractions of Bark of *T. arjuna* Evaluated by Three Methods

0.05% (w/v) *T. arjuna* Sohxlet Extracts	FRAP Assay	DPPH Assay	Ferrylmyoglobin/ABTS Assay
		Antioxidant Activity (TEAC) mmol/l	
Aqueous	0.28 ± 0.02	0.24 ± 0.02	0.18 ± 0.001
Methanolic-HCl	0.31 ± 0.04	0.70 ± 0.05	0.10 ± 0.05
Methanolic	0.37 ± 0.03	0.35 ± 0.02	0.34 ± 0.04
Acetone	0.26 ± 0.02	0.18 ± 0.03	0.08 ± 0.01
Chloroform	0.01 ± 0.01	0.07 ± 0.01	0.01 ± 0.002

Note: Data presented is mean ± S.E. from four different experiments.

and skin cancer include lipid peroxidation as a main source of membrane damage.[25] The systems used for inducing oxidative damage *in vitro* is ascorbate-Fe^{2+}, a system relevant to endogenous oxidative damage in rat liver mitochondria.

Table 24.2 presents data on antioxidant activities of Sohxlet extracts of bark of *T. arjuna*. In the case of the FRAP assay, the total antioxidant activity of methanolic fraction of *T. arjuna* was equivalent to that of 0.37 mmol/l of Trolox and was the highest among other fractions. Methanolic-HCl (TEAC = 0.31) and aqueous (TEAC = 0.28) are second and third in reducing properties. TEAC of the methanolic-HCl fraction was the highest, i.e., 0.70 mmol/l as calculated experimentally by the DPPH method. The antioxidant activity of methanolic extract (TEAC = 0.35) was half of that of methanolic-HCl. In the ferrylmyoglobin/ABTS assay, methanolic extract of *T. arjuna* gave the maximum antioxidant activity (TEAC = 0.34) among all other fractions. This was followed by aqueous (TEAC = 0.18) and methanolic-HCl (TEAC = 0.10).

Figure 24.2a and Figure 24.2b present the data on effect of *T. arjuna* extracts on lipid peroxidation in rat liver mitochondria induced by AAPH and γ-radiation, respectively. In the case of AAPH-induced lipid peroxidation, the methanolic-HCl fraction is the most potent inhibitor of damage, whereas, in γ-radiation-induced TBARS formation, methanolic, methanolic-HCl, and aqueous fractions are equally effective. Figure 24.2c shows the protection afforded by baicalein in rat cardiac homogenate against AAPH-induced damage. The inhibition of TBARS formation was found to be in a concentration-dependent manner. At a concentration as low as 5 to 10 μM, baicalein gives significant protection of about 40%. In these experiments, the methanolic fraction of *T. arjuna* shows the highest antioxidant potential. It was found by HPLC determination that baicalein is present in the methanolic fraction (data not shown). Hence, the antioxidant ability can be attributed to baicalein.

Table 24.3 presents the data on bimolecular rate constants of plumbagin and baicalein on reaction with biologically relevant free radicals as studied by pulse radiolysis. In the case of the hydroxyl radical reaction, both the compounds give

a)

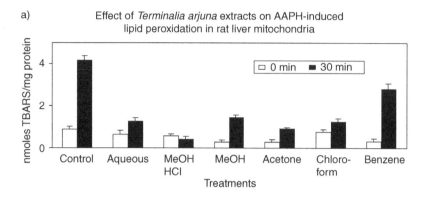

b)

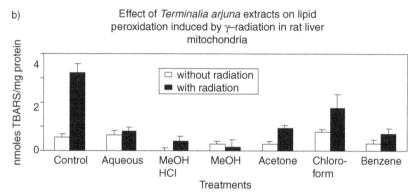

c)

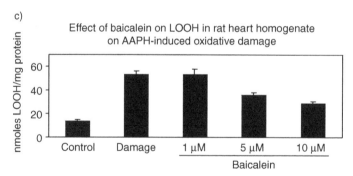

FIGURE 24.2 (a) Effect of Sohxlet extracts of *T. arjuna* bark on AAPH-induced lipid peroxidation (30-min incubation) in rat liver mitochondria. (b) Data on effect of these extracts on γ-radiation-induced lipid peroxidaiton in rat liver mitochondria, when present during irradiation. (c) Effect of baicalein; lipid peroxidation is induced by AAPH in rat heart homogenate. The formation of TBARS was estimated as a measure of lipid peroxidation. The values are ± SEM from four experiments.

high rate constants due to the very high oxidation potential of •OH. The transient absorption spectra suggest that radical formation in the reaction of •OH with these two compounds are due to the one electron oxidation. On reaction with model

TABLE 24.3
Bimolecular Rate Constants of the Pure
Phytochemicals from Indian Medicinal Plants as
Measured by Pulse Radiolysis Technique

	Bimolecular Rate Constants (dm^3 mol^{-1} sec^{-1})	
Free Radicals	Baicalein	Plumbagin
Hydroxyl radical ($^\bullet OH$)	3.7×10^9	2.03×10^9
Trichloroperoxyl radical ($CCl_3OO^\bullet$)	1.34×10^9	1.1×10^9
Linoleic acid peroxyl radical ($LOO^\bullet$)	8.5×10^7	6.7×10^7

peroxyl radical, trichloroperoxyl radical ($CCl_3OO^\bullet$), and linoleic acid peroxyl radical ($LOO^\bullet$), baicalein has higher rate constants than plumbagin. This can be due to the three hydroxyl groups present in its structure, which forms a phenoxyl type of transient radical. Thus, the ability of these compounds to react with ROS with high rate constants may partly explain their efficacy in medicinal preparations.

Mushrooms are functional foods and are traditionally used in folk medicine and several systems of medicine. Medicinal mushrooms, possessing antioxidant properties, in the diet would be potentially useful to help reduce oxidative damage in the human body. The results of the investigation (Figure 24.3) reveal that extracts of *P. rimosus, P. florida, P. sajor-caju,* and *G. lucidum* are effective in lowering the formation of LOOH in rat liver mitochondria induced by γ-radiation. Our studies reveal that all the four mushrooms show high radical-scavenging activities and potent reducing power using standard assays. Our results indicate that among the samples examined, the extract of *G. lucidum* is the most effective in rendering antioxidant protection.

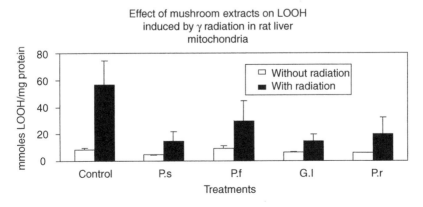

FIGURE 24.3 Effect of mushroom extracts on γ-radiation-induced lipid hydroperoxide formation in rat liver mitochondria. The values are ± SEM from four experiments. (From Lakshmi, B. et al., *Indian Acad Sci,* 2005; 88(3): 484–488. With permission.)

The results obtained showed that the various extracts and active ingredients from these medicinal plants have significant antioxidant effects to varying extents in the relevant model systems at different levels of protection. These antioxidant abilities may at least in part explain the reported therapeutic effects of these medicinal plants.

REFERENCES

1. Halliwell B, Gutteridge JMC. *Free Radicals in Biology and Medicine*. Oxford University Press: Oxford, 1997.
2. Sies H. *Antioxidants in Disease, Mechanisms and Therapy*. Academic Press: New York, 1996.
3. Agarwal SS, Singh VK. Immunomodulators: A review of studies on Indian medicinal plants and synthetic peptides. Part I: Medicinal plants. *Proc Indian Natl Sci Acad B* 1999; 65: 179–204.
4. Devasagayam TPA, Sainis KB. Immune system and antioxidants, especially those derived from Indian medicinal plants. *Indian J Expt Biol* 2001; 40: 639–655.
5. Kirtikar KR, Basu BD. *Indian Medicinal Plants*. Vol. 1 and Vol. 2. Blatter E, Cauis JR, Mhaskar KS, Basu LM, Eds. Allahabad, India, 1984.
6. Thakur RS, Puri HS, Husain A. *Major Medicinal Plants of India*, Central Institute of Medicinal and Aromatic Plants: Lucknow, India, 1989.
7. Sharma I, Gusain D, Devi PU. Hypolipidaemic and antiatherosclerotic effects of plumbagin in rabbits. *Indian J Physiol Pharmacol* 1991; 35: 10–14.
8. Gupta MM, Verma RK, Uniyal GC, Jain SP. Determination of plumbagin by normal phase high performance liquid chromatography. *J Chromatogr* 1993; 637: 209–212.
9. Itoigawa M, Takeya K, Furukawa H. Cardiotonic action of plumbagin on guinea-pig papillary muscle. *Planta Med* 1991; 57: 317–319.
10. Sugie S, Okamoto K, Rahman KM. Inhibitory effect of plumbagin and juglone on azoxymethane-induced intestinal carcinogenesis in rats. *Cancer Lett* 1998; 127: 177–183.
11. Dastur JF. *Medicinal Plants of India and Pakistan*, D. B. Taraporwala Sons & Co. Pvt. Ltd.: Bombay, 2001.
12. Kaur S, Grover IS, Kumar S. Antimutagenic potential of ellagic acid isolated from *Terminalia arjuna*. *Indian J Exp Biol* 1997; 35: 478–482.
13. Gauthaman K, Maulik M, Kumari R, Manchanda SC, Dinda AK, Maulik SK. Effect of chronic treatment of bark of *Terminalia arjuna*: a study on the isolated ischemic reperfused heart. *J Ethnopharmacol* 2001; 75: 197–201.
14. Gupta R, Singhal S, Goyle A, Sharma VN. Antioxidant and hypocholesterolaemic effects of *Terminalia arjuna* tree bark powder: a randomised placebo-controlled trial. *J Assoc Physicians India* 2001; 49: 231–235.
15. Sumitra M, Manikandan P, Kumar DA, Arutselvan N, Balakrishna K, Manohar BM, Puvanakrishnan R. Experimental myocardial necrosis in rats: Role of arjunolic acid on platelet aggregation, coagulation and antioxidant status. *Mol Cell Biol* 2001; 224: 135–142.
16. Munasinghe JTC, Senevirante CK, Thabrew MI, Abeysekara AM. Antiradical and antilipoperoxidative effects of some plant extracts used by Sri Lankan traditional medical practitioners for cardioprotection. *Phytother Res PTR* 2001; 15: 519–523.

17. Jones S, Janardhanan KK. Antioxidant and antitumor activity of *G. lucidum* (Curt: Fr) P. Karst–Reishi (Aphyllophoromycetideae) from South India. *Int J Med Mushrooms* 2000; 2: 195–200.

18. Ajith TA, Janardhanan KK. Antioxidant and anti-inflammatory activities of methanolic extract of *Phellinus rimosus* (Berk) pilat. *Indian J Expt Biol* 2001; 39: 1166–1169.

19. Mau JL, Chao GR, Wu KT. Antioxidant properties of methanolic extracts from several ear mushrooms. *J Agric Food Chem* 2001; 49: 5461–5467.

20. Ekanem EO, Ubengama VS. Chemical composition, anti-nutritional factors and shelf-life of oyster mushroom (*Pleurotus ostreatus*). *J Food Sci Technol* 2002; 39: 635–638.

21. Ajith TA, Janardhanan KK. Antioxidant and antihepatotoxic activities of *Phellinus rimosus* (Berk) Pilat. *J Ethnopharmacol* 2002; 81: 387–391.

22. Aquino R, Morelli S, Lauro MR, Abdo S, Saija A, Tomaino A. Phenolic constituents and antioxidant activity of an extract of *Anthurium versicolor* leaves. *J Natl Prod* 2001; 64: 1019–1023.

23. Alzoreky N, Nakahara N. Antioxidant activity of some edible Yemeni plants evaluated by Ferrylmyoglobin/ABTS assays. *Food Sci Technol Res* 2001; 7: 141–144.

24. Devasagayam TPA, Kamat JP, Mohan H, Kesavan PC. Caffeine as an antioxidant: Inhibition of lipid peroxidation induced by reactive oxygen species in rat liver microsomes. *Biochim Biophys Acta* 1996; 1282: 63–70.

25. Yoshikawa T, Toyokuni S, Yamamoto Y, Naito Y. *Free Radicals in Chemistry Biology and Medicine*, OICA International: London, 2000.

26. Giles GI, Tasker KM, Jacob C. Hypothesis: The role of reactive sulfur species in oxidative stress. *Free Radic Biol Med* 2001; 31: 1279–1283.

27. Gil MA, Tomas-Barberan FA, Hess-Pierce B, Holcraft DM, Kader AA. Antioxidant activity of pomegranate juice and its relationship with phenolic composition and processing. *J Agric Food Chem* 2000; 48: 4581–4589.

28. Benzie IFF, Strain JJ. The ferric reducing ability of plasma (FRAP) as a measure of "Antioxidant power" The FRAP assay. *Anal Biochem* 1996; 239: 70–76.

29. Pulido R, Bravo L, Saura-Calixto F. Antioxidant activity of dietary polyphenols as determined by a modified ferric reducing/antioxidant power assay. *J Agric Food Chem* 2000; 48: 3396–3402.

30. Kamat JP, Boloor KK, Devasagayam TPA, Jayashree B and Kesavan PC. Differential modification of oxygen-dependent and -independent effects of γ-irradiation in rat liver mitochondria by caffeine. *Int J Radiat Biol* 2000; 76: 1281–1288.

31. Kamat JP, Devasagayam TPA. Tocotrienols from palm oil as potent inhibitors of lipid peroxidation and protein oxidation in rat brain mitochondria. *Neurosci Lett* 1995; 195: 1–4.

32. Mukherjee T, Ahmad SA. *Atomic, Molecular and Cluster Physics*, Narosa, New Delhi, 1997, pp. 299–316.

25 Antioxidant Activity of Tendai-Uyaku Extract

Toshiki Masumizu
JEOL, Ltd., Tokyo, Japan

Takao Kaneyuki
Kurasiki Sakuyo University, Kurasiki, Japan

Akitane Mori
Okayama University, Okayama, Japan

CONTENTS

25.1 INTRODUCTION

Tendai-uyaku (uyaku) is the dried root of *Lindera strychnifolia (Sieb. et Zucc.) F. villar* of the Lauraceae family (1–3). Uyaku has been used as a folk drug for maintaining good health and for the treatment of diseases such as those of the stomach and the kidney, neuralgia, and, in some districts including Shingu in the Wakayama prefecture, to cure rheumatism. We first found that uyaku roots and leaves showed markedly higher superoxide dismutase (SOD)-like activities compared with several other natural extracts (4). The extract of uyaku roots and leaves have potent scavenging activity not only against superoxide anion radical but also against peroxynitrite and effectively inhibited lipid peroxidation and protein carbonyl formation, suggesting beneficial effects for health, especially for prevention or treatment of free-radical-related diseases (5). Recently, an effective preparation technique for uyaku raw material has been developed, which involves using a special apparatus for powdering developed by Sakai Canning Co., Ltd. (Wakayama, Japan; Japanese Patent No.: HEI 2-135149). The extract preparations for drinking available commercially are called *Jo-Fuku-no-sei* (the spirit of Jo-Fuku) after Jo-Fuku (*Xu Fu* in Chinese). According to a traditional story, the first emperor of the Qin dynasty ordered Jo-Fuku to look for a drug effective for longevity. After a long trip by sea, he and his colleagues arrived at Shingu in Wakayama, Japan, and found Uyaku in 219 BC. In this study, the antioxidant activities of uyaku extract Jo-Fuku-no-sei were estimated and compared with our previous experimental results.

25.2 MATERIALS AND METHODS

25.2.1 CHEMICALS

6-Hydroxypurine (hypoxanthine) was purchased from Sigma-Aldrich Chemical Co. (St. Louis, MO). Xanthine oxidase (XOD: 20 units/ml from cow milk) was from Roche Diagnostics GmbH (Germany). The spin trap, 5,5-dimethyl-1-pyrroline-N-oxide (DMPO) was from Labotec Co., Ltd. (Tokyo, Japan). Metal chelating agent diethylenetriamine N,N,N',N'',N'''-pentaacetic acid (DETAPAC) was from Dojiindo Laboratory (Kumamoto, Japan). Ferrous sulfate and sodium hydrate were from Kanto Kagaku Co., Ltd. (Tokyo, Japan). All chemicals were of analytical quality. Pure water was used throughout.

25.2.2 UYAKU EXTRACT

Uyaku leaf powder was supplied from Sakai Canning Co., Ltd. One gram of its raw material (600-mesh powder) is equivalent to 4 g of uyaku leaves. To obtain the extract, the powder (10 g) was added to 200 ml of water (1:20 w/v), and the sample was extracted with boiling water for 10 min. After cooling to room temperature, the suspension was filtered (Watman No. 2), and the recovered filtrate was used for the experiment.

25.2.3 MEASUREMENT OF SUPEROXIDE-ANION-RADICAL-SCAVENGING ACTIVITY

Fifteen microliters of DMPO (8.8 M) was first added to a test tube, followed by 5 mM hypoxanthine (50 μl), sample solution (50 μl), 3.5 mM DETPA solution (50 μl), and 0.4 units/ml XOD (50 μl). After vortex mixing for 10 sec at room temperature, the sample solution was transferred immediately into a quartz flat cell. Electron spin resonance (ESR) recording was started exactly 60 sec after the addition of XOD. ESR measurement JES-FA200 (JEOL, Ltd., Tokyo) was used. Conditions for measurements in this study were as follows: microwave frequency = 9414.12–9415.80 MHz, microwave power = 4.00 mW, field center 335.309–335.45 mT, width $\pm$ = 5.00 mT, modulation: frequency = 100.00 kHz/ width = 0.0800 mT, time constant = 0.1–0.3 sec, sweep time = 1.0–2 min., amplitude 500–1500. ESR spectra were measured at room temperature. Scavenging activities were estimated using the relative peak height of ESR spectra of both DMPO spin adducts and 3rd manganese (Mn) signal as standard samples of the cavity.

25.2.4 ESR Direct Measurement of Uyaku Powder at 77 K

Uyaku powder was analyzed directly using a JES FA-200 ESR spectrometer with a liquid N_2 vacuum flask (UCD3X: JEOL, Ltd., Tokyo). ESR conditions for measurements in this study were as follows: microwave frequency = 9050.676–9063.737 MHz, microwave power = 4.00 mW, field center 300.00 mT with width $\pm$ = 250.00 mT, modulation frequency = 100.00 kHz with width = 0.250 mT, time constant = 0.1–0.3 sec, sweep time = 4.0 min, and amplitude = 250.

25.2.5 Hydrogen-Peroxide-Induced Lipid Peroxidation in Rat Brain Homogenate

Male Sprague-Dawley rats (350-g body weight) were anesthetized with ether and perfused through the heart with 0.9% saline at 4°C. The cerebral cortex was removed, and was homogenized in ice-cold 20-mM Tris-HCl buffer, pH 7.4 (1:10 w/v). The homogenate was incubated with or without 5 mM H_2O_2 for 60 min at 37°C. Uyaku was used in combination with H_2O_2 (5 mM) to test for its antioxidant activity *in vitro*. After incubation, the reaction was stopped by placing the homogenates in ice-cold water for 10 min, and then the homogenates were centrifuged at 15,000 $\times$ g for 10 min at 4°C. The level of malondialdehyde plus 4-hydroxyalkenals was assayed in the supernatant as an index of lipid peroxidation using an LPO-586™ Kit (OXIS International, Inc., Portland, OR).

25.2.6 Statistics

All data were expressed as mean $\pm$ SEM. Statistical analysis was performed using the Student's t-test.

25.3 RESULTS AND DISCUSSION

Superoxide-anion-radical-scavenging activity of the newly prepared uyaku leaf extract, Jo-Fuku-no-sei, was examined by an ESR method using the spin-trapping reagent, DMPO (Figure 25.1, Figure 25.2). Uyaku leaf extract showed potent scavenging activity against superoxide anion radical, i.e., scavenging activity of

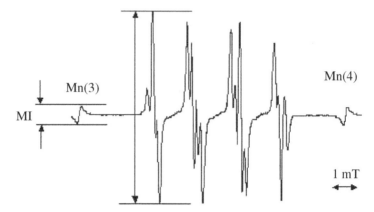

FIGURE 25.1 ESR spectra of DMPO-$O_2^{\cdot}$. The marker is manganese (Mn) inserted into the cavity for ESR spectra. MI: marker intensity, I: intensity of DMPO- $O_2^{\cdot}$, RI: relative intensity I/MI.

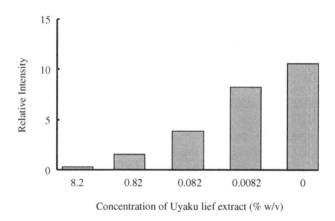

Concentration of Uyaku lief extract (% w/v)

FIGURE 25.2 Relative intensities of superoxide scavenging activities in various concentrations of uyaku leaf extract.

uyaku leaf extract was calculated as 5.40×10^3 SOD equivalent units/mg by comparing with the standard SOD curve. It was 60 times higher compared with our previous experimental results (90 SOD equivalent units/mg), in which the extract was obtained using the same analyzing technique (4).

These results demonstrated the effectiveness of the newly improved method for preparing the raw material, i.e., this difference is probably due to preparation technique used for making dried uyaku leaves to a fine powder (600 mesh) before extraction using hot water; they were crudely cut and extracted in the previous experiment (4). More interestingly, the superoxide-anion-radical-scavenging activity

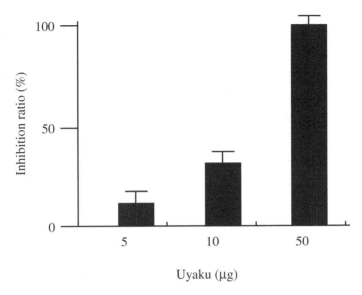

FIGURE 25.3 Inhibitory effect of uyaku leaf extract on hydrogen-peroxide-induced lipid peroxidation in rat brain homogenate.

of uyaku leaf extract was 2.5 times higher than that of green tea extract, which is known to be one of the strongest natural antioxidant substances (4).

Uyaku leaf extract markedly inhibited H_2O_2-induced lipid peroxidation in the rat brain homogenate during incubation (Figure 25.3). The ID_{50} of uyaku leaf extract was calculated as 35.0 µg/ml from the dose–response curve shown in Figure 25.3. This value was compared with those of pomegranate fruit extract and L-ascorbic acid 2-[3,4-dihydro-2,5,7,8-tetramethl-2,4,8,12-trimethyltridecyl]-2H-1-benzopyran-6-yl-hydrogen phosphate]potassium salt (EPC-K1), which were estimated using the same method (6). EPC-K1 is a synthetic phosphate diester of vitamin E and vitamin C, and is used for a standard material of hydroxyl-radical-scavenging activity (7). In this study, it was used for a control of an antioxidant activity because of its higher water solubility (7,8). Inhibitory activity of uyaku was found to be at almost the same level as that of EPC-K1, although a little weaker when compared with pomegranate fruit extract, which is known to be an excellent natural antioxidant (6) (Table 25.1).

Meanwhile, specific ESR spectra of Mn^{2+} were observed using liquid N_2 at 77 K (Figure 25.4). Manganese is commonly known to be a cofactor of hydrolase, decarboxylase, and transferase enzymes, and to be an active center of mitochondrial superoxide dismutase. The role of manganese in uyaku is unknown at the present. It may be useful for product control because the ESR signal of Mn^{2+} in uyaku preparation was very stable both in powder and liquid states.

Studies on the components of uyaku, mainly roots, started in the 1920s. Linderol, linderane, linderene, and linderaic acid were first isolated by Kondo and Sanada (9). Then, many sesquiterpens having a furan ring, e.g., linderane

TABLE 25.1
Inhibitory Effect of Uyaku Leaf
Extract on Lipid Peroxidation
in the Rat Brain Homogenate

Comparison with Other Reported Substances
by ID_{50} (μg/ml)

Uyaku leaf extract	35.0
Pomegranate extract	23.6
EPC-Kl	30.0

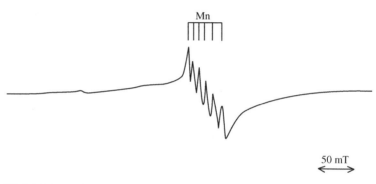

FIGURE 25.4 Detection of ESR spectra of Mn^{2+} in uyaku extract. Specific ESR spectra of Mn^{2+} were observed at 77 K.

(10), linderene (10,11,12,13), lideralactone, and isolinderalactone (14), were isolated and their chemical structures were identified. Recently, strychnistenolide and its acetate, and strychnilacetone were isolated from uyaku root by Kouno et al. (15). Uyaku roots are known to contain azulens, e.g., liderazulene and lindazulen (16,17), and alkaloids such as laurolistine (18), boldine, and reticuline. (19). However, at present almost no information is available about uyak leaves, and the relationship between the antioxidant activities and the components of uyaku still unknown and remain to be clarified.

25.4 CONCLUSION

The uyaku leaf extract showed potent scavenging activity against superoxide anion radical and effectively inhibited lipid peroxidation in the rat brain. These results demonstrated the effectiveness of the new method for preparing the extract from the raw material. Meanwhile, specific ESR spectra of Mn^{2+} were observed at 77 K. These ESR spectra may be useful as indicators for product control.

ACKNOWLEDGMENT

This research was supported in part by a subsidy from Shingu City.

REFERENCES

1. Makino T. *Makino's New Illustrated Flora of Japan*, Revised edition, Hokuryukan: Tokyo, 1989, p.126.
2. Hsu H-Y, Chen Y-P, Shen S-J, Hsu C-S, Chen C-C, and Chang H-C. *Oriental Materia Medica: A Concise Guide*, Oriental Healing Arts Institute: Long Beach, 1986, p. 420.
3. Bensky D and Gamble A. *Chinese Herbal Medicine*, Eastland Press: Seattle, 1986, p. 431.
4. Noda Y, Mori A, Anzai K, and Packer L. Superoxide anion radical scavenging activity of uyaku (*Lindera strychnifolia*), a natural extract used in traditional medicine. In *Antioxidant Food Supplementations in Human Health*, Packer L, Hiramatsu M, and Yoshikawa T. Eds., Academic Press: San Diego, 1999, pp. 471–479.
5. Noda Y, Mori A, and Packer L. Antioxidant activity of uyaku (*Lindera strychnifolia*), a natural extract used in traditional medicine. American Aging Association 32nd Annual Meeting, Baltimore, MD, June 6–9, 2003.
6. Noda Y, Kaneyuki T, Mori A, and Packer L. Antioxidant activities of pomegranate fruit extract and its anthocyanidines: delphinidin, cyanidin, and pelargonidin. *J Agric Food Chem* 2002; 50: 166–171.
7. Mori A, Edamatsu R, Kohno M, and Ohmori S. A new hydroxyl radical scavenger: EPC-K1. *Neurosciences* 1989; 15: 371–376.
8. Noda Y, Kohno M, Mori M, and Packer L. Automated electron spin resonance free radical detector assays for antioxidant activity in natural extracts. *Method Enzymol* 1999; 299: 28–34.
9. Kondo H and Sanada T. Ueber die bestandteile der wurzel von *Lindera strychnifolia* Vill. (I. Mitteilung). *Yakugaku Zasshi* 1925; 40: 1047–057. [In Japanese].
10. Suzuki H. Ueber die bestandteile der wurzel von *Lindera strychnifolia* Vill. (II. Mitteilung) *Yakugaku Zasshi* 1930; 50: 714–720. [In Japanese].
11. Takeda K and Shimada T. Components of the root of *Lindera strychnifolia* Vill. IV. Structure of linderene (1). *Yakugaku Zasshi* 1944; 64: 154–161. [In Japanese].
12. Takeda K, Components of the root of *Lindera strychnifolia* Vill. VI Structure of linderene (2). *Chem Pharm Bull (Jpn)* 1953; 1: 244–251.
13. Takeda K, Ikuta M, and Miyawaki M. Components of the root of *Lindera strychnifolia* Vill. X. Structure of linderene. *Tetrahedron* 1964; 20: 2991–2997.
14. Takeda K, Minato H, and Ishikawa M. Components of the root of *Lindera strychnifolia* Vill. Part VII. Structures of linderalactone and isolinderalactone. *J Chem Soc (C)* 1964; 4578–4582.
15. Kouno I, Hirai A, Fukushige A, Jiang Z-H, and Tanaka T. New eudesmane sesquiterpenes from the root of *Lindera strychnifolia*. *J Nat Prod* 2001; 64: 286–288.
16. Kondo H and Takeda K. Components of *Lindera strychnifolia* Vill. III. *Yakugaku Zasshi* 1939; 59: 504–519.

17. Takeda K and Nagata W. Components of the root of *Lindera strychnifolia* Vill. V. Azulenes isolated from linderene by zinc-dust distillation. *Chem Pharm Bull (Jpn)* 1953; 1: 164–169.

18. Tomita M, Sawada T, Kozuka M, Hamano D, and Yoshimura K. The alkaloids of *Lindera strychnifolia* (Sieb. et Zucc.) F. Vill. and *Lindear umbellata* Thumb. *Yakugaku Zasshi* 1969; 89: 737–740. [In Japanese].

19. Kozuka M, Yoshikawa M, and Sawada T. Alkaloids from *Lindera strychnifolia*. *J Nat Prod* 1984; 47: 1063.

26 Detection of Active Oxygen Species in the Decomposition Process of Peroxovanadium (V) Complexes

Seiichi Matsugo and Shingo Hayakawa
University of Yamanashi, Kofu, Japan

Chie Mihara and Kan Kanamori
Toyama University, Toyama, Japan

CONTENTS

26.1 OVERVIEW

Peroxovanadium(V) complexes (pVs) display insulin mimetic effect *in vivo*. By synthesizing some pVs, we have studied the decomposition process precisely. As a result, active oxygen species were generated during the decomposition of pVs. In this chapter, we report the formation of active oxygen species produced during the decomposition of pVs.

26.2 INTRODUCTION

Peroxovanadium(V) complexes (pVs) display insulin mimetic effect and may undergo various oxidation reactions (1). However, the precise decomposition mechanism of pVs is still unclear. Therefore, we have synthesized several pVs, including cmhist (N-carboxymethyl-D,L-histidinate) and dpot (2-oxo-1,3-diaminopropane-N,N,N',N'-tetraacetate), and investigated the behavior of pVs in aqueous solution. We previously reported that these pVs are quite stable in neutral pH conditions; however in acidic pH conditions (2), the decomposition of pVs takes place spontaneously, accompanying the reduction of vanadium(V) to vanadium(IV). The reduction of vanadium(V) is due to the decomposition of the peroxide.

26.3 MATERIALS AND METHODS

Peroxovanadium complexes $K[VO(O_2)(cmhist)] \cdot H_2O$ and $Cs_3[(VO)_2(O_2)_2(dpot)] \cdot 3H_2O$ were synthesized according to the method previously reported [1].

26.3.1 DETECTION OF SUPEROXIDE ANION BY THE WST-1 METHOD

Five mM solutions of pVs (cmhist complex, dpot complex) and 10 mM solution of WST-1 were prepared in 50 mM sodium phosphate buffer (pH 7.4). Both solutions were mixed in the same buffer. The final concentration of pVs and WST-1 were adjusted at 500 μM. The reaction mixtures were left standing for 4 h (dpot complex system) or 48 h (cmhist complex system) at 37°C. Subsequently, the absorption at 438 nm corresponding to the formation of WST-1 formazan ($\varepsilon = 37,000$) from the reaction mixtures were quantified by UV/visible spectrometer (PerkinElmer UV/VIS Spectrometer: Lambda Bio/20). Both pVs solution and WST-1 solution at 500 μM were also left standing for the same time as the control. The concentration of the decomposed complex was calculated by the decrease of the peroxo CT band intensity.

26.3.2 DETECTION OF HYDROXYL RADICAL BY THE HYDROXYLATION OF BENZOIC ACID

The pVs were dissolved in 50 mM sodium phosphate buffer (pH 7.4) and adjusted at 10 mM solution. A 100 mM of benzoic acid solution was prepared using 0.1 M NaOH solution. Then, both solutions were mixed. The final concentration of pVs and benzoic acid were adjusted at 2 mM and 20 mM (cmhist complex system), 40 mM (dpot complex system), respectively. The formation of hydroxylated benzoic acid was measured using HPLC after standing the reaction mixture for 144 h (cmhist complex system) and 24 h (dpot complex system) at 37°C, respectively. The mobile phase used in these experiments were 10 mM sodium phosphate buffer (pH 6.0). Elution was performed at a flow rate of 1.0 ml/min. Hydroxybenzoic acid was detected by UV absorption (240 nm). The concentration of the decomposed complex was calculated from the decrease of the peroxo CT band intensity using the reaction mixture.

26.3.3 EVALUATION OF OXIDIZING ABILITY TO PROTEIN BY THE THIOL OXIDATION ASSAY

Briefly, BSA (2 mg/ml) was reacted with the defined concentration (50 to 150 μM) of dpot complex in 50 mM phosphate buffer (pH 7.4) at 37°C for 1 h. Subsequently, 200 μl of 2 M HCl was added to the reaction mixture, followed by the addition of 700 μl of 3% SDS aqueous solution. After mixing, the reaction mixture was placed at room temperature for 5 min. BSA was deposited by the addition of 2 ml of ethanol and 2 ml of hexane, respectively. The solution was centrifuged at 3000 rpm for 20 min, and then precipitate (BSA pellet) was collected. The BSA pellet was washed twice with a solution of ethyl alcohol and ethyl acetate (1:1). The BSA pellet was dried at 37°C for 90 min, and then dissolved in 1 ml of 50 mM sodium phosphate buffered solution containing 2% of SDS (pH 8.0). DTNB (5,5′-dithiobis(2-nitrobenzoic acid)) was added to the BSA solution and was left standing for 30 min. The amount of the sulfhydryl group remained in the BSA solution was determined using DTNB assay ($\varepsilon_{412\,nm} = 13600$ $M^{-1}cm^{-1}$). On the other hand, an aliquot (50 μl) of the modified BSA solution was used for the quantification of the remaining protein amounts.

26.4 RESULTS AND DISCUSSION

In the WST-1 assay, the relative amounts of the superoxide generated in the self-decomposition of cmhist complex and dpot complex were calculated to be *ca.* 6.6 and 3.8%, respectively. In our previous work, we succeeded to detect the hydroxyl radical from the decomposition of pVs by ESR spin trapping technique using DMPO as a spin trapping reagent (3). According to HPLC assay for the formation of hydroxylated benzoic acids, all hydroxybenzoic acids (HBA: *o*-hydroxybenzoic acid, *m*-hydroxybenzoic acid, and *p*-hydroxybenzoic acid) clearly showed the different retention times as shown in Figure 26.1. The total amount of hydroxyl radical from the decomposition of cmhist complex and dpot complex based on the formation of hydroxylated benzoic acids were calculated to be *ca.* 20.5 and 55.6%, respectively.

In the protein sulfhydryl assay, the sulfhydryl amount of BSA decreased significantly to that of the control sample with increase in the concentration of the dpot complex as shown in Figure 26.2. The sulfhydryl residues in BSA were oxidized *ca.* 35, 57, and 70% by 50, 100, and 150 μM dpot complex, respectively.

In this study, we succeeded in detecting superoxide anion and hydroxyl radical produced in the self-decomposition process of pVs. The amount of hydroxyl radical was much more than that of the superoxide. These results indicate that the major pathway for the decomposition of pVs passes through the formation of hydroxyl radical. The formation of hydroxyl radical can be explained by

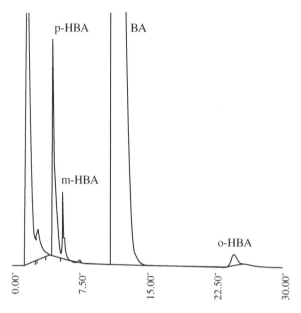

FIGURE 26.1 HPLC chromatogram corresponding to reaction mixture. [cmhist]$_0$ = 2 mM, [BA]$_0$ = 20 mM, pH 7.4, 144 h, 37°C.

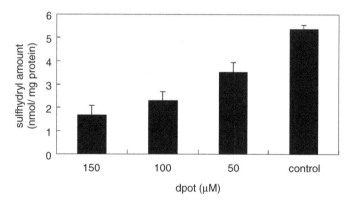

FIGURE 26.2 Sulfhydryl concentration of protein. [dpot]$_0$ = 150, 100, 50 µM, [BSA]$_0$ = 2 mg/ml, pH 7.4, 1 h, 37°C ($p < .01$;**: Dunnett).

considering the electron transfer mediated decomposition of vanadium(V) complexes (3). We previously confirmed that pVs oxidize BSA to produce protein carbonyl formation in high yield. In the protein sulfhydryl assay, pVs also oxidize sulfhydryl groups in BSA. More than 50% sulfhydryl group in BSA was oxidized by at 100 µM of pVs, which also indicates the formation of highly reactive oxygen radicals such as hydroxyl radical in the decomposition of pVs. Further studies are now in progress to elucidate the precise reaction mechanism for the decomposition of pVs.

REFERENCES

1. Butler A, Clague MJ, and Meister GE. Vanadium peroxide complexes. *Chem Rev* 1994; 94: 625–638.
2. Kanamori K, Nishida K, Miyata N, Okamoto K, Miyoshi Y, Tamura A, and Sakurai H. Syntheses, structures, stability, and insulin-like activities of peroxovanadium(V) complexes with a heteroligand. *J Inorg Biochem* 2001; 86: 649–656.
3. Kanamori K, Nishida K, Miyata N, Shimoyama T, Hata K, Mihara C, Okamoto K, Abe Y, Hayakawa S, and Matsugo S. Mononuclear and dinuclear monoperoxovanadium (V) complexes with a heteroligand. Part 1: Self-decomposition reaction, detection of reactive oxygen species, and their oxidizing ability. [Submitted to *Inorganic Chem*].

27 Structural Determination and Antioxidative Activity of Oxidation Products of Flavonoids with Reaction Oxygen Species

Yuko Hirose, Toshiyuki Washizu, Yuzo Uchida, and Seiichi Matsugo
University of Yamanashi, Kofu, Japan

CONTENTS

27.1 OVERVIEW

Flavonoids may be useful for the prevention or treatment of free-radical-associated diseases; hence, much attention has been paid to the physiological functions of flavonoids as antioxidants. However, there was a lack of information on the molecular mechanism in which flavonoids play the antioxidative role. It has been reported that their function is strongly dependent on their structural properties. We have already studied the oxidation of quercetin with hydrogen peroxide and sodium hypochlorite in alcoholic aqueous solution and determined the structure of the two major oxidation products (Figure 27.1). Through the structural analysis of the oxidation products, it was suggested that the hydroxyl group at the C-3 position in the C ring plays an important role in the antioxidative action. In this study, we oxidized morin by sodium hypochlorite and determined the structure of the oxidation products. Additionally, we compared the antioxidative activity of two flavonoids with that of their oxidation products and discussed the structure-antioxidant activity relationships between these flavonoids. The result indicated

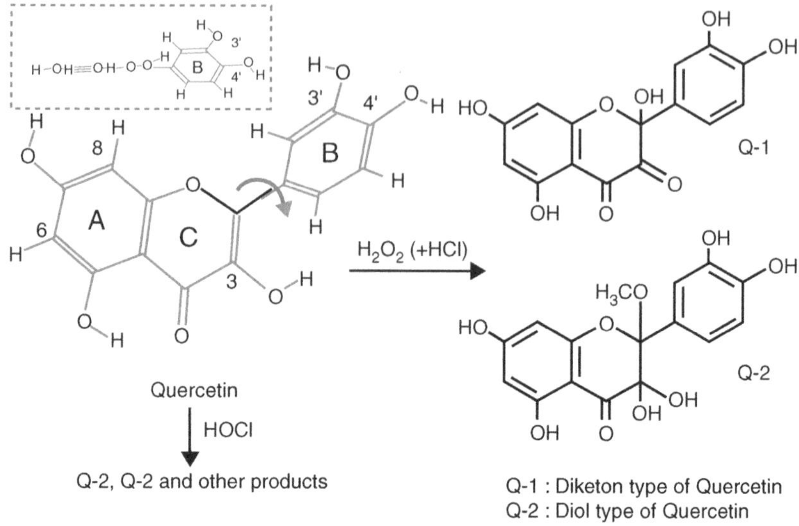

FIGURE 27.1 The structure of quercetin and their oxidation products. The geometry of quercetin was optimization using the UHF formalism as implemented in the Gaussian 98 program. The decline between the A and C rings and the B ring is shown in the inset.

that the positions of hydroxyl groups in the B ring affect the electronic state and structural geometry of flavonoids.

27.2 INTRODUCTION

Flavonoids are phenolic compounds that are found in dietary plants and are also known to act as antioxidants with strong scavenging ability against reactive oxygen species (ROS). In recent years, it was indicated that some diseases such as cancer and heart disease are induced by oxidative stress, and hence the antioxidative activity of some flavonoids were determined in many experimental conditions [1,2]. It has been demonstrated that flavonoids play a role as hydrogen-donating free radical scavengers, and their function is strongly related to their structural properties, of which there are three key features: (1) the catechol group in the B ring, (2) 2,3-double bond with the 4-oxo group in the C ring, and (3) 3-(and 5-)hydroxyl group of the C (and A) ring [3]. However, the precise oxidation mechanism of flavonoids is still a subject of debate. To determine the chromophore responsible for the antioxidative action, it is quite meaningful to identify the oxidation products of flavonoids. Quercetin, one of the most abundant flavonoids in fruits and vegetables, satisfies all these structural criteria and shows strong antioxidative activity. We have already proved that both the catechol chromophore in the B ring and the enol moiety in the C ring of quercetin are essential for the antioxidative activity during the lipid peroxidation [4], whereas the enol moiety plays an important role in the antioxidative action in alcoholic aqueous solution against hydrogen peroxide and sodium hypochlorite [5,6]. These results suggested that all the three aforementioned functional-group features are not always required for antioxidative action. In our study, morin, which is a quercetin-analog-possessing resorcinol structure in the B ring, was oxidized, and the structures of their oxidation products were determined based on spectral analyses such as UV, IR, MS, and NMR. Moreover, the antioxidative activities of the oxidation products were evaluated by ESR spin-trapping technique using the DMPO spin-trapping reagent.

27.3 MATERIALS AND METHODS

Reagents: Flavonoids and DMPO were purchased from Sigma Chemical Co. (St. Louis, MO), and all other reagents were purchased from Kanto Chemical Co. Ltd. (Tokyo, Japan). All reagents used were of the highest grade and used without further purification.

Oxidation of morin with ROS: Morin dissolved in MeOH was oxidized with hydrogen peroxide or sodium hypochlorite in water. The reactants were extracted with ethyl acetate to give the reaction products. The products were dissolved in MeOH, injected into preparative HPLC, using ODS column (YMC D-ODS-5, 20 mm × 250 mm), and detected at 290 nm.

Structural determination of oxidation products: Structures of the products were determined based on the spectral analyses of UV, MS, IR, and NMR. NMR spectra were obtained in CD_3OD operating at 400 MHz (^1H) and 100 MHz (^{13}C) and using TMS as an internal standard.

2,2',4',5,7,-Pentahydroxyflavan-3,4-dione (M-1): ^1H-NMR: δ 8.23 (d, $J = 9.1$ Hz), 6.46 (dd, $J = 9.1, 2.3$ Hz), 6.37 (d, $J = 2.3$ Hz), 6.10 (s), 6.09 (s), ^{13}C-NMR: δ 195.0 (C3), 190.9 (C4), 172.9 (C9), 170.0 (C7), 167.9 (C2'), 166.6 (C4'), 159.4 (C5), 136.1 (C6'), 110.9 (C1'), 109.2 (C5'), 105.1 (C2), 103.6 (C3'), 101.5 (C10), 98.9 (C6), 92.2 (C8).

2',3,3,4',5,7-Hexahydroxy-2-methoxyflavan-4-one (M-2): ^1H-NMR: δ 7.34 (d, $J = 8.2$ Hz), 6.54 (dd, $J = 8.2, 2.1$ Hz), 6.38 (d, $J = 2.1$ Hz), 6.03 (s), 6.03 (s), 3.57 (s), ^{13}C-NMR: δ 188.8 (C4), 169.3 (C7), 166.0 (C5), 161.9 (9), 161.8 (C2'), 160.4 (C4'), 126.0 (C6'), 117.3 (C1'), 110.3 (C5'), 109.3 (C2), 100.8 (C10), 100.5 (C3), 99.6 (C3'), 97.7 (C6), 96.3 (C8), 53.2 (OMe).

8-Chloromorin (Mc-1): ^1H-NMR: δ 7.29 (d, $J = 8.4$ Hz), 6.43 (d, $J = 2.2$ Hz), 6.39 (dd, $J = 8.4, 2.2$ Hz), 6.41 (s), ^{13}C-NMR: δ 176.3 (C4), 160.9 (C4'), 159.5 (C7), 159.0 (C5), 157.2 (C2'), 152.1 (C9), 149.6 (C2), 136.7 (C3), 131.9 (C6'), 109.2 (C1'), 107.2 (C5'), 104.4 (C10), 103.2 (C3'), 98.4 (C6), 97.2 (C8).

3,8-Dichloromorin (Mc-2): ^1H-NMR: δ 7.26 (d, $J = 8.6$ Hz), 6.64 (d, $J = 8.6$ Hz), 6.42 (s), ^{13}C-NMR: δ 176.2 (C4), 159.3 (C7), 158.7 (C5), 156.2 (C4'), 152.5 (C2'), 152.0 (C9), 147.6 (C2), 136.9 (C3), 128.8 (C6'), 107.3 (C5'), 110.2 (C3'), 108.6 (C1'), 104.2 (C10), 98.1 (C6), 97.0 (C8).

6,8-Dichloromorin (Mc-3): ^{1}H-NMR: δ 7.31 (d, *J* = 8.4 Hz), 6.43 (d, *J* = 2.2 Hz), 6.40 (dd, *J* = 8.4, 2.2 Hz), ^{13}C-NMR: δ 175.7 (C4), 160.8 (C4'), 156.9 (C2'), 154.9 (C7), 154.4 (C5), 150.0 (C2), 149.8 (C9), 136.5 (C3), 131.7 (C6'), 108.6 (C1'), 106.9 (C5'), 104.0 (C10), 103.1 (C3'), 102.3 (C6), 98.6 (C8).

3,6,8-TrichloromorinMc-4: ^{1}H-NMR: δ 7.27 (d, *J* = 8.62 Hz), 6.64 (d, *J* = 8.63 Hz), ^{13}C-NMR: δ 176.0 (C4), 156.4 (C4'), 154.4 (C5), 154.3 (C7), 152.6 (C2'), 149.9 (C2), 148.4 (C2), 137.1 (C3), 128.8 (C6'), 110.0 (C1'), 108.6 (C3'), 107.3 (C5'), 104.2 (C10), 103.1 (C6), 98.7 (C8).

Antioxidative activity: The antioxidative activity toward hydroxyl radical produced by Fenton reaction was evaluated by ESR spin-trapping techniques.

27.4 RESULTS AND DISCUSSION

The oxidation of morin with hydrogen peroxide gave two products, M-1 and M-2. From the results of spectral analyses, the structures of M-1 and M-2 were determined as 2,2',4',5,7,-pentahydroxyflavan-3,4-dione and 2',3,3,4',5,7-hexahydroxy-2-methoxyflavan-4-one, respectively. On the other hand, the oxidation of morin with sodium hypochlorite gave four compounds, Mc-1, Mc-2, Mc-3, and Mc-4, as major oxidation products, and they were determined to be 8-chloromorin, 3',8-dichloromorin, 6,8-dichloromorin and 3',6,8-trichloromorin, respectively, as shown in Figure 27.2. These products suggested that the oxidation reaction of morin with hypochlorous acid is an electrophilic aromatic substitution by chlorination. In the oxidaiton of quercetin with sodium hypochorite, diketone-type (Q-1) and diol-type flavanone (Q-2), which were the same oxidation products obtained by the reaction with hydrogen peroxide, were obtained as major products (Figure 27.1), and the chloroinated derivatives were not major products [5,6]. On the other hand, diketone-type (M-1) and diol-type flavanone (M-2) were not detected in the oxidation of morin with sodium hypochlorite. Therefore, the antioxidation mechanism of morin is different from that of quercetin, and it was suggested that the position of the hydroxyl group were responsible for the antioxidation mechanism.

The antioxidative activity of these chlorinated flavonoids toward hydroxyl radical were evaluated using ESR spin-trapping technique using DMPO as a spin-trapping regent. As a result, chlorinated morins showed almost the same hydroxyl-radical-scavenging activity as that of morin (data not provided in this chapter). Further, for the analysis of the electronic state of these flavonoids, we carried out the calculation of the optimized structures of these compounds using the Gaussian 98 program (the optimized structures of quercetin and morin are shown in Figure 27.1 and Figure 27.2, respectively). The calculation indicated that the positions of the hydroxyl groups in the B ring affect the electronic state and structural geometry of flavonoids. Further calculations are required for the elucidation on the antioxidation mechanism of flavonoids. These are now in progress in our laboratory.

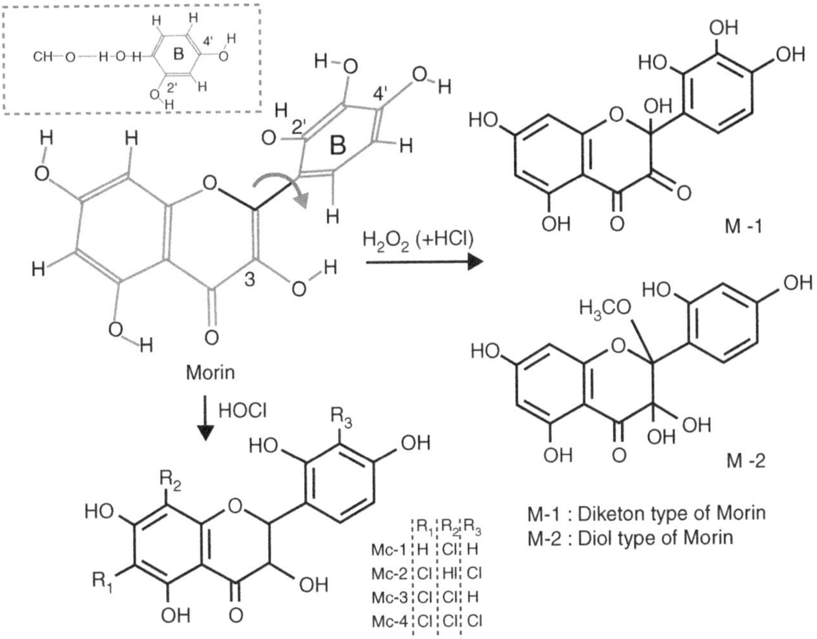

FIGURE 27.2 The structure of morin and the oxidation products. The geometry of morin was optimization using the UHF formalism as implemented in the Gaussian 98 program. The decline between the A and C rings and the B ring is shown in the inset.

REFERENCES

1. Formica JV and Regelson W. Review of the biology of quercetin and related bioflavonoids. *Food Chem Toxicol* 1995, 33: 1061–1080.
2. Sestili P, Guidarelli A, Dacha M, and Cantoni O. Quercetin prevents DNA single strand breakage and cytotoxicity caused by tert-butylhydroperoxide: Free radical scavenging versus iron chelating mechanism. *Free Radic Biol Med* 1998; 25: 196–200.
3. Birda S and Oleszek W. Antioxidant and antiradical activities of flavonoids. *J Agric Food Chem* 2001; 49: 2774–2779.
4. Hirose Y, Fujita T, and Nakayama M. Structure of doubly-linked oxidative product of quercetin in lipid peroxidation. *Chem Lett* 1999; 775–776.
5. Hirose Y, Kakita M, and Matsugo S. Structure determination of the oxidation products of quercetin with hypochlorous acid. 33th Symposium on Chemical and Biological Oxidation, Kanazawa, Japan, P-4, November 2000.
6. Hirose Y and Matsugo S. Structural determination of the oxidative reaction products of quercetin with hydrogen peroxide. 79th Annual Conference of Chemical Society of Japan, Kobe, Japan, March 3G108, 2001.

28 Reactivity of Anthocyanin toward Reactive Oxygen and Reactive Nitrogen Species

Takashi Ichiyanagi and Yoshihiko Hatano
NUPALS, Niigata, Japan

Seiichi Matsugo
University of Yamanashi, Kofu, Japan

Tetsuya Konishi
NUPALS, Niigata, Japan

CONTENTS

28.1 INTRODUCTION

Anthocyanin, a reddish pigment, is a family of flavonoids distributed in many colored food materials, for example, blueberry, blackcurrant, strawberry, and eggplant, and we ingest large amounts of these in our daily diets. Recently, many physiological functions were reported, such as improvement of vision [1], α-glucosidase inhibitory activity [2], induction of apoptosis [3], anticancer activity [4], and antioxidant activity [5]. In our study, we focused on the antioxidant activity of anthocyanin [6]. However, isolation of anthocyanin is difficult compared to other flavonoids such as catechin because of their unstable nature [7]. Hence, the structure–activity relationship of anthocyanin, except for a few purified

FIGURE 28.1 Structure of anthocyanins.

Anthocyanins	R_1	R_2	R_3
Dephinidin 3-O-b-D-glucopyranoside	H	OH	glucoside
Dephinidin 3-O-b-D-galactopyranoside	H	OH	galactoside
Dephinidin 3-O-a-L-arabinopyranoside	H	OH	arabinoside
Cyanidin 3-O-b-D-glucopyranoside	H	H	glucoside
Cyanidin 3-O-b-D-galactopyranoside	H	H	galactoside
Cyanidin 3-O-a-L-arabinopyranoside	H	H	arabinoside
Petunidin 3-O-b-D-glucopyranoside	H	OCH$_3$	glucoside
Petunidin 3-O-b-D-galactopyranoside	H	OCH$_3$	galactoside
Petunidin 3-O-a-L-arabinopyranoside	H	OCH$_3$	arabinoside
Peonidin 3-O-b-D-glucopyranoside	CH$_3$	H	glucoside
Peonidin 3-O-b-D-galactopyranoside	CH$_3$	H	galactoside
Peonidin 3-O-a-L-arabinopyranoside	CH$_3$	H	arabinoside
Malvidin 3-O-b-D-glucopyranoside	CH$_3$	OCH$_3$	glucoside
Malvidin 3-O-b-D-galactopyranoside	CH$_3$	OCH$_3$	galactoside
Malvidin 3-O-a-L-arabinopyranoside	CH$_3$	OCH$_3$	arabinoside

anthocyanins [8], has not been sufficiently studied. In bilberry (a wild-type blueberry), 15 anthocyanins exist (Figure 28.1). Thus, bilberry is good sample for studying the structure–reactivity relationship of anthocyanin toward reactive oxygen species (ROS) and reactive nitrogen species (RNS). To discuss the reaction of anthocyanin, we first established the capillary zone electrophoretic (CZE) method for simultaneous separation of bilberry anthocyanins [9–11]. Using this method, we evaluated the structure–reactivity relationship of anthocyanins found in the bilberry extract in reactions toward ROS *in vitro* including hydroxyl radical ('OH), superoxide anion (O_2^-), hydroperoxide (H_2O_2), and singlet oxygen ($^1O_2$) [12,13]. Reactions of anthocyanins towards RNS such as nitric oxide (NO) and peroxynitrite (ONOO') were also studied [14]. To discuss the *in vivo* antioxidant activity of anthocyanins, it is important to understand their absorption and distribution mechanisms. Hence, the uptake and tissue distribution of bilberry anthocyanin after oral administration were further studied in rats.

28.2 RESULTS AND DISCUSSION

Reactivity of anthocyanin toward ROS and RNS was different depending on the anthocyanin structure and the type of ROS and RNS studied. The type of conjugated sugar did not significantly affect the reactivity of anthocyanins, but the aglycon structure strongly affected them. Reactivity of anthocyanins toward ROS and RNS was found to be classified into two general categories: In group-1 anthocyanins, the reactivity towards ROS and RNS was mainly determined by the number of hydroxyl group on aglycon B ring, whereas the group-2 reaction was not affected by the aglycon structure.

Figure 28.2 shows anthocyanin reactivity towards O_2^-. The reactions of anthocyanin with O_2^- and $^1O_2$ were classified as group-1 reactions; delphinidin-glycosides

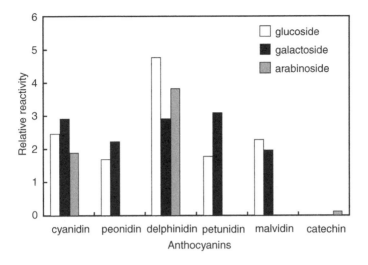

FIGURE 28.2 Reactivity of anthocyanin toward O_2^-.

having pyrogallol structure on aglycon B ring were the most reactive, followed by cyanidin-glycosides having catechol structure. Moreover, methylation of anthocyanins decreased the reactivity toward these species. Figure 28.3 shows the reaction of anthocyanins toward ˙OH. The reactions of anthocyanin toward ˙OH and NO were not dependent on anthocyanin structure (group-2 reaction). However, in the reactivity of anthocyanin toward ONOO˙, delphinidin-glycosides showed specifically high reactivity compared to other anthocyanins, which showed little difference in their reactivity (Figure 28.4).

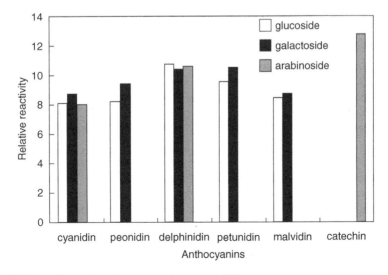

FIGURE 28.3 Reactivity of anthocyanin toward ˙OH.

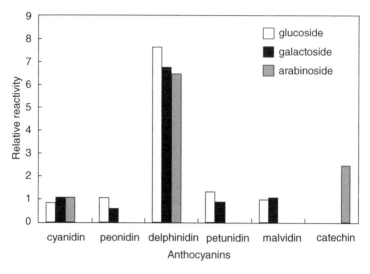

FIGURE 28.4 Reactivity of anthocyanin toward ONOO⁻.

We demonstrated that the scavenging potential of anthocyanins toward ROS and RNS *in vitro* largely differed depending on the anthocyanin structure and also the type of ROS and RNS. However, to discuss further the *in vivo* antioxidant potential of these anthocyanins, it is important to understand their absorption and tissue distribution. Bilberry extract as an anthocyanin mixture was orally administrated to rats, and their plasma and tissue distribution were studied in detail. The uptake profile of anthocyanins in the plasma was significantly affected by their aglycon structure and also their conjugated sugar type. Figure 28.5 shows the HPLC chromatogram of anthocyanins in bilberry extract (A) and anthocyanins extracted from plasma (B). In plasma, delphinidins and cyanidins were found distributed higher than that of methylated anthocyanins such as petunidins (3′-O-methyl delphinidin), peonidins (3′-O-methyl cyanidin), and malvidins (3′,5′-O-methyl delphinidin). The plasma level of anthocyanins having the same aglycon structure was in the following order: galactoside > glucoside > arabinoside (Figure 28.6). On the other hand, in tissues, anthocyanins' distribution pattern was completely different from that in plasma. Figure 28.7 shows a chromatogram of anthocyanins extracted from liver (A) and kidneys (B). In these tissues, major anthocyanins distributed were methylated anthocyanins and only a trace amount of cyanidins and delphinidins were detected. From this observation, it was predicted that orally administrated anthocyanins will show different physiological functions in the tissues in which they are distributed. This finding is thus important for further investigations on the antioxidant role of anthocyanin in food materials.

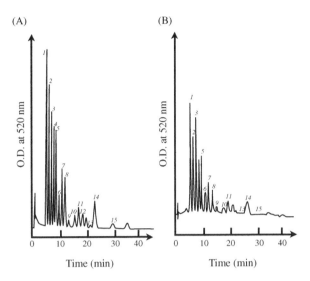

FIGURE 28.5 (A) Typical HPLC chromatogram of bilberry extract and (B) anthocyanins extracted from rat plasma after 15 min of oral administration. Peak numbers: 1: delphinidin 3-galactopyranoside, 2: delphinidin 3-glucopyranoside, 3: cyanidin 3-galactopyranoside, 4: delphinidin 3-arabinopyranoside, 5: cyanidin 3-glucopyranoside, 6: petunidin 3-galacto-pyranoside, 7: cyanidin 3-arabinopyranoside, 8: petunidin 3-glucopyranoside, 9: peonidin 3-galactopyranoside, 10: petunidin 3-arabinopyranoside, 11: peonidin 3-glucopyranoside, 12: malvidin 3-galactopyranoside, 13: peonidin 3-arabinopyranoside, 14: malvidin 3-gluco-pyranoside, and 15: malvidin 3-arabinopyranoside.

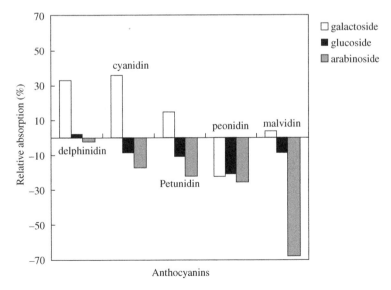

FIGURE 28.6 Relative plasma concentration of anthocyanins after 30 min of oral administration.

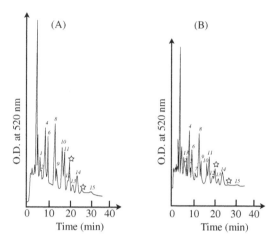

FIGURE 28.7 Typical HPLC chromatogram of anthocyanins extracted from rat tissues after 15 min of oral administration. (A) liver, (B) kidneys. Peak numbers are as shown in Figure 28.5.

REFERENCES

1. Matsumoto H, Nakamura Y, Tachibanaki S, Kawamura S, and Hirayama M. Stimulatory effect of cyanidin 3-glycosides on the regeneration of rhodopsin. *J Agric Food Chem* 2003; 51: 3560–3563.
2. Matsui T, Ueda T, Oki T, Sugita K, Terahara N, and Matsumoto K. α-Glucosidase inhibitory action of natural acylated anthocyanins 2. α-Glucosidase inhibition by isolated acylated anthocyanins. *J Agric Food Chem* 2001; 49: 1952–1956.
3. Katsube N, Iwashita K, Tsushida T, Yamaki K, and Kobori M. Induction of apoptosis in cancer cells by Billberry (Vaccinium myrtillus) and the anthocyanins. *J Agric Food Chem* 2003; 51: 68–75.
4. Hou DX. Potential mechanisms of cancer chemoprevention by anthocyanins. *Curr Mol Med* 2003; 3: 149–159.
5. Serraino I, Dugo I, Dugo P, Mondello I, Mazzon E, Dugo G, Caputi AP, and Cuzzocrea S. Protective effects of cyanidin 3-O-glucoside from blackberry extract against peroxynitrite-induced endothelial dysfunction and vascular failure. *Life Sci* 2003; 73: 1097–1114.
6. Ichikawa H, Ichiyanagi T, Xu B, Yoshii Y, Nakajima M, and Konishi T. Antioxidant activity of anthocyanin extract from purple black rice. *J Med Food* 2002; 4: 211–218.
7. Brouillard R. Chemical structures of anthocyanins, in *Anthocyanins as Food Colors.* 1982; 1–40.
8. Kahkonen MP and Heinonen M. Antioxidant activity of anthocyanins and their aglycons. *J Agric Food Chem* 2003; 51: 628–633.
9. Ichiyanagi T, Tateyama C, Oikawa K, and Konishi T. Anthocyanin distribution in different blueberry sources by capillary zone electrophoresis. *Biol Pharm Bull* 2000; 23: 492–497.

10. Ichiyanagi T, Kashiwada K, Ikeshiro Y, Hatano Y, Shida Y, Matsugo S, and Konishi T. Complete assignment of bilberry (*Vaccinium myrtillus* L.) anthocyanins in capillary zone electrophoresis. *Chem Pharm Bull* 2004; 52: 226–229.

11. Ichiyanagi T, Oikawa K, Tateyama C, and Konishi T. Acid hydrolysis of blueberry anthocyanins. *Chem Pharm Bull* 2001; 49: 114–117.

12. Ichiyanagi T, Hatano Y, Matsugo S, and Konishi T. Kinetic comparison of anthocyanin reactivity towards hydroxyl radicals, superoxide anion and singlet oxygen. *ITE Lett* 2003; 4: 788–793.

13. Ichiyanagi T, Hatano Y, Matsugo S, and Konishi T. Kinetic comparison of anthocyanin reactivity towards AAPH radicals and hydroperoxides by capillary zone electrophoresis. *Chem Pharm Bull* 2004; 52: 434–438.

14. Ichiyanagi T, Hatano Y, Matsugo S, and Konishi T. Kinetic comparison of anthocyanin reactivity towards reactive nitrogen species by capillary zone electrophoresis. *Chem Pharm Bull* 2004; 52: 1312–1315.

Index

T - #0033 - 111024 - C0 - 229/152/21 - PB - 9780367391683 - Gloss Lamination